Two Manuals Combined into One Book

Public Affairs Tactics, Techniques and Procedures

And

Military Role in Civil Disturbance Operations

Department of Defense

Pentagon
Publishing

Preface

The mission of Army Public Affairs (PA) is to fulfill the Army's obligation to keep the American people and the Army informed, and to help establish the conditions that lead to confidence in America's Army and its readiness to conduct operations in peacetime, conflict and war. PA is a critical battlefield function in today's global information environment since every aspect of an Army operation is subject to instantaneous scrutiny.

This field manual (FM) sets forth tactics, techniques and procedures (TTPs) for conducting PA operations in accordance with the doctrinal principles contained in FM 3-0 (100-5), *Operations* and FM 3-61 (46-1), *Public Affairs Operations*. Although this manual is primarily designed to be used by public affairs officers, noncommissioned officers and civilians, it presents TTPs that all leaders conducting operations in the information age should be familiar with.

FM 3-61-1 (46-1-1) is applicable to units and individuals in both the active and reserve components. It serves as a foundation for integrating PA into Army doctrine, training, leader development, organization, materiel and soldier initiatives. In conjunction with the Army Training and Evaluation Program (ARTEP) and other training guidance, it should also be used to plan, integrate and execute individual and collective PA training in units throughout the Army.

The proponent for this manual is the Office of the Chief of Public Affairs. Send comments and recommendations on DA Form 2028 to Director, Army Public Affairs Center, ATTN: SAPA-PA, Fort Meade, MD 20755-5650.

Unless this publication states otherwise, masculine nouns or pronouns do not refer exclusively to men.

This document is approved for public release; distribution is unlimited.

Introduction

With the extremely sophisticated communication technologies of the global information environment (GIE), the nature of media coverage has a significant impact on the conduct of war and stability and support operations at the strategic, operational and tactical levels. Civilian and military news media coverage influences the perceptions of soldiers, family members, the public and political leaders, and affects the direction, range and duration of operations. It has a direct relation to the confidence these key audiences have in the Army and its execution of operations.

Effective PA operations are critical to successful Army operations in the information age. They assist the commander in monitoring and understanding public opinion, explaining the situational context of events and communicating the Army's perspective clearly and without filters. They enable the commander to interpret the perceptions of external and internal audiences and influence the way in which discussion of the operation is framed.

Synchronized, well-planned and actively executed PA tactics, techniques and procedures significantly clear the fog of war and impact the morale and effectiveness of the force. They reduce distractions, misinformation, uncertainty, confusion and other factors that cause stress and undermine efficient operations. They enhance understanding, acceptance and support. Effective PA operations contribute to soldier confidence, discipline, will to win, and unit cohesion.

FM 3-61-1 (46-1-1), *Public Affairs Tactics, Techniques and Procedures*, builds on the doctrinal foundation of FM 46-1. It translates the PA fundamentals and principles into detailed guidance for the planning, coordination and execution of PA operations. It provides what is required for the information age -- a sophisticated approach to conducting PA operations. It is the TTP that brings Army Public Affairs into the "information age."

FM 3-61 (46-1), *Public Affairs Operations*, addresses fundamental PA concepts in depth, and provides the linkage between PA and the Army's keystone doctrine, FM 3-0 (100-5), *Operations*. It recognizes that a refocused, restructured military will conduct operations in an information environment in which detailed, graphic, and live coverage of events are transmitted around the world. It builds from the understanding that information availability will influence strategic decisions and the direction, range and duration of operations.

FM 3-61 (46-1) also examines PA operations at the different levels of war and across the range of operations. It discusses PA operations with respect to the Principles of War and the Tenets of Army Operations. It analyzes the PA contributions to build and sustain combat power, defines the PA mission, and establishes strategic PA goals, fundamental PA principles, and underlying considerations for planning integrated information strategies.

Public affairs frequently deals in intangibles -- perceptions and implications -- that are not easily quantifiable or qualifiable, but are essential to commanders. The PA objectives, processes and methods presented in FM 3-61-1 (46-1-1) will assist Army leaders and PA professionals to develop solutions to the complex PA issues they will confront.

FM 3-61.1

Field Manual
No. 3-61.1

Headquarters
Department of the Army
Washington, DC, 1 October 2000

Public Affairs Tactics, Techniques and Procedures

Contents

Page

PREFACE ... iv

INTRODUCTION ... v

Chapter 1 PUBLIC AFFAIRS FUNDAMENTALS ... 1-1

PA Mission ... 1-1

PA Force ... 1-2

PA Units ... 1-3

Chapter 2 PA FUNCTIONS AND RESPONSIBILITIES ... 2-1

Responsibilities ... 2-2

Communications ... 2-4

Logistics and Support ... 2-5

Chapter 3 PLANNING ... 3-1

Types of Plans ... 3-2

PA Estimate of Situation ... 3-9

PA Guidance ... 3-9

Annex ... 3-10

Chapter 4 MEDIA FACILITATION ... 4-1

Media Center Responsibilities ... 4-2

MOC Operations ... 4-4

Media Pools ... 4-8

DISTRIBUTION RESTRICTION:. Approved for public release; distribution is unlimited.

		Page
Chapter 5	**INFORMATION STRATEGIES**	5-1
	Responsibilities	5-3
	Information Strategy Process	5-4
	Information Program Evaluation	5-8
Chapter 6	**TRAINING**	6-1
	Training Objectives	6-1
	PA training for non-PA personnel	6-4
	Staff Section and Unit Training	6-5
Chapter 7	**COMMUNITY RELATIONS**	7-1
	Community Relations Activities	7-2
	Community Assistance	7-9
	Readiness Enhancement	7-10
Chapter 8	**PA ORGANIZATIONS**	8-1
	Brigade PAO	8-1
	Division	8-3
	Tactical Command Post PA Section	8-5
Chapter 9	**PA AND INFORMATION OPERATIONS**	9-1
	IO Staff Organization	9-2
	PA Estimate and IO Planning	9-3
	IO Campaign Cycle	9-4
Appendix A	**DOD PRINCIPLES OF INFORMATION**	A-1
Appendix B	**DOD POLICY FOR COVERAGE OF COMBAT OPERATIONS**	B-1
Appendix C	**SAMPLE PA ESTIMATE**	C-1
Appendix D	**SAMPLE PA ANNEX**	D-1
Appendix E	**SAMPLE PA GUIDANCE**	E-1
Appendix F	**FORMAT FOR PA OPERATIONS BRIEFING**	F-1
Appendix G	**PA AREA STUDY FORMAT**	G-1
Appendix H	**NEWS BRIEFING AND PRESS CONFERENCES**	H-1
Appendix I	**WAIVER OF LIABILITY STATEMENT**	I-1
Appendix J	**PREDEPLOYMENT CHECKLIST**	J-1
Appendix K	**SAMPLE PUBLIC AFFAIRS SECTION SOP**	K-1
Appendix L	**MEDIA OPERATIONS CENTER ORGANIZATION**	L-1

Appendix M	MEDIA QUERY	M-1
Appendix N	GUIDE FOR MEDIA INTERVIEWS	N-1
Appendix O	MEDIA RESEARCH AND ANALYSIS	O-1
Appendix P	INFORMATION PROGRAM EVALUATION	P-1
Appendix Q	PRINCIPLES OF PA SERVICE AND STANDARDS	Q-1
Appendix R	PRIVACY ACT/FREEDOM OF INFORMATION ACT	R-1
Appendix S	SAMPLE PA GUIDANCE FOR TERRORISM COUNTERACTION	S-1
Appendix T	AFRTS BROADCAST OPERATIONS PLANNING CONSIDERATIONS	T-1
Appendix U	COMMUNITY SURVEY	U-1
Appendix V	AUDIENCE SURVEY	V-1
Appendix W	PA LESSONS LEARNED	W-1
Appendix X	MEDIA GROUND RULES	X-1
	GLOSSARY	Glossary-1
	BIBLIOGRAPHY	Bibliography-1
	INDEX	Index-1

Chapter 1
Public Affairs Fundamentals

Public Affairs fulfills the Army's obligation to keep the American people and the Army informed and helps establish the conditions that lead to confidence in America's Army and its readiness to conduct operations in peacetime, conflict and war.

FM 3-61 (46-1), Public Affairs Operations

THE PA MISSION

1-1. The American public, internal Army audiences, allies, adversaries and other critical audiences have access to an ever expanding array of public and military media. Newspapers, magazines, radio, television and electronic media are independent conduits of information. They provide news, analysis, interpretation and commentary and serve as a forum for ideas, opinions and public debate. What appears in the media, both civilian and military, shapes perceptions, attitudes and opinions, and can have a direct impact on mission success.

1-2. The vast majority of both civilian and military media representatives are committed to providing responsible, accurate, balanced coverage. Although there are exceptions, most media representatives, even when editorializing, are focused on achieving a credible presentation. To accomplish this, media representatives investigate issues, ask tough, challenging questions, and pursue verifiable answers. They seek information, interpretation and perspective. Army leaders at all levels need to educate media representatives and facilitate their efforts to provide an accurate, balanced and credible presentation of timely information.

1-3. Army leaders do this by integrating public affairs into the planning process and synchronizing PA operations with other facets of the operation. Integrating and synchronizing public affairs issues allows commanders to communicate their perspective and achieve a balanced, accurate, credible information presentation.

1-4. The PA mission identifies the essential contribution that PA makes to America's Army. The mission and the strategic goals derived from it provide the foundation on which public affairs operations are built. Strategies, which are developed from the perspective that every aspect of every operation could become an issue of interest in the global information environment, are the most successful. Developing such strategies requires that PA personnel access, analyze and anticipate potential issues by conducting a thorough mission analysis.

1-5. The challenge for commanders, and personnel supporting them, is to plan and execute operations, which accomplish this mission and support these goals. To do this, PA must be integrated into the planning and decision-making process from receipt of the mission.

1-6. The need to integrate and synchronize PA early derives from the fact that in most situations media representatives will be present in an area of operations before the arrival of Army forces. They will know the area of operations and because they are covering the story as it evolves, will have an understanding of, and opinion about, the situation.

1-7. Media interest will normally be the most intense at the onset of operations. Media representatives will cover the deployment of Army forces, their arrival in the area and their initial conduct. To support the commander and the force in their interactions with media representatives during these early stages, public affairs personnel should be deployed in the first days, if not hours, of the operation. Dealing with a large international press corps constitutes the most immediate public affairs challenge facing the commander during contingency.

PUBLIC AFFAIRS FORCE

1-8. The changing information environment in which the Army conducts war and stability and support operations makes it necessary for PA officers, NCOs and specialists to respond to increasingly complex,demanding challenges. They must be prepared to support the commander with a wide range of knowledge about and understanding of the communication process, the global information environment (GIE) and its potential impact on operations.

1-9. PA personnel must also thoroughly understand the fundamentals of Army operations and the strategic context within which the Army conducts operations. As the GIE compresses the strategic, operational and tactical levels of operations, PA personnel need to appreciate the linkage between public opinion, political decision-making and the national security strategy. They must understand the Army's approach to fighting, influencing events in operations other than war and deterring actions detrimental to national interests.

1-10. PA leaders must, therefore, be among the most informed people in the command. They must be thoroughly aware of all aspects of the operation. They need to know what is being reported about the operation in the global news media and how internal and external audiences are reacting to that information.

1-11. The Public Affairs Officer. The PAO's primary mission is to assess the PA situation, advise the commander on PA issues, assist him in making the best possible decisions, and translate his decision into effective PA operations. PAOs employ the decision-making process to plan, coordinate and supervise the implementation of a PA strategy that helps the commander meet his obligation to communicate with the American public, soldiers, home station communities and the Department of the Army community. PAOs analyze the situation, anticipate issues, assess implications, and develop comprehensive operations, which meet the news and information needs of internal and external audiences and facilitate media operations.

1-12. The Public Affairs Non-Commissioned Officer. The strength of the Public Affairs functional area is its non-commissioned officer corps. PA NCOs are experts on the global information environment, media operations, information strategies, and PA training. They are integral to all facets of the PA planning and decision-making process and provide the essential functional area expertise and continuity required for successful PA operations.

1-13. PA NCOs work closely with the PAO, and in many situations, a PA NCO is the commander's senior PA advisor. Therefore, PA NCOs are fully prepared to assess the PA situation, develop, synchronize and coordinate a PA strategy, implement and monitor PA operations, and measure and evaluate the success of the PA effort.

1-14. The Public Affairs Specialist. In addition to learning traditional soldier skills, PA specialists are trained to support the entire spectrum of PA operations conducted in the global information environment. They are trained on news media operations, news media facilitation, information strategies and information provision. They register media representatives, gather information, develop information products, support news media briefings, respond to news media inquiries and requests for assistance, and track and monitor news media activities throughout the area of operations. They work with news media representatives to gather accurate information and provide timely, balanced coverage of the operation.

1-15. The Department of the Army Public Affairs Civilian. Civilian PA practitioners assigned to Tables of Distribution and Allowances (TDAs) have the same skills as military PA personnel. They provide critical support during war and non-combat operations by providing a vital link between deployed forces and the home station community, and in many situations, may be called upon to deploy with the units they support, or as individual augmentees.

PUBLIC AFFAIRS ELEMENTS

1-16. Battlefield commanders have two sources for tactical PA support. The first is the PA section organic to a warfighting headquarters. The second is the PA unit, which is attached to a headquarters to augment the command's PA capability.

1-17. Conducting PA planning, facilitating news media operations on the battlefield, providing news and information, and executing PA training and support operations is manpower intensive. The austerely staffed PA sections organic to warfighting headquarters will nearly always be overwhelmed trying to meet PA requirements in war and other operations. PA staff sections, therefore, rely on early augmentation by PA units, or individual augmentation if appropriate, to accomplish the battlefield PA mission.

1-18. **Organic Public Affairs Sections.** Organic PA sections are found in warfighting headquarters at various levels including brigades, divisions and corps throughout the Army. Army PA personnel are also assigned to the organic PA sections of joint and combined headquarters.

1-19. In headquarters without organic PA sections, the commander is responsible for PA and must plan and execute PA operations or assign responsibility for PA operations as a special or additional duty to an officer or senior NCO in the command.

1-20. Regardless of the echelon, the PA staff section's primary responsibility is to assist the commander in accomplishing his mission. The staff:

- provides PA information expertise and advice
- conducts PA assessments
- provides analysis of the information environment
- conducts PA planning
- develops information strategies and guidance
- implements PA operations
- measures the effectiveness of the PA effort
- conducts PA training

1-21. The PA staff element controls augmenting PA units. It determines requirements, defines priorities and assigns missions to the augmenting unit. In conjunction with the augmenting unit commander, the staff element task organizes the unit, allocating personnel and equipment to accomplish objectives. If the PA staff element is a Public Affairs Operations Center or a Task Force Headquarters, it will coordinate Armed Forces Radio and Television Service (AFRTS) support activities for the command.

1-22. **Public Affairs Units.** PA units are fully deployable TOE organizations designed to augment the PA staff sections of warfighting units, although they can operate independently in certain limited situations. When a PA unit is deployed to augment a PA staff section, the personnel in the unit cannot be reassigned as replacements or employed as individual fillers for other public affairs elements.

1-23. PA units depend upon the unit they augment for personnel administration, finance, legal and health services, communications, food service, unit maintenance, and supplemental transportation support. PA units operating at corps and below must have the capability to transport all of their TOE equipment in a single lift using authorized organic vehicles.

1-24. There are currently four types of PA units:

- Public Affairs Detachment (PAD)
- Mobile Public Affairs Detachment (MPAD)
- Public Affairs Operations Center (PAOC)
- Broadcast Operations Detachment (BOD)

1-25. **Public Affairs Detachment (SRC 45500AA00).** The smallest of the PA units, the Public Affairs Detachment (PAD) (Figure 1-1) is commanded by a captain and includes seven PA soldiers..

Public Affairs Detachment

CDR	CPT	46A
PA NCO	SSG	46Q
JOURN	SGT	46Q
JOURN(x2)	SPC	46Q
BR JOURN	SPC	46R
JOURN	PFC	46Q
BR JOURN	PFC	46R

Figure 1-1. PAD

1-26. The PAD normally augments a division, separate brigade and armored cavalry regiments and deploys in support of combined, unified or joint operations.

1-27. The PAD commander assumes responsibilities as the PAO or deputy PAO, and the PAD PA personnel are integrated into the supported command's PA section based on operational requirements.

Historical Perspective

The 10th Mountain Division, Fort Drum, N.Y., arrived in Somalia with hundreds of reporters already there, and absolutely no public affairs personnel accompanying them. The first public affairs support arrived at the 2nd Brigade, 10th Mountain Division, 10 days later and only because the JIB dispatched one of its own PADs.

(PA After Action Report, Operation Restore Hope, 10th Mountain Division, Dec 1992)

1-28. Because of the size of the unit, the PAD provides limited:

- Personnel and equipment for digital imagery and audio acquisition.
- Personnel for media escort within the supported unit's area of operations
- Coordination of an information product distribution system that can relay news and information products to members of the supported unit, higher echelons and home station.
- Planning, developing and implementing strategy to support civilian news media and facilitate news gathering efforts throughout the supported unit's area of operations
- Personnel and equipment to coordinate, assist or conduct press conferences and briefings

- Personnel to train, advise and assist leaders and soldiers interacting with or supporting civilian news media within the supported unit's area of operations.

1-29. **Mobile Public Affairs Detachment (SRC 45413A000).** The workhorse of PA units, the Mobile Public Affairs Detachment (MPAD) is a modular, task organizable unit, which provides the full range of PA services. (Figure 1-2). The MPAD is commanded by a major and includes 20 soldiers.

Mobile Public Affairs Det

Command Cell

CDR	MAJ	46A
PA SPVR	MSG	46Z
CLERK	SPC	71L

MEDIA TEAM*

TM OIC	CPT	46A
NCOIC	SFC	46Q
JOURN	SGT	46Q
BR JOURN	SGT	46R
JOURN	PFC	46Q
BR JOURN	PFC	46R

BROADCAST TEAM*

TM OIC	CPT	46A
NCOIC	SFC	46R
BR JOURN	SSG	46R
BR JOURN	SPC	46R
BR JOURN	SPC	46R
BR JOURN	PFC	46R

PRINT TEAM*

TM OIC	CPT	46A
NCOIC	SSG	46Q
JOURN(x2)	SPC	46Q
JOURN	PFC	46Q

*This is one possible configuration. Paragraph 2 members can be configured as one, two or teams of various strengths, based on mission.

Figure 1-2. MPAD

1-30. The MPAD normally augments a Corps PA section or a Public Affairs Operations Center (PAOC). In support of a PAOC, it provides manpower and equipment to establish and operate a media center at Theater Army, TAACOM and Corps.

1-31. It may also be deployed to directly support a joint service task force or non-DoD governmental agency conducting disaster relief, humanitarian assistance, counter drug, peacekeeping, or other contingency operations.

1-32. MPADs in direct support of a gaining command PAO provide acquisition capability for print, audio and video.

1-33. Additional MPADs in direct support can expand the media escort capability of the supported PAO, augment divisions and other elements in theater and provide the PAO with staff augmentation. MPADs may be deployed forward to establish media centers or a sub-JIB in a joint environment.

1-34. MPADs have the capability to:

- Monitor and assess the perceptions of external audiences through access to civilian commercial news sources.
- Conduct assessments of the information environment, to include development of a PA estimate of the situation, as the initial part of operational planning.
- Assist the PAO in operational planning and policy and ground rules for media, coordination for logistical support to PA, and coordination of PA operations with higher and subordinate headquarters.
- Plan and develop information products, which will be produced through contracted services and/or the use of organic equipment and facilities.
- Acquire, produce and transmit information products throughout the theater, between the theater and home station, and between the theater and HQDA.
- Create and disseminate print, photographic, audio and video products for external release directly to civilian media who do not have representatives within the theater of operations. Conduct media facilitation and develop information strategies.
- Prepare commanders, staff personnel and other command members for interviews, press conferences, and similar media interaction.

1-35. **Public Affairs Operations Center (SRC 45423A000).** The Public Affairs Operations Center (PAOC) consists of command, media facilitation and post-production sections (Figure 1-3). It is commanded by a lieutenant colonel and includes 32 soldiers.

1-36. The PAOC normally augments the PA staff section at echelons above division to establish and operate a media center in support of civilian and military media representatives working in the theater.

1-37. The PAOC commander serves as the media center commander but works under the control of the PAO of the supported unit or task force.

1-38. In joint or combined operations, a PAOC serves as the Army element of the joint media operations center.

1-39. When the PAOC functions as the Army element of a joint or combined media center, the PAOC commander works for the joint or combined media center commander.

1-40. For major operations in which there is a significant media interest, the PAOC is augmented by up to three MPADs. The MPADs are either integrated into the main media center operation or tasked to operate subordinate media centers at outlying locations. A PAOC can support up to 100 news media representatives. When augmented by three MPADs, the PAOC can support up to 300 news media representatives.

Public Affairs Operations Center

Command Cell

CDR	LTC	46A
XO	MAJ	46A
PAO	MAJ	46A
CH PA NCO	SGM	46Z
PA NCO	SFC	46Q
SPLY SGT	SGT	92Y
MECHANIC	SPC	63B
ADM SPC	SPC	71L
PERSONNEL	SPC	75B
ADM CLERK	PFC	71L

ESCORT TEAM

CHIEF	CPT	46A
PRESS OFF (x2)	CPT	46A
JOURN (x2)	SGT	46Q
JOURN (x4)	SPC	46Q
JOURN (x4)	PFC	46Q

BRIEFING SECTION

CHIEF	MAJ	46A
BRIEF OFF (x2)	CPT	46A
BR JOURN	SGT	46R
BR JOURN	SPC	46R
BR JOURN (x2)	PFC	46R

A/V PRODUCTION

BR NCO	SSG	46R
PROD SPC	SPC	25V
BR JOURN	SPC	46Q

Figure 1-3. PAOC

1-41. PAOCs are modular, task organizable units having the capability to:

- Implement the theater or corps strategy to support civilian news media and facilitate news gathering efforts in theater.
- Coordinate and provide services to registered civilian news media sponsored by the command.

- Provide the personnel and equipment to coordinate and conduct media support within the theater of operations.
- Provide personnel and equipment to plan and conduct daily news media briefings.
- Provide personnel to train, advise and assist leaders and soldiers interacting with or supporting civilian news media representatives.

1-42. **Broadcast Operations Detachment (SRC 45607A00).** The BOD consists of a command element, two broadcast teams and a maintenance team. (Figure 1-4). It is commanded by a major and includes 26 soldiers.

1-43. The BOD augments a fixed or field expedient AFRTS facility under the control of a senior AFRTS facility commander, or it can establish and operate a separate radio and/or television broadcast facility to support theater level operations.

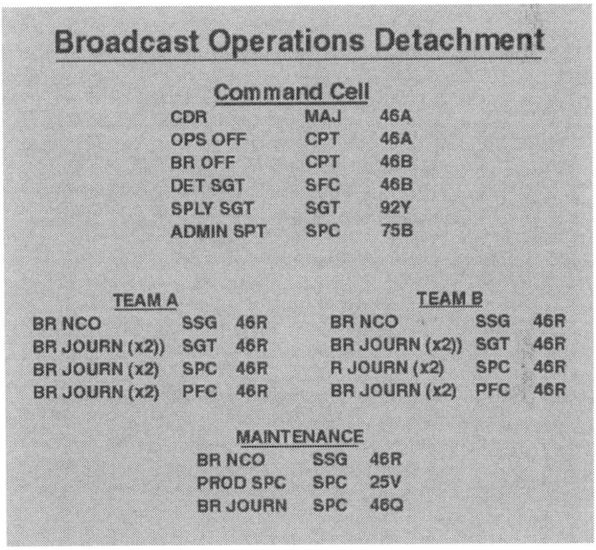

Figure 1-4. BOD

1-44. The BODs must be authorized and equipped by AFRTS to perform this mission. When deployed to perform this mission, the BODs are assigned to the PAOC supporting the command.

1-45. BODs have the capability to:

- Provide on-air broadcasters recorded materials and satellite down links to operate a 24-hour a day radio outlet.

- Provide on-air broadcasters, recorded materials and satellite down links to operate a television station.
- Originate audio and video news, feature and entertainment programming from within theater.
- Provide post production services for audio and video news and feature material supplied in unedited format.
- Provide limited audio and video materials to other public affairs operations for dissemination outside the theater.
- Acquire audio and video electronic newsgathering coverage of operations in the theater for use in internal and external information programs.
- Perform field maintenance and repair above operator level to broadcast equipment organic to the unit.
- Provide the commander with an alternate means of communications when tactical communications are not adequate or not available.

1-46. Public Affairs organizations are built around a force of soldiers who are selected and trained to articulate the goals and missions of the Army.

1-47. More than 65 percent of the total public affairs force and 85 percent of the deployable TOE unit structure is positioned in the U.S. Army Reserve and Army National Guard. These reserve units and personnel must be seamlessly integrated with the active component and focused on supporting the overall Army goals and objectives.

Chapter 2

Public Affairs Functions and Responsibilities

THE CHALLENGE

2-1. The global information environment and continually evolving information communication technologies make it imperative that information and messages be consistent at all levels. The personal comments made by a deployed soldier in a remote area of operations and the official statements released by DoD at the Pentagon must be mutually supporting. The information targeted to internal audiences must parallel the information released through the news media to the American public and other external audiences. The Army's need for security, and the soldier's and family member's right to privacy must be balanced with the Army's obligation to provide timely, accurate, complete information to internal and external audiences. The commander's information strategy must ensure that the information available in the public domain, regardless of the source, does not conflict, contradict or otherwise undermine the credibility of the command or the operation.

Historical Perspective

Civilian news coverage contributed greatly to maintaining soldier morale during Desert Storm. The coverage was generally positive; the American people were behind the operation and soldiers felt this impact. Problems arose when the coverage created rumors, and command information was not consistent with what the soldiers were seeing or hearing in the world media. Family members and non-deployed soldiers were greatly affected by news coverage, often creating problems for rear commanders and detracting from their credibility.

(After Action Report, Desert Storm 1990, Center for Army Lessons Learned)

2-2. Accomplishing this presents unique command and control challenges for commanders, PA practitioners at all levels and others involved in using information to help accomplish the mission in the most effective, efficient manner. It requires careful coordination between staff elements and necessitates continual liaison between levels of command from the tactical through the operational to the strategic.

2-3. Further complicating PA command and control challenges are PA force structure realities. The small size of the PA staff sections organic to war-fighting headquarters necessitates augmentation, especially for operations with a high level of visibility. The availability of augmenting PA units, the majority of which are located in the reserve components, and the difficulties inherent in deploying PA civilians result in heavy dependence on augmentation by individuals. This leads to the creation of ad hoc, unequipped PA elements, which have not trained together or developed relationships with other staff sections or commands, and do not have established internal or external operating procedures.

2-4. For PA personnel therefore, the critical challenge is to rapidly define command and control channels, establish lines of communication and develop operating procedures. The responsibility for doing this usually lies with the Corps PAO who normally leads the commander's PA effort. He identifies requirements, assesses resources and plans, organizes, directs, coordinates and controls the PA operation.

PUBLIC AFFAIRS RESPONSIBILITIES

2-5. Effective PA command and control establishes a public affairs organization based on analysis of mission, enemy, terrain, troops, time available and civilians (METT-TC), tailored to the situation, which reflects the commander's concept of the operation. It ensures that there are sufficient, experienced, PA personnel at each echelon to provide the commander and his force with the most effective and efficient support possible.

2-6. PA command and control begins at the DoD level. The Office of the Assistant to the Secretary of Defense, Public Affairs (OASD(PA)), retains primary responsibility for the development and consistent implementation of DoD information policies and determines who should serve as the initial source of information about operations. Although (OASD(PA)) delegates PA release authority to the combatant commander as soon as practical, it retains responsibility for approving Public Affairs Guidance (PAG), establishes public affairs policy, and coordinates and approves PA strategies and plans.

2-7. The Office of the Chief of Public Affairs (OCPA) for the Army is responsible for Army PA resources. OCPA develops PA doctrine, designs PA organizations, determines training and leader development requirements, identifies materiel needs, and manages PA personnel to ensure that sufficient assets are available, qualified and ready to conduct successful PA operations in support of any assigned mission.

2-8. Commanders supported by their PA staff personnel, plan PA operations for their assigned missions based on the situation, published in DoD directives, instructions, doctrine and guidance, and in coordination with OASD(PA). The CINCs prescribe the chain of command, organize and employ forces, give authoritative direction, assign tasks and designate objectives through component commanders, subordinate unified commanders, commanders of joint task forces and other subordinate commanders. The commander establishes responsive PA structures and ensures that they are provided with the personnel, facilities, equipment, transportation and communications assets necessary to provide adequate PA support. A failure to establish these structures results in a duplication of effort and a waste of resources. The commander is responsible for the full range of PA activities -- PA planning, media facilitation, information strategies and PA training and at sustaining base, community relations. He is also responsible for establishing, resourcing and guiding the operations of Joint Media Operation

Centers and planning all AFRTS radio and television support operations in the area of operations.

> **Historical Perspective**
>
> The establishment of the Joint Information Center under the auspices of the DOT Presidential Task Force on September 1, 1992 was vital to a coordinated and successful Joint Public Affairs effort. The JIC was an "umbrella" organization that served as the clearing house for dissemination of hurricane relief information to the news media. More than 10 federal agencies involved in relief operations had public affairs representatives at the JIC. Daily meetings and consistent interaction among the agencies involved resulted in a coordinated federal information effort.
>
> (Public Affairs Lessons Learned Library, Joint Information Center, Hurricane Andrew, 1992)

2-9. Within the Army, the Corps is usually the hub for PA operations. The Corps commander, supported by his PA staff element, is responsible for the development and coordination of PA strategies, the implementation of information campaigns and the execution of PA operations based on METT-TC, the information environment, and guidance and policy received from the combatant CINC.

2-10. The Corps PAO is the principal PA advisor to the Corps commander and deploys with the lead element of the Corps headquarters. When fully deployed, the Corps PAO section operates from the Corps main command post, with a liaison officer/NCO located at the Corps plans cell. It also is responsible for establishing coordination with the PA elements of higher, lower and adjacent commands.

2-11. The Corps, through the PAO, controls the employment of augmenting Army PA units deployed in support of the operation. Up to one PAOC and six MPADs are normally allocated to augment the Corps PA section, although the size and scope of the mission will determine the actual augmentation required for each operation. The Corps PAO and his staff task organize the personnel and organizations available and allocate the equipment, communications support and facilities. When augmented, the Corps PAO operates a media operations center and establishes satellite centers as required.

2-12. Below the Corps level, the PA staff section organic to a war fighting headquarters is extremely austere. The mission of the PA section below Corps is to advise the commander by providing immediate planning expertise and guidance on issues with critical PA implications. The PA section deploys with the lead elements of the headquarters, and operates from the command's main CP.

2-13. PA units deployed to augment the staff sections organic to a headquarters are normally placed under the control of the supported PAO, who assigns the augmenting PA unit missions and tasks. He will do so in conjunction with the augmenting PA unit commander, who will retain command of his unit and ensure that his unit's personnel are not employed as individual fillers. Whenever possible, augmenting PA units should be linked with the supported command headquarters at that command's home station prior to deployment to facilitate establishment of command and control relationships.

PUBLIC AFFAIRS COMMUNICATIONS

2-14. Reliable, survivable, flexible communications are essential for effective PA command and control. In today's global information environment, information must flow to and from users, up and down the chain of command, and horizontally across the battlefield. Technology has compressed time and space and forward-deployed PA sections can be in direct communication with officials at DoD working PA strategies. The challenge is to ensure coordination and interoperability so that all elements have the communications capability necessary to effectively carry out their assigned mission, especially in today's joint, combined or interagency environment.

2-15. Deliberate, detailed planning can prevent communications shortfalls. PA planners assess their information transmission and reception needs and requirements. They then identify the communications capabilities they need access to, and determine the communications support they will need from command signal organizations. Through close coordination with the staff signal section, the identified PA communications requirements are integrated into the overall communication architecture.

INFORMATION OPERATIONS

2-16. The three central defining characteristics of the global information environment -- the facility of information acquisition and transmission, the speed of information communication and the breadth of information saturation -- combine to increase information availability. The American public, internal audiences, allies and adversaries have ready access to information. Information security is transitory and it is critical that information operations at every echelon are mutually supporting and directed at a clearly defined, decisive and attainable objective.

2-17. Credibility is essential for successful information operations. If an information source is not perceived as believable, then the desired effect of that communication cannot be achieved. Regardless of the source, target or objective of an information effort, in the GIE, credibility is founded in truth and enhanced by validation, corroboration, and consistency.

2-18. Commanders require integrated, coordinated, synchronized information operations. PA operations, which occur at, and impact on, the strategic, operational and tactical levels -- often simultaneously -- are a critical element of these operations. News media coverage of conflicting messages and information communicated by different elements of the command compromises credibility.

2-19. Integrating, coordinating and synchronizing every element of the commander's information operation -- Public Affairs, Psychological Operations, Civil Affairs, Combat Camera, Operations Security and others -- results in a synergistic information strategy. It minimizes the possibility of conflicting messages, which undermine credibility, jeopardize operations and endanger mission accomplishment.

LOGISTICS SUPPORT

2-20. Logistics is critical at all levels of command for Public Affairs mission success, during any phase of combat or garrison operations. Commanders must ensure their Joint Table of Allowances (JTA), Modified Table of Equipment (MTOE), Table of Distribution and Allowances (TDA) or Common Table of Allowances (CTA) reflects appropriate equipment levels to maintain a PA staff and media support under field and garrison conditions. Maintenance also plays an important role in Public Affairs operations. A Public Affairs element that has all its equipment cannot function properly if its equipment is inoperative, broken or deadlined. Each Public Affairs element must develop its own internal SOP in regard to logistics. (See Appendix K.)

2-21. Public Affairs staff members must be trained in the areas of supply, budget, property book, ordering, class A procurement, etc. Public Affairs must be an integral player in all mission and operational planning sessions to ensure logistical requirements are identified and resourced.

2-22. Responsibilities: The Public Affairs staff has the responsibility to identify to its resource manager, property book manager and ordering officer all fiscal and logistical requirements for field operations and home station support. Requisitions for equipment, supplies, services and allowances will be ordered and processed in accordance with appropriate Army Regulations, AR 710-2 *Unit Supply Update* and budgetary guidelines.

2-23. Requesting supplies: Commanders must ensure that equipment and components authorized by JTAs, CTAs, MTOEs, or TDAs are on hand or requested. The organization's supply operation is responsible for identifying, acquiring, accounting, controlling, storing and properly disposing of materiel authorized to conduct the mission of the unit and maintain the soldier. The organization is the foundation of the supply system. Exceptions and procedures are outlined in AR 710-2.

2-24. The Direct Support and General Support Activities provide class 1, 2, 3 (packaged and bulk), 4, 5, 7, 8 and 9 supplies directly to the using units on a customer support basis. These supplies are routinely procured through the unit supply rooms. In the event a Public Affairs element is operationally attached outside its assigned organization, it should coordinate before deployment for logistical support when possible. If prior coordination is not possible, contact for support should be made upon arrival into the theater of operations through the C-4, J-4, G-4, or S-4.

2-25. Accountability: All property acquired by the Army, regardless of source or whether paid for or not, must be accounted for, in accordance with applicable Army regulations and AR 710-2.

- Nonexpendable property is personal property that is consumed in use and that retains its original identity during the period of use. It requires formal property book accountability throughout the life of the item. It will be accounted for at the using unit level using property book procedures. Examples are desks, computers, file cabinets, chairs.
- Expendable items are property which is consumed in use or that loses its identity in use, and all items not consumed in use with a unit price

of less than $100 and not otherwise classified as nonexpendable or durable. It requires no formal accounting after issue to the user. The following classes or types of property will be classified as expendable.

- Supplies consumed in the maintenance and upkeep of the public service. Examples are oil, paint, fuel and cleaning and preserving materials.
- Supplies that lose their identity when used to repair or complete other items. Examples are assemblies, repair parts, and accessories.
- Office supplies and equipment (such as paper, staplers and hole punchers) with a unit price of less than $100.
- Durable property is personal property that is not consumed in use, does not require property book accountability, but because of its unique characteristics requires hand receipt control when issued to the user. Examples are hammers, lawnmowers, audiovisual production material and books.

2-26. Conservation of resources and property accountability is ultimately a supervisory responsibility. Property responsibility must be assigned and acknowledged in writing using hand receipts and property books as outlined in AR 735-5, *Policies and Procedures for Property Accountability* and AR 710-2.

2-27. Property book: Effective supply support at the using element or property book level requires timely and accurate processing of supply requests and receipts, accurate accounting records and adequate property control. Turn in, transfer, substitutions, hand receipt, etc., are accomplished in accordance with Army Regulations, AR 710-2 and logistical SOPs.

2-28. Budget: Budgets must be programmed in advanced. Organizations plan their budget in the previous fiscal year They must be established and managed with the appropriate command resource manager/budget analyst. When allocating funds consideration must be given to equipment replacement and upgrades, recurring supply needs TDYs, maintenance, contracts, etc. Normal operating funds are allocated/dispersed by a public affairs element operational headquarters; however, during deployments for exercises/operations funds may be available from the tasking headquarters up front or on a recuperative basis.

2-29. Maintenance: Public Affairs elements must maintain their equipment in a deployment ready state. Preventive Maintenance Checks and Services (PMCS) is an important part of the maintenance program, and is the user's responsibility. The Public Affairs element's operational headquarters provides maintenance support. For example, an embedded Public Affairs section assigned to the Headquarters Company of a separate brigade would seek maintenance support from the Headquarters Company, then the brigade maintenance section. The company/brigade's maintenance SOP would be followed for 1st, 2nd, 3rd, etc., echelons of maintenance. Maintenance for communications and data processing (computers) equipment is coordinated through the G6 (DOIM).

2-30. Local Purchase: Local purchases may be an option for procurement provided the action is in the best interest of the government in terms of timeliness, quality and cost. Local purchase requests must be made in accordance with AR 710-2. Approval for local purchase of nondevelopmental items starts at the first level of command authority and is accomplished in accordance with local policies and Army regulations. Nondevelopmental items is a generic term that covers materiel available from a variety of sources with little or no development effort from the Army. Sources include commercial items which fully meet an approved need, items being used by other U.S. services or agencies or items used by military or other agencies of foreign government. Most of these purchases are covered under the IMPAC Card program. Local guidance covers implementation.

2-31. Class A Procurement/Credit Card: Class A agents and government credit card holders must be identified and trained prior to their ability to accomplish those functions. Training is routinely accomplished at the installation level. Purchases for other than national-stock-numbered items are routinely accomplished using Class A agents and ordering officers, and the U.S. Government Credit Card.

2-32. SSSC: Self Service Supply Centers are managed at the installation or theater level. Users are required to have a valid SSSC account. Accounts are identified by DODAAC or UIC. Subaccounts are authorized IAW AR 710-2. Field resupply centers are often established at divisional-level logistics bases.

2-33. The best means of ensuring supply discipline is to be proactive and not reactive in supply operations. Enforcing compliance with regulations requires constant emphasis.

Chapter 3

Public Affairs Planning

During Operation Just Cause, PA planning and integration were inadequate. Commanders at all levels failed to involve public affairs officers in planning from fear of OPSEC leaks. The results were insufficient PA guidance provided to soldiers, family members and commanders; sometimes miscommunication to and confusion within Army family elements; and misuse of PA assets. A matter of urgent concern was the failure to plan for and use Reserve Component PA assets to relieve the pressure on an already small active PA force so that it could better handle both internal and external communication.

- Public Affairs After Action Report, TRADOC LLC, Phase II, Sept. 90

COMMUNICATING INFORMATION

3-1. Public Affairs operations assist the commander in communicating information and messages about his force and the operation to internal and external audiences. Like other operations, PA operations are conducted to bring about clearly specified, meaningful objectives, which support the commander's intent and contribute to mission success. Those objectives are defined in terms of the effect the PA operations are intended to have on target audiences -- the impact on target audience behavior that is desired -- and are measurable.

3-2. Once PA objectives are defined, PA operations are planned and executed to achieve those objectives. PA operations focus on the communication process -- an on going, dynamic, ever-changing process. The communication process is composed of elements involved in receiving, collecting, analyzing and interpreting data, identifying and analyzing audiences and formulating and transferring messages. This process is used to bring about a specified objective, while measuring and analyzing the outcome and effectiveness of the effort.

3-3. To support the commander's effort to communicate, PA professionals concentrate on five basic functions or core processes — planning, media facilitation, information provision, force training and community relations.

3-4. This chapter focuses on the systematic process for Public Affairs planning and decision-making. It addresses the information environment and the impact of information at the strategic, operational and tactical levels across the range of operations that requires public affairs considerations be totally integrated into the planning and decision-making process. Doing so enables PA personnel to prepare for potential situations, to synchronize efforts with other agencies that manage information communication, and to more successfully influence the coverage, interpretation and understanding of events. It limits the need for reactive, defensive attempts to buy time or control damage.

3-5. PA planning prepared in support of the CINC's theater campaign plan requires a series of decisions related to policy at the national level and the techniques at the tactical level. From policy to techniques, however, basic planning considerations are the same: What should the PA objective accomplish? With what audience? When? How? PA planning must not only be done at all echelons and within national policy but also within the limits of operational plans and capabilities.

TYPES OF PLANS

3-6. The amount of time available significantly influences the planning process. Two different methods of planning are described in the JCS-published Joint Operations Planning and Execution System (JOPES).

3-7. **Deliberate or Peacetime Planning** is the process used when time permits the total participation of commanders and staffs. Development of the plan, coordination among supporting commanders and agencies, reviews by staffs, planning conferences and development of proposed public affairs guidance can take many months. Deliberate or Peacetime Plans are prepared in prescribed formats--the complete operational plan (OPLAN) or the conceptual operational plan (CONPLAN).

3-8. **Time-Sensitive or Crisis Action Planning** (CAP) is conducted in response to crisis where U.S. interests are threatened and a military response is being considered. Crisis Action Planning is carried out in response to specific situations as they occur and that often develop very rapidly.

3-9. It is within the CAP process that established, working relationships between the PAO and operational planning staffs are crucial to the inclusion of PA considerations into OPLANS and OPORDS.

3-10. Both deliberate and crisis action planning are conducted within JOPES. Joint Pub 5-03.2, JOPES Volume II, describes detailed administrative and format requirements for documenting the annexes, appendixes, etc. of operational plans, and conceptual plans, the products of deliberate planning.

3-11. The purpose of JOPES is to bring both deliberate and crisis action planning into a single architecture to reduce the time required to complete deliberate or crisis action planning. This makes the refined results more readily accessible to planners, and makes it a more manageable plan during execution.

3-12. The overall procedures are the same, at all echelons, for both deliberate and crisis action planning.

- Receive and analyze the task to be accomplished
- Review the situation and begin to collect necessary intelligence
- Develop and compare alternative courses of action
- Select the best alternative
- Develop and get approval for its concept
- Prepare a plan
- Document the plan

Operation Just Cause

The basic problem---planning. No discussion concerning use of personnel can proceed without an understanding of the planning problem. Public Affairs in general was not sufficiently planned for by leaders or public affairs officers for Operation Just Cause. PAOs were not given time to plan. Only outstanding unit mission accomplishment, American public support and the hard work of public affairs personnel, prevented major PA failings in Panama. A longer duration and less popular action could have turned into a public affairs disaster.

The SOUTHCOM, XVIII Airborne Corps, and Army Special Operations Command PAOs were not informed that the operation would occur until 17 December and then they were given instructions not to discuss it with key persons on their staff. The 82d PAO did not learn of the operation until 18 December, 7th ID PAO, 19 December and OCPA **one hour** before H hour.

More critical however, was the absence of joint coordination. OASD-PA received a PA plan from SOUTHCOM public affairs in November, but the plan was never coordinated due to worries concerning possible security leaks. The XVIII Airborne Corps PAO indicated that he also knew of the operation in general terms in November but was unable to coordinate planning with the Director of Public Affairs SOUTHCOM. Because the plan was not staffed, OCPA was caught unaware. Divisional public affairs officers all indicated that they were not sufficiently drawn into planning. Some PAOs said they never saw a plan. It is obvious that sufficient public affairs planning did not occur **at any level**.

There was also a ripple effect downward caused by the lack of staffing by the Joint/OASD-PA. In addition, senior leaders at division level and above did not draw PA officers into planning to maximize the limited planning time that was available. Commanders strongly complained about poor balance of coverage in the media, inability to send command information at the same pace as civilian media reporting, and lack of sufficient public affairs guidance; yet senior leaders are reluctant to draw PAOs into the planning process to prevent these problems from occurring. Army leaders must come to grips with this dilemma.

(Public Affairs After Action Report TRADOC Lessons Learned Collection Phase II. Sept 1990)

3-13. An OPLAN is a complete, detailed plan. It includes a description of the concept of operations from the commander's perspective and presents additional annexes provided by various staff sections which identify specific functional area requirements, restrictions, limitations, or considerations. The inclusion of a public affairs annex is essential to successful integration of PA principles and guidance into the OPLAN.

3-14. A CONPLAN is an abbreviated operational plan, which requires considerable expansion or alteration to convert it into an OPLAN or OPORD. Detailed support requirements are not included. The commander determines what annexes will be included to complete the CONPLAN.

3-15. A Public Affairs Estimate is an assessment of a specific mission from a Public Affairs perspective. It is an examination of critical Public Affairs factors, their influence on the planning and execution of operations, and their potential impact on mission success. The senior PAO at each echelon is responsible for consolidating information and preparing the PA Estimate. A sample PA Estimate is included at Appendix C.

3-16. The Public Affairs Annexes to OPLANS or CONPLANs provide the details and instructions necessary to implement Public Affairs media facilitation, news and information provision, and force training operations. It is coordinated with all staff agencies, especially those that significantly impact the information environment -- Psychological Operations, Civil Affairs, Signal, and Military Intelligence -- to ensure that Public Affairs activities are synchronized with other activities.

DELIBERATE PLANNING

3-17. A commander continually faces situations involving uncertainties, questionable or incomplete data, or several possible alternatives. As the primary decision maker, the commander, with the assistance of the staff, must not only decide what to do and how to do it, the commander must also recognize if and when to act. How the commander arrives at a decision is a matter of personal determination. However, superior decisions (those, which offer the best solution, decisively, at precisely the correct time,) result from the commander's thorough, clear, and unemotional analysis of facts and supported assumptions. This is done through the "deliberate planning process."

3-18. To support the commander's decisions and command objectives, the PAO must develop a thorough, clear, comprehensive public affairs strategy. This strategy allows the PA to link public affairs considerations into planning for contingency, future and current operations. With the PA strategy, the PAO defines the public affairs perspective of the operation, and identifies how the Army public affairs involvement in this operation supports strategic goals. It provides the intent for PA operations and the Army approach to meeting the information needs of critical internal and external audiences. It is the framework for defining and developing the PA scheme of operations.

3-19. Based on the PA strategy, PA plans are developed for integration into OPLANs. A PA plan is produced by the operational commander's PAO, and it details the media facilitation, news and information provision, and force training and support procedures which will be employed in support of the operation.

3-20. The first crucial step in fulfilling the PA strategy requires the PA Plans officer/NCO to establish and maintain a routine, ongoing relationship with operational planners within the organization. The PA plan is coordinated with key staff agencies, integrated into the OPLAN as a PA Annex. Synchronization with these other activities ensures services and support required by the PAO is provided and multiplies the impact of the PA plan. This process is followed at subordinate echelons as planning guidance is communicated down the operational chain of command.

3-21. There are five phases in the deliberate planning process. Items in parenthesis identify PA actions performed within each phase:

- **Phase I** - Initiation. The task assigning directive outlines the major combat forces available for planning, gives general planning instructions, lists assumptions for planning, and specifies the product document such as an OPLAN, CONPLAN. (PA planners begin assessing the information environment, its impact on operations and the PA requirements to operate within a specific arena.)

- **Phase II** - Concept Development. (Using the supported CINC's mission statement and concept of envisioned operations, the supporting PA planners analyze the mission, formulate tentative courses of actions and develop the PA Estimate for the operational scenario and requirements.)
- **Phase III** - Plan Development. Subordinate commanders use the CINC's concept and the allocated major combat forces as the basis to determine the necessary support, including forces and sustaining supplies for the operation. (The PA planners provide the CINC with recommendations for public affairs assets required, phasing of PA forces and support into the theater of operations, and perform a transportation analysis of their movement to the destination to ensure that the PA segment of the entire plan can feasibly be executed as envisioned. For the supported CINC's PA requirements, above those organic to the tasked major combat elements, the supporting commands [force providers] of each service, as much as possible, identify real-world PA assets to take part in the plan and sustainment to meet requirements. The supporting command identifies PA requirements in OPLANs, OPORDs, and taskings, through operational channels, to major subordinate commands.)
- **Phase IV** - Plan Review. The review process is more than a single phase in deliberate planning. The Joint Staff performs or coordinates a final review of operations plans submitted by the combatant CINCs. It is a formal review of the entire operation plan. Approval of the plan is the signal to subordinate and supporting commands to develop their plans in support of the CINC's concept. (PA planners do not wait until the plan is approved before beginning to develop their supporting plans; they have been involved in doing this, while coordinating with their command's planning staff. In the meantime the CINC has been building the overall plan.)
- **Phase V** - Supporting Plans. The emphasis in the Supporting Plans Phase shifts to the subordinate and supporting commanders. (This is the phase in which PA planners begin to concentrate on how to meet tasks identified in the approved operation plan by preparing Public Affairs Annexes to supporting plans. This input outlines the actions and relationships of assigned and augmenting PA assets.)

INFORMATION ENVIRONMENT ASSESSMENT AND PA ESTIMATE

3-22. Planning fosters effective application of knowledge, logic, and judgment. Analysis of the information environment (IE) starts the process used to develop an estimate of the situation.

3-23. The IE analysis provides the basis for the development of all PA operational plans and is a channel for integration of strategic, operational and tactical planning guidance. It is a method of identifying factors within the information environment that have potential implications for the planning and execution of Army operations. PA planners study and evaluate the dynamics of the area information environment to identify specific public affairs operational considerations.

3-24. Analysis of the information environment focuses on research into the following areas, which will be put into the "Situation and Considerations" portion of the PA estimate:

- Information infrastructure
- Media presence
- Media capabilities
- Media content analysis
- Public opinion assessment
- Information needs assessment
- Impact assessment/courses of action (COA)

3-25. An analysis of the IE using this approach builds a complete picture of the conditions facing commanders and their PA forces, providing them the tools necessary to anticipate trends, actions, issues, and conflicts. The PA staff officer or NCO conducts research and assessment for the estimate then evaluates, prioritizes, and suggests courses of action that public affairs can best support, while considering the information environment. To acquire all the information necessary for an accurate picture of the operational environment, the staff officer or NCO must work closely with intelligence, civil affairs, psychological operations, military police, visual information and other staff sections involved with information gathering.

3-26. The PA assessment must include those aspects under the control of the commander, as well as those the commander cannot control. This can only be achieved with a thorough integration of PA planning at all stages and into all aspects of the planning and decision-making process. Although a variety of techniques may be used in the analysis of the IE, the PA assessment should address the following primary categories.

- **Information Channels and Infrastructure**. This element focuses on an assessment of the information infrastructure. It addresses the resources, communications facilities, organizations, and official and unofficial information channels available within the area of responsibility (AOR). It addresses the means to transmit and receive unofficial information. It addresses specific requirements for American Forces Radio Television Service (AFRTS) information services and the availability of assets to meet theater requirements. It identifies the availability of host nation telephone service for voice and data transmission, the accessibility of audio/video channels, the prevalence of private communications devices such as cellular telephones, facsimiles, computers with modems, radios and televisions, and the nature of the information available through these information channels. It addresses alternate means of voice and data communications, whether military or government contracted, for use in the absence of host nation information channels and infrastructure. Much of the information required for this category may be obtained through civil affairs or psychological operations elements assigned or attached to the command and U.S.Information Service offices supporting consulates or the embassy within the area of operations.

- **Media Presence**. This is an assessment of the media presence in the area of operations prior to the introduction of American forces and an assessment of the expected level of media presence commanders should anticipate once deployment begins. It includes a description of the type of media (print or broadcast), the visibility of the media (local, national, or international; American or foreign), and the focus of the news media present (news or entertainment) covering the operation. The assessment of the media presence should address the authority under which media representatives are operating (open or closed borders, and free press or controlled press) and the reporters' degree of access to the theater of operations.
- **Media Capabilities**. This element is an assessment of the media's information collection, production, transmission and communication capabilities in the AOR. This element analyzes the technological capabilities of the media representatives present within the AOR. It describes their level of sophistication (if they must transport products out of the area of operations for transmission to parent media or do they have self-contained interactive satellite telecommunications access). It also addresses the media's level of logistics support and its potential impact on Army commanders who are required to provide the media free and open access to the AOR. It includes information about their transportation assets, resupply channels, and equipment maintenance requirements. Additionally, the media's general ability to provide their own security should be assessed.
- **Media Content Analysis**. Media content analysis is an assessment of news coverage, the media's agendas and an analysis and prioritization of the potential strategic and operational issues confronting the command. Media content analysis assesses what is being said, by whom, and how it is being presented. It is a constant process that must begin well before planning for a specific operation begins and continues through Mobilization, Deployment, Employment, Sustainment, and Re-deployment. Content analysis reveals the meaning, tone, and accuracy of messages, how the information was presented, and the cumulative affect of the information. A media content analysis will provide an evaluation of the quantity of coverage, both in and out of theater, and the nature of that coverage. This will assist the commander to understand the strategic context, the measure of success and the definition of an end-state for the operation as viewed from outside the command and the Army itself. It will also be an essential element of friendly information (EEFI), as explained in FM 3-13 (100-6), in determining objectives and strategies for communicating the Army perspective, and for working to achieve a balanced, fair and credible flow of information.
- The specific methods for conducting a media content analysis are explained in Appendix O.
- **Public Opinion**. A public opinion assessment surveys the national and international attitude about the operation and the command, leaders and soldiers conducting it. This assessment looks at the perceptions held by major audience and coalition groups, and the relative solidity or strength of those attitudes. It addresses the perceptions held by international

audiences: those traditionally allied with the United States and those traditionally considered to be adversaries of the US. The public opinion assessment should include as a minimum, consideration of the following groups:

- American public (general)
- Civilian political leadership
- Coalition and allied forces and their general population
- Host nation citizens
- International public
- Internal command audience
- Home station community
- Specific special interest groups (if needed)
  - In determining the effects of the media on public opinion, there are three general types of evidence which explain behavior response: direct indicators; indirect indicators; and post-event sampling. Direct indicators are evidence that provide a direct link between the information received by the public and the behavioral response. These indicators include but are not limited to: personal interviews and surveys to estimate awareness and understanding of an issue; dissident group marches, meetings, advertising and other activities; monitoring internal and external law, order and discipline activity; and chain of command after action reports, staff journals and duty logs. Indirect indicators are evidence that identifies behavioral response generated by separate events or activities which appear to be the result of reception of media information. These indicators include: cause-effect estimates from information products and sources other than the military or civilian commercial media; interest level in news media products; shifts in social or economic trends; shifts in political support.
  - Post-event sampling considers the qualitative and quantitative statistical evidence that identifies the level of and nature of awareness and behavioral response to information. This includes the results of surveys, interviews, group observation, probability and non-probability samples, which will identify if and how the public was influenced by information products or messages.

- Information Needs. This is an assessment of the information needs and requirements of the previously identified key publics. It analyzes and prioritizes key external and internal audiences and assesses their news and information expectations. It identifies the types of information that should be made available to soldiers, their family members, other home station community audiences, the American public, and the host nation local populace. It will identify other audiences, such as allied or adversary leaders and publics that will be interested in available "cross-border" information.

PA ESTIMATE AND PA GUIDANCE COORDINATION

3-27. The purpose of the PA Estimate is to determine whether the mission can be accomplished and to determine which COA can best be supported by public affairs. In preparing its estimate, the Public Affairs staff:

- Reviews the overall mission and situation from the public affairs and information environment perspective.
- Examines all public affairs factors impacting on or impacted by the mission.
- Analyzes each COA from the public affairs perspective.
- Compares each COA based on the public affairs functional analysis.
- Concludes whether the mission can be supported by public affairs, and from the public affairs perspective, which COA can best be supported.

3-28. The Public Affairs Estimate summarizes the information environment, prioritizes the major issues confronting the command and predicts anticipated outcomes in detail. It measures the effectiveness of previous and current information strategies, and based on this evaluation, identifies possible courses of action to support command PA objectives. The PA Estimate also contributes to the development of Public Affairs Guidance (PAG) for specific operations or missions. PAG is a primary tool that guides commanders and PA leaders in the application of doctrine and policy during operations. PAG provides the PA force at all echelons standard operating procedures.

3-29. But to be effective, PAG must be developed with the needs of the front-line PA force in mind. PA planners must be able to "see" and "feel" the battlefield. They must have an understanding of the information environment and how it will change throughout the operational continuum. They must be aware that all the resources available at the planning headquarters may not be available or feasible in the theater of operations. Issues that need to be addressed include information release authority restrictions (national, theater or local). These restrictions often place the PA leader in a difficult situation -- one in which an overwhelming number of news media on the scene will seek answers to legitimate questions about unfolding events -- activities that the PA leader cannot discuss. The result is a loss of credibility for the Army.

3-30. DOD policy requires that proposed PAG be provided to the Assistant to the Secretary of Defense for Public Affairs (OTASD-PA) by the unified, specified and other major commands for all operations. This requirement includes major joint training exercises that could attract national and international attention. Subordinate command PA leaders should conduct continuing PA assessments as a part of mission training for theater-specific contingencies in anticipation of PAG development requirements there.

PAG DEVELOPMENT

3-31. Upon receipt of a "warning order," the commander, through the PA staff, will begin development of proposed PAG. In reality this warning order may be preceded by a telephonic "heads up" call from a PA planner at a higher headquarters that allows PA planners to begin working on proposed PAG before the hard copy of the warning order arrives. This proposed PAG should be based on the warning order or other planning guidance, the proposed command operations plan (OPLAN), and the PA Estimate. Once the proposed PAG is developed, it is staffed through command staff. Once approved by the commander, it is forwarded through major command and Unified/Specified command PA channels to DoD.

3-32. Commanders of Unified/Specified commands should ensure the proposed PAG is coordinated with appropriate elements and functional staffs within the theater of operations. This includes governmental and non-governmental organizations such as the State Department and its embassies, civil affairs, country assessment teams, host governments, allied force public affairs teams, the four U.S.military services and their subordinate commands.

3-33. PA leaders at all levels, specifically major command and above, should work to approve PAG as quickly as possible in order to provide subordinate PA leaders the opportunity to develop and implement PA strategies to support their commands. The format for PAG is included as an appendix to this manual at Appendix E.

PA ANNEX DEVELOPMENT

3-34. Once the PA estimate and proposed PAG are completed, and the other staff officers have completed their estimates, the commander selects a course of action. The commander then outlines it to the staff. The commander may select one of the proposed COAs, a combination of two or more, or a completely new one. The PA staff must then be prepared to enter the plan development phase (Phase III) which requires development of a PA annex. A format for a PA Annex is included at Appendix D.

3-35. The operation-specific approach to conducting public affairs activities is called a PA scheme of maneuver. This PA scheme summarizes the commander's PA intentions, and details the media facilitation, news and information provision, and force training and support procedures, which will be employed to support a particular operation.

3-36. The PA scheme consists of the PA estimate of the situation, higher command PA guidance, and the selected course of action. It is coordinated with key staff agencies, integrated into the operation plan through the development of a PA annex, and synchronized with the other activities to be executed as part of the basic plan.

3-37. The PA scheme, when included in the PA annex, should not only identify public affairs force requirements for the operation, but more importantly, it must provide the commander a visual picture of how public affairs will support the commander's concept of operation as outlined in the plan.

3-38. The PA activities addressed in the PA scheme of maneuver are:

- **Media Facilitation**. Media Facilitation is activities executed to support news media efforts to cover the operation, facilitate the timely, accurate, balanced provision of information which communicates the Army perspective, and minimizes the media disruption of operations or endangerment of mission accomplishment. Media facilitation is accomplished by the early establishment of a media center as the focal point for media representatives seeking to cover the operation. Normal media center operations include scheduling briefings, coordination for interviews; responding to media queries; coordinating unit visits and media escort requirements; and resolving media - military incidents. To prepare for encounters with the media, commanders must accept and understand the role of the news organizations and the journalists in the theater, and their capabilities in getting information from the battlefield or area of operations. Commanders must provide media access to the force, keeping in mind the impact their technology will have on operations security. Commanders must identify and provide support and resources to assist the media in their mission.

- **Information Strategies**. Activities executed to fill the news and information needs and expectations of internal and external audiences. Proliferation of personal computers, the World Wide Web, the Internet, on-line services, fax machines, E-mail, cable television, direct broadcast satellites, copy machines, cellular and wireless communication and many other information technologies have created an endless stream of data and information that flow into a world filled with images, symbols, words, and sounds. Public affairs specialists acquire information using a variety of sources. Because of the volume of information and the vast number of potential distribution mediums, the PA staff uses a systematic acquisition strategy. They acquire information from participants, leaders, developed sources, the media, research and development, intelligence, culture at large, and subject matter experts. Print, video, audio and electronic information products are provided to deployed soldiers, home station audiences such as family members and the home station community and news media representatives using contracted services and organic military assets. They communicate the Army perspective and contribute to timely, balanced coverage of the operation.
- **Force Training and Support**. Activities executed to assist members of the DA community in interacting with media representatives. Force training and support are conducted to educate soldiers, family members and DA civilian employees on their rights and responsibilities with respect to news media representatives attempting to provide coverage of an operation and related issues. It focuses on helping them to respond when they encounter news media representatives seeking interviews, photo opportunities, responses, reactions, interpretations or comments on an operation, policies or events. The intent of force training and support is to assist members of the community and media representatives in approaching each other with mutual respect. Training for public affairs personnel expands on soldier and unit leader training. It stresses individual as well as collective tasks with an aim of developing units fully prepared to accomplish the range of public affairs missions. It integrates public affairs into the battle staff and trains PA planners to assess the operation environment from a public affairs perspective, produce a PA Estimate, develop the PA Annex and PA Guidance.

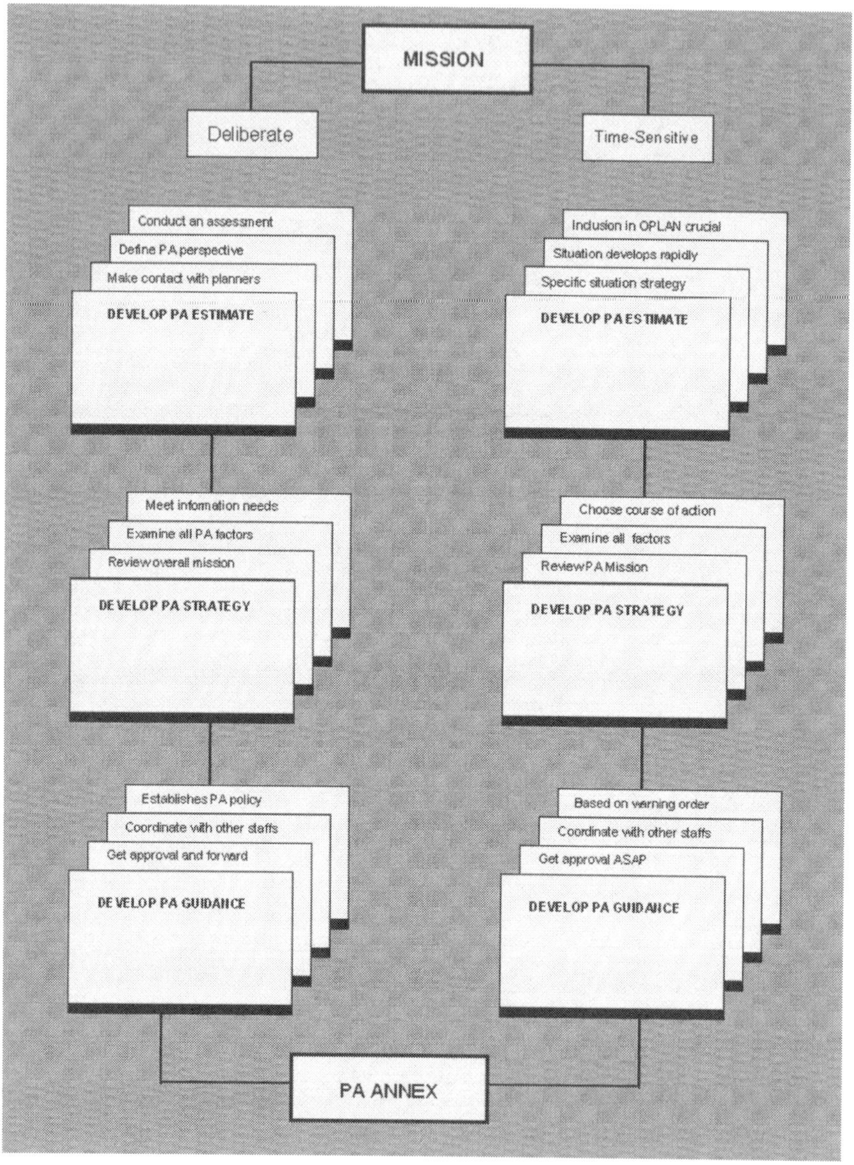

Figure 3-1. PA Planning

PA INTEGRATION INTO STAFF PLANNING

3-39. Concurrent with formulation of the PA staff estimate, PAG development, and production of the PA Annex to the OPLAN, PA planners must be an integral part of the staff planning process, especially on the following matters.

3-40. **Force Planning.** In force planning, the PA staff works with the J3/G3/S3 staff. Force planning consists of PA force requirements determination, force list development and refinements in light of PA force availability and PA force shortfall identification and resolution. In force list development, the PA assets needed to meet the mission are identified. Force availability is considered based on the strength and readiness of organic PA units, their personnel and equipment. Identification of PA force shortfalls addresses the lack of organic or mission-capable PA assets and the additional requirements and augmentations for PA units and personnel needed to accomplish the CINC's concept of operations. All taskings for unit or personnel augmentation must be validated and requested through the J3/G3/S3 operational channels. Tasking authority does not exist PA staff to PA staff or PA staff to subordinate unit.

3-41. **Support Planning.** To plan for logistical support of PA units and personnel assigned to carry out the CINC's concept of operations, the PA staff coordinates with and identifies support requirements to the J4/G4/S4. Specific logistical areas of concern include support in all classes of supply to the PA force, availability and authorized levels of support to civilian media, local purchase and contract support, property accountability, and vehicle transportation and maintenance support as tasked for through the J3/G3/S3.

3-42. **Transportation Planning.** PA forces move from their home station to a specified destination in the theater, either as part of their parent organization or a task-configured PA unit. This movement involves planning by several echelons of command, possibly stops at several intermediate locations en route, and a schedule constrained by a variety of operational requirements and priorities. Key staff for the PA planner to interact with include the command's transportation officer, movements control officer, and staff officers within the J3/G3/S3 and J4/G4/S4 that have staff supervision in this area. Key items PA planners need to track in this process are:

3-43. **Time-Phased Force and Deployment Data (TPFDD).** The TPFDD is the JOPES database portion of an operations plan. It contains time-phased force data, non-unit related cargo and personnel data, and movement data for the operation plan. The Appendix 1 to Annex A of the operation plan is the Time-Phased Force and Deployment List (TPFDL) which identifies types and/or actual units required to support the operation plan and indicates origin and port of debarkation or ocean area. It may also be generated as a computer listing from the TPFDD. PA planners must ensure that the TPFDD/TPFDL contains their unit line numbers (ULNs) for units, personnel, or cargo. Assets not listed on the TPFDD do not deploy. It is crucial to the planning process that the PA staff closely coordinate with the J3/G3/S3 and J4/G4/S4 to ensure that PA assets are reflected on the TPFDD or included as organic/attached assets to parent units with validated ULNs on the TPFDD.

3-44. **Destination (DEST)** - the geographic location where the force is to be deployed/employed.

3-45. The distances between the port of debarkation (POD) within the theater of operations to the destination (DEST), to the port of support (POS), to the marshaling area or assembly area. Where troops land at the APOD (Aerial Port of Debarkation) or SPOD (Seaport of Debarkation) they may be substantial distances from the port where the PA element's equipment arrives in theater and operations begin.

3-46. Transport of equipment must be planned for it to be available at the earliest possible date within the theater of operations. Thus, the PA planner must set a realistic, achievable required delivery date (RDD). This must be aligned with the CINC's required date (CRD). Planners begin with the RDD to establish two interim dates, the earliest arrival date (EAD) and the latest arrival date (LAD). Once these dates are established, then the ready to load date (RLD) and the available to load date (ALD) are established at home station to meet the earliest departure date (EDD).

3-47. **Communications/Automation Planning**. Key planners on the coordinating staff for communications and automation planning are the J6/G6, G3/S3 and CE officer. Specific concerns include priorities for radio/telephone communications, satellite uplinks and downlinks, number of telephone links/trunks allocated to PA requirements, E-mail access, and inclusion into the Communications Electronics Operating Instructions (CEOI). In addition, consider possible development of web pages or sites, like BosniaLink, the Task Force Eagle Homepage, the Desert Voice in Kuwait or Task Force Falcon in Kosovo.

3-48. **Information Environment**. When formulating PA plans and coordinating on the overall plan with the commander's staff, PA planners work closely with the staff element having supervising responsibility for each separate issue.

- Operational Security - G3/S3
- Psychological Operations - G3/S3
- Civil Military Operations - G5/S3
- Combat Camera Operations - G3/S3
- Armed Forces Radio and Television Operations - G3 and Armed Forces Information Service

APPLYING METT-TC

3-49. To function as part of a deployed or deployable organization the PAO and PA NCO must think and state requirements in terms that the rest of the organization can understand. Moreover, the PAO must fit the operational PA requirements into the operational planning procedure of the organization about to deploy. Operators think in terms of mission, enemy, terrain and weather, troops, time available and civilian considerations or METT-TC. METT-TC is used to envision how the operation will occur, to identify potential risks or hazards, and to define troop and equipment requirements.

3-50. **Mission** - alert, marshall, deploy, operate (internal information, media facilitation, information provision) redeploy

3-51. **Enemy** - rumors, disinformation, propaganda, OPSEC

3-52. **Terrain and weather** - theater of operation, theater of war, area of responsibility, intermediate staging base, homestation and weather condition

3-53. **Troops** - embedded assets, units (PADs, MPADs, BODs, PAOCs) AFRTS, HTNRs, ABS, NBS, JIBs, Star and Stripes, stringers, surrogate PAOs, Adjutants, and S1s. All other AG services, signal, USIS, DoS and homestation forces and audiences

3-54. **Time available** - timeline, transition, reports

3-55. Essentially, the information required for a METT-TC analysis is provided by the PA estimate of the situation, which contains the selected course of action and detailed descriptions of PA actions to be performed. These requirements are then translated into the command's planning language and format, resulting in the PA Annex to the OPLAN or OPORD.

3-56. At theater level and above, the PA annex is normally Annex F to the OPLAN. At corps and below, commanders can tailor their plans to fit specific needs or preferences, so the PA annex may fall in another location among the annexes. Regardless of where it is located, the PA Annex is used to provide information about the conduct and execution of public affairs operations in support of the basic OPLAN. The PA annex outlines the situation, identifies the specific PA mission and explains the concept of the operation. It also provides detailed information and guidance PA personnel need to conduct successful PA operations at the operator level. A sample PA Annex format is included in this manual at Appendix D.

3-57. Phase IV, Plan Review, consists of staff coordination and plan adjustment or correction.

3-58. The final phase of the planning process, Phase V, Supporting Plans, follows the same course as the first three phases, with attention aimed at the specific aspects of the overall plan. These supporting plans focus on conducting specific operations, which must be successful in order to guarantee success of the larger mission. PA support to these supporting plans is as important as PA coordination and input to the main campaign plan.

STANDING OPERATING PROCEDURES

3-59. Integral to the operational effectiveness of PA sections are their standing combat operating procedures. These routine procedures ensure that all members of the section are working in concert toward the same PA objectives and that PA activities are easily blended into the actions of the command's staff.

3-60. PA SOPs differ from PA plans and PA annexes to OPLANS in that they specifically detail and describe how PA is conducted within a certain command or unit. They are routine procedures and actions that apply to each section or unit.

3-61. The senior PA NCO prepares the staff section or unit PA SOP. PA units designated to support or augment specific commands in the execution of contingency missions should use SOPs from these supported commands.

3-62. SOPs should address:

- Preparation for combat. Stockage, prepackaging, and maintenance of vehicles, equipment, and expendable and nonexpendable supplies.

- Vehicle load plans.
- Alert and mobilization actions, routines and procedures.
- Composition of quartering and/or advance parties and rear echelons.
- Organization for combat, including detailed delineation of duties for each individual, shift compositions, and plans for reconstitution in the event of combat losses.
- Operations center and media center layouts (theater, corps and/or division main/rear CPs).
- Procedures for preparing, disseminating and disposing of records, reports, estimates and orders.
- Physical, document, and tactical security.
- Communications procedures. These steps include radio/telephone operating procedures unique to the command, message routing and preparation formats, and operation of communications and data transmission equipment.
- Movement and displacement.
- Operations under NBC conditions.
- Field Maintenance.
- Personal hygiene, rest, and morale, welfare and recreation requirements and procedures during deployment.
- Post-operations and reconstitution procedures. Maintenance, restocking and packaging composition of advance and rear parties; disposition of records, and preparation of after-action reports are included.
- A PA SOP outline is included in this manual at Appendix K.

POST-MISSION PLANNING CONSIDERATIONS

3-63. During mission planning and preparation, Public Affairs planners should consider debriefing and other post-mission activities. These activities normally include:

- Collective debriefing of the operational element on all aspects of mission execution, including lessons learned.
- Collection of maps, notebooks, logbooks, plans, annexes, duty officer/NCO logs, serious incident reports, news releases, tapes and transcripts of news briefings and conferences, and all other information products pertinent to the mission after action report.
- Maintenance and storage of unit and personal equipment.
- Individual debriefing of key personnel.
- Other reconstitution measures as required.

3-64. Upon completion of these activities, the operational element begins pre-mission sustainment training or prepares for its next mission. The planning staff begins review of lessons learned for integration into future plans. See Appendix W for information on producing PA Lessons Learned for the Center for Army Lessons Learned.

Chapter 4

Media Facilitation

"It is likely that small pools of news media will be assigned directly to operational units to cover all facets of activity. With few exceptions, there will be no security review of media copy or audiovisual products. The policy will be to maintain security at the source. It is important to support the efforts of the media and our dealings with them should not be confrontational, but professional and courteous."

- GEN Binford Peay
Commander, U.S. Central Command [1994]
FM 3-61 (46-1), Public Affairs Operations

PRINCIPLES OF INFORMATION

4-1. In the past 20 years the Army has undergone a fundamental shift in our approach to dealing with the news media. In response to the perceived treatment by the press during the Vietnam War, we have gone from adopting an exclusionary tactic for the conduct of the invasion of Grenada in 1983, to managing the controversial pool system for covering the initial stages of the Gulf War in 1991, and more recently evolving into an almost completely open access policy in Somalia (1993), Haiti (1994) and Bosnia (1995) operations.

4-2. The results of this policy evolution are the "DoD Principles of Information," which form the foundation for the PA function of media facilitation. The basic approach that DoD and the Army take to media facilitation is contained in Appendix A, The DoD Principles of Information and Appendix B, The Guidelines for Coverage of DoD Combat Operations.

4-3. Simply stated, media facilitation is providing assistance to civilian and military news media representatives covering an operation. The objective of media facilitation is to support news media efforts. This includes providing accurate, timely, balanced, credible coverage of the force and the operation, while minimizing the possibility that media activities will disrupt the operation.

4-4. Media facilitation includes assisting media entry into the area of operations, registering media representatives, orienting them on the ground rules for coverage and ensuring that they understand the security policies and constraints under which they must agree to operate if they desire Army support.

4-5. Media facilitation also involves arranging interviews and briefings, coordinating unit visits and escorts, and assisting media representatives with transportation, messing, billeting, communication support, safety and equipment. Media facilitation involves the early establishment of a media center as a focal point for media wishing to cover an operation, for Army personnel seeking assistance with media representatives in their area, and for resolution of problems or incidents resulting from media-military interaction.

4-6. A primary strategic goal of any Public Affairs staff is to support an operational commander in achieving a constant flow of complete, accurate and timely information about the mission and U.S. forces.

4-7. The PA staff accomplishes this goal by making information fully and readily available within the constraints of national security and OPSEC, and by facilitating inclusion of civilian and military news media representatives in military units whenever possible.

PA STAFF AND MEDIA CENTER RESPONSIBILITIES

4-8. **PA Staff Sections.** The prime focus of the PA staff is staff coordination. The staff will be the element tasked with executing the media facilitation strategy.

4-9. The staff provides PA planning and operational guidance to the PAO. They ensure leaders within the command understand the commander's media relations policies, and serves as the command ombudsman in the settlement of conflicts between the media and the military.

4-10. As an active participant in the command's information planning element, the staff coordinates with G2, G3, G5, PSYOP, U.S. Information Service, and other staff elements in developing the commander's information strategy to ensure synergy and to reduce the probability of conflicting messages.

4-11. **Media Operations Centers.** Currently, joint, combined and Army media centers fulfill the requirement for a focal point for the news media during military operations. In essence, the media operations center (MOC) is a command post for media support efforts. It serves as both the primary information source, and as a logistical support and coordination base for commercial news organizations covering the operation.

4-12. Media centers are organized when large numbers of news media representatives are anticipated to cover military activities. Media centers may be formed for all types of operations or for any stage within an operation.

4-13. When operated by unified/specified commands, these media centers may be called a Joint Information Bureau (JIB). At the combined commands, they are called an Allied Press Information Center (APIC), Coalition Press Information Centre or Combined Information Bureau (CIB). At theater level and below, they are simply referred to as Media Operations Centers.

4-14. MOCs support the commander and are subordinate to the command's PAO. They provide the commander a professional, immediately available, fully trained organization designed to respond to national and international civilian media interest in American military operations.

4-15. In addition, the media operations center provides the following functions:

- A single point of contact and information source for media within the theater
- Briefings and enforcement of media guidelines and ground rules
- Primary information release authority for the senior PAO
- Coordination of news media coverage with corps, divisions, brigades, etc.
- Coordination with all service branches for each service, agency or country
- Identification and communication of host-nation sensitivities to all personnel in theater
- Preparation for and conducting press briefings and news conferences
- Registration of news media personnel
- Media Operations Center Staffing and Organization

4-16. Organization and personnel staffing of media operations centers are determined by the responsible command in coordination with the PAO and his staff. APIC staffs should be a proportionate representation of the forces, with representation from all services involved in the operation. This will be determined by CPA at the unified command. Regardless of the echelon establishing a media operations center, the organizational model is functionally designed and remains relatively the same.

4-17. MOCs normally consist of two major elements: a Headquarters Group and a Media Operations Center Group.

4-18. The headquarters is made up of the command group and support staff. The command group contains the commander, deputy commander and/or executive officer, and the sergeant major. The support staff is normally task organized to support tailored forward deployed MOC teams or sub-MOCs when the APIC operates as other than a single element. The support sections provide administrative support, conduct lease and purchase contracting, setup, operate and maintain the unit's equipment, and conduct the day-to-day operation of the MOC. The support sections are responsible for the execution of MOC communications, supply operations, administration support, vehicle maintenance, security and other support functions as required.

4-19. The Media Operations Center consists of a Plans Section, a Media Support Section and an Information Operations Section.

4-20. The plans section is responsible for all MOC media planning. It establishes MOC requirements and determines operating procedures and policies. It maintains channels of communication with OASD(PA) and the JPAO (or senior command PAO). It is responsible for recommending and assisting in the development and dissemination of PA Guidance. It monitors available major U.S., international and local television and radio broadcasts and print publications providing coverage of the operations, conducts news media analysis and evaluates the effectiveness of MOC operations.

4-21. The Media Support Section (MSS) is the primary point of contact for news media representatives (NMRs) in an area of operation seeking information or assistance in covering the force and the operation. The MSS receives and registers NMRs, briefs NMRs on the media ground rules and security procedures or concerns, and orients them on the force, the operation and other pertinent issues (special safety or host nation considerations).

4-22. The MSS orchestrates the command's news briefings and coordinate for subject matter experts to explain and discuss operations and capabilities. The MSS is also responsible for coordinating for appropriate, knowledgeable escorts, unit visits, and service member interviews. It assists the Joint Force or other senior PAO in preparing service members for interaction with the news media. Finally, it provides support to Joint Force elements and service component PA elements seeking assistance with NMRs.

4-23. The Information Operations Section is responsible for monitoring plans and operations from within the command's operation center and assessing the PA implications of events occurring throughout the area. It ensures that the MOC has current situation information, is aware of issues of potential media interest, and can obtain any operational information necessary for the development of responses to media inquiries in a timely manner.

4-24. The IO section ensures that PA operations are synchronized with other combat functions and promote early coordination of PA, CA and PSYOP functions.

4-25. **MOC Staffing**. Currently, media operations center staffing requires augmentation, either by PA-trained individual fillers or by Army PA units. As fully independent units, the Public Affairs Operations Center (PAOC) (SRC 45423A000) and Mobile PA Detachment (MPAD) (SRC 45413A000) are currently organized, trained, and prepared to fill this role.

4-26. In fact, these Army PA detachments are specifically designed to function as an Army media operation center in theater, corps, or division-controlled operations. MPADs can be combined to form media sub-centers in forward battle areas.

4-27. PA personnel from non-deployed commands and installation PA sections may be called upon to augment news media centers however, requests for individual augmentation should be coordinated through operational channels. Reserve and Guard unit personnel can be used to augment on a voluntary basis.

4-28. An example of a media operations center is included at Appendix L.

MOC OPERATIONS

4-29. In major operations -- actions conducted by unified commands -- a Joint Information Bureau will usually be the first to deploy for this purpose. A JIB will be staffed by public affairs personnel from the services represented in the joint force; participating services may establish their own media centers subordinate to the JIB to disseminate information about their particular missions.

4-30. As the operation unfolds, the Army plans for and contributes to a replacement PA organization for the JIB which consists of individual PA personnel from each of the services and Army PA detachments.

4-31. Media Center operations will be based on five primary assumptions:

- Accurate information is available in a timely manner and adheres to the DoD Principles of Information in Appendix A.
- Current trends in communications technologies within the information environment will continue to reduce the news media's reliance on military support and assistance when covering operations and will continue to increase the availability of information to a worldwide audience.
- Media representatives will be in an area of operations at the start of, and in most cases, before an operation begins.
- Media interest and coverage in non combat operations may be higher at the outset, and barring a significant event which renews national or international attention or interest, will taper off over time. During a high-intensity conflict, media interest could remain high.
- Military PA elements require access to complete information, state-of-the-art communication equipment, and must possess sophisticated coordination channels in order to pre-empt speculative, inaccurate or biased reporting.

4-32. Media centers will support and be responsible to the senior commander of the operation on a 24-hour basis. Media centers are usually established by unified command CINCs to support the news media in an area of operation.

4-33. During the first 24 hours after arrival in a new theater of operation, a media center can provide limited media support services.

4-34. Within this first operational day, the MOC must:

- Establish a "hasty media center" as the initial focal point for the news media until additional media support forces arrive.
- Establish communication with OASD (PA), each service's PA chain of command, and with units operating within the theater.
- Request operational information release authority within the theater.
- Establish command structure/lines of authority.
- Coordinate with appropriate authority for leasing and purchasing contracts.
- Begin to register news media personnel in the area
- Provide basic media support (coordination of media access to subordinate units and media escort as resources permit).
- Assist or conduct command news briefings and conferences.
- Coordinate Subject Matter Expert (SME) interviews.
- Be capable of assisting in the transmission of media products.

4-35. Media Support -- Initially, the media operations center will need to provide varying degrees of support to news media personnel including specialized equipment (flak vest, NBC gear, helmets), transmission of media products, etc.

4-36. This support may include but not be limited to:

- Coordinate media contact with units or individuals to include SME interviews
- Provide a single point of contact for information on operational issues
- Provide news releases, fact sheets, copies of transcripts for news briefings/conferences and copies of archival file products
- When other means are not available, the media center may provide coordination for transportation (to and from interview sources), transmission of media products and food and billeting
- Provide limited media escort within the area
- (SOPs for MOCs should be pre-established for each theater of operation and used for media operations within that theater.)

REGISTRATION OF MEDIA

4-37. Principle to supporting the commander's information strategy is the inclusion of news media representatives (NMR) within Army units from the earliest pre-deployment stages of all operations. The personal safety of media representatives, as acknowledged by the media themselves, is not a reason for excluding them from operations.

4-38. However, all media requesting support or access to units to cover Army operations must be registered. This includes freelance journalist, military media representatives, such as those who are assigned to Armed Forces Radio and Television Service, Stars & Stripes newspaper and other Armed Forces Information Service (AFIS) personnel who are not supporting units on the battlefield.

4-39. **Registration versus Accreditation**. Accreditation is the verification and validation that a person represents a legitimate commercial news organization. This means that accrediting governments or military organizations will physically verify the affiliation of an applicant with a specific news organization.

4-40. This is difficult to perform amidst an ongoing operation, especially when deployed far away from CONUS. It is generally accepted that, when overseas, the decision to accredit news media is made by the host nation's government in coordination with the combined or unified commander.

4-41. When accreditation isn't required by the host nation, responsibility for this determination is held by the combined or unified commander. Accreditation is normally performed at Corps level or higher.

4-42. Accreditation is a major problem for many commands because they are forced to determine the legitimacy of smaller, lessor-known news organizations and freelance journalists without news organization affiliation.

4-43. Unless it is absolutely required by host nations, the American military will attempt to avoid accreditation.

4-44. Registration, however, is merely an accounting tool, which provides PAOs the ability to know what media are represented in the theater, where they are located, and their movement around the theater. This information is helpful in planning and conducting media logistical support and transportation, and in preparing subordinate commands for media encounters. It is also helpful to commanders who might want to provide newsworthy events to the media.

4-45. Registration also identifies which news media have asked for military assistance and access, and have agreed to the command's media ground rules.

4-46. **Registration Requirements.** The registration process is conducted in five basic steps:

- Verify the identity of the media representative (including checking for valid passport/visa, professional media organization membership card, media ID card, other military press credentials, etc.).
- Have them sign an agreement to abide by the established media ground rules for the operation in exchange for granting support, access to units, information and other privileges. If required, revoke credentials for those who violate the ground rules. (Enforcement of this requirement is essential.)
- Have NMR agree to and sign a liability waiver that frees the military of responsibility if the NMR is killed or injured as a result of covering the operation. (An example of a waiver of liability is at Annex I).
- Give NMRs proof of registration (memorandum, press badge or other identification).
- Maintain a roster of registered NMRs and monitor their movements during the time they are receiving military support.

4-47. NMRs who refuse to agree to the military ground rules and who are not registered will receive only the support and information assistance as provided to the general public.

4-48. NMRs should be informed that registration and acceptance of media ground rules will entitle them to better access to units and subject matter experts, and provision of military ground and air transportation when possible.

MEDIA GROUND RULES

4-49. Media ground rules will assist inprotecting the security and the safety of the troops involved while allowing you the greatest permissible freedom and access in covering the story. All interviews with news media representatives will be on the record.

4-50. Security at the source will be the policy. (An example of media ground rules is in Appendix X.)

4-51. The following categories of information are releasable:

- Arrival of major U.S. units when officially announced by a U.S. spokesperson. Mode of travel (sea or air) and date of departure from home station

- Approximate friendly force strength figures, after review by host nation government
- Approximate enemy casualty and POW figures for each action operation
- Non-sensitive, unclassified information regarding U.S. air, ground and sea operations (past and present).
- Friendly force size in an action or operation will be announced using general terms such as multi-battalion or naval task force
- Specific force/unit identification/designation may be released when it has become public knowledge and no longer warrants security protection
- Identification and location of military targets and objectives previously under attack
- Generic origin of air operations such as land or carrier based.
- Date/time/location of previous conventional military missions and actions as well as mission results
- Types of ordnance expended will be released in general terms rather than specific amounts
- Weather and climate conditions
- Allied participation by type of operation (ships, aircraft, ground units, etc)

4-52. Information Not Releasable

- Information about future military plans, activities or operations
- Vulnerabilities or weaknesses on command, control, personnel or the operation
- Friendly unit and command strengths, on-hand equipment or supplies; the presence, activities and methods of operation of specifically designated units or equipment
- Information on friendly force security and deception measures and countermeasures, and intelligence collection activities
- Specific information on friendly force current operations and movements, deployments and dispositions
- Information on in-progress operations against hostile targets
- Information on nuclear, biological or chemical weapons, equipment or training

MEDIA POOLS

4-53. Journalists, as a group, are strongly opposed to media pools in any form. The media pool is seen as a restriction placed on the media representatives and their ability to provide coverage of the news. They are grudgingly tolerated, and should be only used as a last resort when space onboard military transportation is limited, access to an area must be controlled, and after all other possibilities have been explored and eliminated. Even under conditions of open coverage, pools may be appropriate for specific events. Both the Army and the news media are in agreement, however, that limited access is better than no access at all.

4-54. When a pool system is required, the military PAO will identify the maximum size of the pool that can be supported. The news media representatives on the scene will select media pool members. A roster of media personnel registered with the Army PAO will be used to identify the media representatives eligible to participate. The pool should consist of, but not be limited to, a minimum of one video crew (camera operator, sound technician and reporter), one still photographer (wire service, newspaper, or magazine), one radio reporter, and one newspaper or wire service reporter. Special consideration must be given to international reporters as well. While this is a fair and representative pool structure, it is the media themselves who must determine the make-up of the pool. Some news events and situations may lend themselves more to print, or conversely television reporting, and the media representatives may choose to select an unbalanced pool.

4-55. All pool members must be willing and able to meet deadlines and supply information products (video, audio, still media, and text) in a timely manner to all media representatives who are entitled to material generated by the pool. The military media center will also have access to this information and will make it available to all other requesting news media organizations.

4-56. Consistent with its capabilities, the military will supply PAOs with facilities to enable timely, secure, compatible transmission of pool material and will make these facilities available whenever possible for filing independent coverage. In cases when government facilities are unavailable, journalists will, as always, file by any other means available. The military will not ban communications systems operated by news organizations, but electromagnetic operational security in battlefield situations may require restrictions on the use of such systems.

4-57. Once a media pool has been selected, the media pool will select a team leader. It is the responsibility of this team leader to ensure that members of the media pool meet their obligation to share information. The Army PAO will not involve himself in settling internal disputes of the media pool.

4-58. Finally, the pool is an option of last resort. It should be disbanded as soon as free and open access to the operational area can be allowed, normally within the first 24 hours of an operation.

THE DOD NATIONAL MEDIA POOL

4-59. The DoD National Media Pool was established to prevent recurrence of problems encountered with media coverage during Operation Urgent Fury in Grenada in 1983. During the first 24 hours of Urgent Fury, more than 600 reporters attempted to gain access to the operation. The large numbers overwhelmed the limited Public Affairs elements available to assist them.

4-60. In 1985, the Secretary of Defense established the DoD National Media Pool, a civilian news element of approximately 16 media representatives from various national news organizations, with the mission of covering an operation from its initial stages until open coverage could be allowed.

4-61. The pool members remain on call in Washington, D.C., and are available for immediate worldwide deployment. Their products are shared by the open news media until the pool is disbanded and access is granted to the entire news community.

4-62. Supported commanders are responsible for providing operational support to the DoD National Media Pool. At a minimum, the pool members will require:

- Daily, comprehensive and unclassified operational news briefings.
- Access to ongoing combat operations. The media are aware of the personal risks involved in covering combat operations. They will not be denied access to them based on risk to their personal safety.
- Reasonable access to key personnel. All information gathered from these personnel is unclassified and on the record.
- An escort -- usually a lieutenant colonel or colonel -- to coordinate pool support and access requirements.
- Transportation and itinerary planning and coordination that will allow media to gain access to the theater of operations and to disperse pool members throughout the operational area.
- In today's global information environment, when news media can report live from almost anywhere in the world in almost any environment, the technological capabilities of most news organizations decreases the importance of the DoD National Media Pool once word of an operation has spread.
- When the DoD media pool is operational, PAOs will attempt to provide the same information support concerning theater operations to all other media in the operational area.
- However, their primary responsibility is to the DoD Media Pool. After the DoD media pool is dissolved, all media in theater will be dealt with in an equitable manner with respect to information and support provided.
- As soon as open access to the operational area can be allowed (normally within the first 24 hours of an operation), the DoD National Media should be disbanded.

NEWS BRIEFINGS

4-63. There are several reasons for holding news briefings, in addition to the daily operational news briefing required at the unified command level.

- Credibility: The physical presence of a briefer and his willingness to meet the issue head on leads to a much more credible presentation
- Uniformity: All media get the same information at the same time.
- Expression of concern: A briefer represents the face of the command, which shows more concern than an impersonal news release, especially in situations where there is loss of life or extensive damage.
- Complexity of material: Where material is technical or complicated, the news briefing makes the subject matter more easily understandable. The question and answer session that accompanies a news briefing saves time in call-backs by news reporters needing clarification.

4-64. News briefings should be done daily during an operation and when important events dictate. They should:

- Get out a specific message
- Explain complex or technical matters
- Reach a large number of media interested in the same subject matter area.

4-65. PA specialists should think about media deadlines and set the time to help the media meet those deadlines. Be sure to invite all media within the area in a timely manner.

4-66. A knowledgeable and articulate spokesperson should be chosen to present the material. This should be the subject matter expert (SME), but may be the PAO or the commander. At the very minimum, a person of prominence within the command should be selected.

4-67. Other SMEs may be in attendance at the briefing to field technical questions. The SME interviews should be at the request of an individual media representative and the time should be set to facilitate the media to meet those deadlines.

4-68. Consideration should be given to the appearance of the presentation, the message, space, lighting, electrical needs, suitable setting, chairs, tables and press packets. All handouts should be reviewed.

4-69. Appendix H provides briefing and press conference formats.

Chapter 5

Information Strategies

The defining trend of the 1990s, from corporate boardrooms to private living rooms, is connecting everyone and everything to everyone and everything else. The American public is faced with many choices: interactive television, cellular phones, modems, faxes, personal digital assistants, an assortment of daily newspapers, online computer information services and other easily accessible information services -- if it can be connected, it seems to be in demand. The computer has invaded the American home. In this decade it has become routine that news and information can now be tailored to fit the individual needs of the consumer using on-line services, distributed electronically and received instantaneously at his or her computer. With the sophistication, power and miniaturization of these technologies improving each year, and the cost dropping at an equally rapid pace, the public will expect access to these devices and services and the information they carry. These emerging technologies have contributed to the refocusing of the Public Affairs mission. It has resulted in shifting the information provision function from an emphasis on producing specific products (such as post and field newspapers and radio/television news programs) to focusing on the processing of our themes and messages and their intended effects -- the function of information communication, rather than the form. This chapter explains the objectives of information strategies, identifies and explains the elements of information strategies, and describes the relative advantages and disadvantages of present day and emerging information communication channels available to PA organizations. Most importantly, it explains how best to use these information channels to satisfy the information needs of the various target audiences as we enter the information age.

INFORMATION STRATEGIES

5-1. Information Strategies is the sum of all actions and activities, which contribute to informing the American public and the Army. The responsibility for this activity is assigned to an element within each PA section, which focuses entirely on accomplishing the information strategy mission. This section is usually called the Public Affairs Information

Services Section. At all echelons, it employs numerous techniques to provide news and information to internal and external audiences. The Army provides an expedited flow of complete, accurate and timely information, which not only communicates the Army perspective, but also attempts to educate audiences and engender support for the force.

5-2. Using a combination of contracted services, organic military assets, and government and commercial communications networks, Public Affairs organizations provide information to news media representatives, deployed soldiers, home station audiences and the American public. The Information Services Section within a Public Affairs organization coordinates information efforts and develops informational products (such as digital text, graphics, and photos, printed publications, audio/video news releases and graphic imagery) into consolidated campaigns designed specifically to present the Army's perspective. This means that Army Public Affairs communicates information to create an informed American public and Army force, assist them in gaining a clear understanding of the strategic, operational and tactical situation.

5-3. To accomplish this, the Information Services Section must develop information objectives or goals during the planning process prior to an operation. These information goals are similar to the PA standards of service and support, which appear in Appendix Q, in that they establish a basis for determining successful information communication operations.

5-4. These information objectives should include:

- Ensuring an understanding of the role of America's Armed Forces in American society.
- Ensuring an accurate perception of the particular military situation or mission.
- Ensuring an understanding of individual and unit roles in mission accomplishment.
- Establishing confidence in America's Army to accomplish the assigned mission in accordance with our national values.
- Establishing confidence in and support for American soldiers.

5-5. By establishing a comprehensive information strategy program, Public Affairs can assist in mission accomplishment by increasing audience understanding of the situation and establish confidence in and support for the force. This contributes to unit cohesion and provides commanders with increased range of action, free of distractions and limitations.

5-6. This is best accomplished by three basic types of information campaigns:

- **MISSION**. Both external and internal publics need to know what the mission is, what they're being asked to do and why. They need to know not only the organization's mission, but also how it fits into the big picture -- the political/strategic-level situation, and why it is important.
- **ROLE.** All military members and civilian employees need to have an understanding of their job and how it relates to mission accomplishment. The general public needs to have an accurate understanding of the military's role and its ability to accomplish

the mission. This understanding results in confidence in the force and demonstrates American unity and resolve.

- **MORALE.** Military members need to have access to news and information about current events and the activities available to them while deployed. They also need to have access to information from civilian commercial news sources. This is important because, in addition to being more credible, it allows the deployed force to see how the operation and their participation in it are being portrayed for the American public. In order to better understand the mission, their role in it, and give it his or her full effort, they have to know what effect the operation is likely to have at the local, regional, national and international levels. The opportunity to involve themselves in educational and other activities is necessary to quality of life and morale. A well-informed service member is more effective.

5-7. The general public is interested in soldiers, their lifestyle, how they are being treated and their ability to accomplish a given mission. Information about these topics provide reassurance, confirming that soldiers maintain professional and ethical values and are being cared for adequately.

Historical Perspective

During Operation Desert Shield/Storm, many commanders developed innovative methods of sharing Command Information products produced in theater and in the rear. The products greatly enhanced morale at both ends. Some commanders and PAOs used the products as issue management tools to dispel rumors in theater and at home station. The products included field newspapers, newsletters, videotapes, audiotapes, etc. They also let soldiers returning from the area meet with family support groups to answer any questions.

(After Action Report 1991)

RESPONSIBILITIES

5-8. Commanders at each echelon are responsible for Public Affairs operations and support. Public Affairs officers and noncommissioned officers at various levels assist commanders in the discharge of these responsibilities. Public Affairs staffs are responsible for accomplishing the Public Affairs information communication mission. This responsibility includes Public Affairs operations in all subordinate, assigned or attached commands.

5-9. PA is only one of many information channels available to a commander. PA information provision cannot substitute for a commander's personal involvement in his "Command Information" program.

5-10. All public affairs practitioners have access to all information that is not classified or violates Operational Security or the Essential Elements of Friendly Information (EEFI) for use in preparing information products. Commanders must ensure the EFFI is up-to-date based on current situations and operational guidance. Public Affairs personnel at all levels must produce and release accurate information packages based on DoD directives and Army policies.

5-11. **Strategic Level Commands** are responsible for providing public affairs guidance to subordinate units. They develop central themes and messages and provide umbrella guidance to subordinate PA staffs. They must also provide subordinate commands with information useful in preparing information products for internal and external release. They are additionally responsible for marketing public affairs information products to subordinate commands, home stations, the Army as a whole, as well as the general public. Strategic level Public Affairs staffs are the primary coordination point for the Armed Forces Radio and Television Service and the geographical manager for radio and television services including personnel, down links, facilities and equipment.

5-12. **Operational Level Commands** are responsible for communicating public affairs guidance to subordinate units. They expand on information campaign themes and messages, and provide additional information products to subordinate command PA staffs. Additionally, they provide subordinate commands with information useful in preparing information products for internal and external release. They are responsible for gathering and producing public affairs information products for release. In the event the Operational Command is the senior command in an area or theater, it assumes the responsibilities of the strategic level command.

5-13. **Tactical Level Commands** are responsible for gathering information products for release through their next higher headquarters to home stations, the Army as a whole, as well as the general public. These commands are also responsible for coordinating the dissemination of information and information products received from senior commands down to subordinate commands. In the event the tactical command is the senior Army command in an area or theater, it assumes the responsibilities of the operational-level command and will be augmented to accomplish these additional functions.

INFORMATION STRATEGY PROCESS

5-14. The Information Services Section uses all available means to gather complete, factual, unbiased information for use in information campaign development. The information is developed, converted into the most appropriate product form based on the information needs/target audience assessment and information communication channel availability, and then transmitted to the intended audiences. This is called the Information Provision Process.

5-15. Although the information strategy process follows a deliberate cycle, it is a continual process. Information campaigns are also conducted simultaneously, with personnel examining external and internal information needs, carrying messages from concept through execution and program evaluation, to accomplish specific PA objectives. The cycle has four phases -- acquire, process, protect and distribute. (See Figure 5-1.) Evaulation is a key component in the cycle. It must be conducted throughout the four phases. This ensures the campaigns are meeting their objectives, and are altered if they do not.

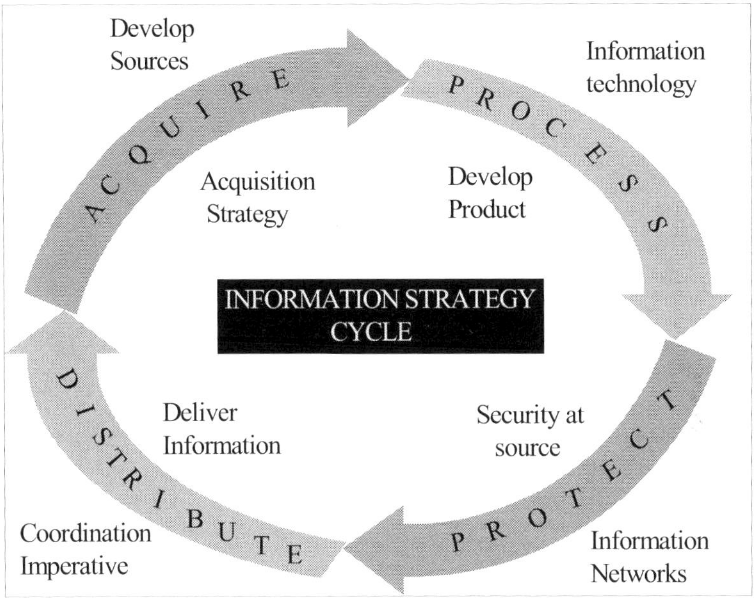

*Evaluation must be conducted throughout each phase

Figure 5-1

ACQUIRE

5-16. During this phase, a ISS identifies and assesses several factors: the situation, the environmental factors, the mission, and the target audiences. They determine the information needs of the various target audiences. They then begin to gather information based on the information requirements of these audiences.

- ***Information Needs Assessment and Audience Analysis.*** Identification of the target audiences and target audience populations and densities must be developed during the planning stages of operations with target audience assessments. Upon assessment of target audiences and consideration of which type of information product will best serve each audience, commanders must ensure adequate public affairs personnel, equipment (to include communications), resources and funds are available and included in OPLANS/OPORDS to achieve mission success. An assessment of soldier information needs is crucial to the information package development and the selection of appropriate products.
- ***Information Gathering.*** Information gathering is the first step in producing packages for release to internal and external audiences. Public affairs personnel gather information through

operational, administrative, logistical, battle, staff, command and support channels. They also acquire information through the media, research, leaders and the culture at large. Information may be obtained via electronic and telecommunication systems in addition to written documents and oral communications. While information often comes from superior and subordinate commands, it may also be obtained laterally.

- Sources of information must be valid and diverse enough to provide a broad overview. Public Affairs staffs must ensure the source does not speculate, nor speak out of his area of expertise. However, soldiers' experiences and personal opinions may lend credibility and provide a "grass-roots" view.

PROCESS

5-17. During this phase, the ISS begins to develop products, prepare them for release and determine methods for distrubition.

- **Product Development.** Development of information products is performed to some degree by PA personnel at all Public Affairs levels and is an on-going process. The term "product," as used in this process, means the message for intended communication, regardless of the format or communication channel proposed. Initial production development may be command directed or initiated by the Public Affairs staff or provided by other Army agencies. The command's resources and the target audience's requirements will determine the product type.
- **Prepare products**. Information packages should be prepared in accordance with Public Affairs Guidance, Soldiers' Manuals, Field Manuals, SOPs and Army Regulations. Ultimately, the Public Affairs staff is responsible for content.
- **Media Forms/Methods.** Soldiers use a variety of technologies to gather information and produce information products. During the gathering process, PA soldiers conduct interviews, attend briefings, withdraw data from government and commercial computer databases, bulletin boards, and e-mail systems. They acquire text, graphics, photography and motion video from government and commercial Internet systems. The nature, distribution, usability and flexibility of public affairs systems are crucial in the processing of information.
- Professional quality systems should be used whenever possible. For printed products, preferred systems include computers, desktop publishing, word processing, laser printing, etc. Reproduction may be Army-contracted, Army-funded or reproduced using the command's assets. Video and audio products intended for release to news media should attempt to be broadcast-industry standard.
- Electronic newsgathering and editing systems should be used when available. Visual products should be generated by modern methods including digital imagery and computer graphics.
- **Print.** Articles released to home station for military publications and family support group publications; for marketing to civilian publications; field publications, e.g. newsletters, with and without photography.

- **Video.** Raw electronic news gathering video and printed news scripts for release to military and civilian outlets.
- **Audio.** Radio interviews; features; internal command information scripts; radio news; news reports for release to military and civilian outlets.
- **Visual.** Digital imagery; photographs; slides, view graphs; graphics for release to military and civilian print and broadcast media.
- **Digital.** Each of the categories described above may be developed and distributed electronically, either through commercial information services directly on the Internet, or by using tactical Army communications systems (SINCGARS and ATCCS). Modern technology in use on the battlefield has made digital transmission the preferred method for all types of products.

PROTECT

5-18. **Security at the source.** No information strategy is complete without a clear cut understanding of how to protect the information. Both sides can benefit from information and use information simultaneously against each other. Pieces of the right information can have a dramatic impact on the outcome of an operation. Public Affairs professionals will continue to protect vital information by practicing security at the source and following established operational security guidelines.

5-19. In addition to protecting the raw and completed information products, public affairs personnel must also take the necessary steps to protect information networks.

DISTRIBUTE

5-20. **Dissemination.** Public Affairs information packages should be released in the format most easily used by the recipient. While this is not always possible, a product stands less chance of being used if it is incompatible with the recipients' equipment. For example, a video product released on Hi-8 to a TV station that works exclusively with Beta SP has less a chance of being aired than a video story in a compatible format.

5-21. Public Affairs information packages must be expedited to the users by the most technologically advanced reliable method. Great consideration must be given to the speed and reliability of the mode of dissemination. This must be included in target audience assessments and conducted during planning stages of operations. Articles and photos may be sent from deployed locations to home stations via computer systems and telephone lines. Yet based on the quality of the telecommunications system, it may be more reliable to use the mail or a courier. As technology improves PA capabilities, Public Affairs will incorporate those improvements into the information gathering and dissemination system to increase its potential to reach an ever-growing, information-hungry public. For example, the emergence of smaller, more powerful satellite link ups can provide PA elements the ability to reach targeted audiences sooner and from more locations.

5-22. Internal information packages must be available to soldiers at all levels of command. Public Affairs must develop and coordinate a distribution scheme with the commander, the general staff, and with signal as the proponent for physical distribution of certain commercial news and information products. An efficient distribution system will also ensure prompt delivery of public affairs products. The public affairs staff must conduct periodic quality control checks and update the distribution scheme as necessary, based on changing population densities or information products. Electronic means are the preferred mode of distribution, however additional methods include, but are not limited to contract delivery, AG distribution and the military postal system.

5-23. **Products for distribution to deployed soldiers.** Publications produced by other military agencies intended for deployed soldiers must be given the same distribution considerations as commercial information packages. For example Stars & Stripes, Soldiers Magazine, Army Trainer and home station post newspapers contracted for delivery to deployed soldiers must be given the same distribution considerations as other publications. However, a separate distribution scheme may be required.

INFORMATION PROGRAM EVALUATION

5-24. The final step in the Information Provision function is the evaluation of our communication efforts. Evaluating communications programs is research, which boils down to a series of questions:

- Did we achieve our objectives?
- Were our policies and programs effectively communicated?
- Was the operation affected positively by our efforts?
- Did the American public support our soldiers? Was unit cohesion enhanced?
- What audiences received our messages? What was the impact of our communication programs on these audiences?
- Research is the foundation of the Information Program Evaluation. Corrections and changes in courses of action should be based on solid factual information. Methods for conducting Information Program research are discussed in greater detail in Appendix P.

NON-HIERARCHICAL AND INTERNET

INFRASTRUCTURE AS A RESOURCE

5-25. Modern technology has provided us with an advanced form of communications structure called non-hierarchical structure. The advantage of non-hierarchical communication structuring is that every "node" in a communications web shares information with every other "node". Each node on a network can identify itself and "find" others in the network in order to communicate specific information. Public Affairs elements act as information nodes, gathering, developing and sharing information vertically and horizontally on the battlefield.

5-26. While this technology was originally intended for command and control, it is essential for other functions on the battlefield as well. Well-coordinated public affairs operations will leverage this capability to move information -- sending messages around the battlefield, to and from home station, and up and down the chain of command.

5-27. The end state of this technology effort is that both organizations and individual soldiers on the battlefield will possess this capability.

TELECOMMUNICATION SYSTEMS

5-28. Information technology provides the means to collect, process, display, develop and disseminate information in an unparalleled manner. This technology has begun to revolutionize our approach to information provision.

5-29. The PA specialist now has the ability to access desired information on a certain issue and tailor and develop this information into a message for dissemination -- all from a personal laptop computer.

5-30. This "from anywhere to anywhere" capability allows the PA specialist at all echelons to accomplish his mission of presenting the Army's perspective -- framing issues and informing targeted audiences.

5-31. There are two telecommunication systems available to Army PA specialists: DoD's internal secure communication network, known as the Defense Data Network (DDN) and the worldwide commercial information network -- Internet. PA specialists must be familiar with both.

DEFENSE DATA NETWORK

5-32. Defense link is an entry point to Internet sites operated by the Office of the Secretary of Defense, the armed services and related defense agencies. It provides a means to search and download Department of Defense directives, obtain transcripts of important speeches and connect to other sites dealing with specific military operations.

5-33. The Department of Defense organizations are extensively represented on the Defense Data Network. Each of the military services has a centralized directory of its Internet sites. Army Link, as its name suggests is a catalog or hotlist of more than 280 Army home pages. The sites are indexed alphabetically and by subject area.

INTERNET

5-34. The Internet is a network of networks. Originally, the Internet was developed by DoD with the goal of building a automation system linking government agencies together which would continue to function in case part of the network ceased to exist. From this inception, the Internet grew, as other agencies and activities noticed the benefit of linked telecommunications. The system now includes organizations from all segments of society: government, defense, business, education, health services, etc. There are more than 30,000 networks participating on the Internet and more than 150 million individual users. And more are connecting everyday.

5-35. Army Public Affairs personnel have access to the Internet through their local Directorate of Information Management (DOIM).

5-36. On the Internet, there are six main activities: "Hypertext" documents with multimedia links on the World Wide Web can be viewed, files can be downloaded with a file transfer program or "FTP," data can be located, communication via e-mail, reading and posting messages to news groups and bulletin boards, and logging in to remote computers using Telnet.

5-37. You can do most of the same things in the Internet that you can do on any of the commercial information services, but you do them in the context of a much larger network that isn't centrally organized or controlled.

5-38. **World Wide Web.** The World Wide Web, also known as WWW, W3, or simply the Web, is one of the most popular activities on the Internet. The Web allows you to view documents that feature graphics, "hypertext" (text which contains highlighted keywords which are linked to other documents or information sources available on Web) and multimedia links. The hypertext and multimedia links are tied to other documents or information forms that might be on the same computer, one across the country, or a machine on the other side of the world.

5-39. The WWW contains thousands of main menus, called "home pages" which identify the various sub-categories at each web site. Web users can search using key words or locations through the lists of home pages for specific web sites.

5-40. **Gophers.** A gopher is a menu driven system that offers text only. Users of the web can access gopher sites and retrieve information. It is a predecessor of the Web.

5-41. **News groups.** They are discussion groups built around a particular topic. Some are managed, others are not. News Groups take two forms: Mailing List: A list of E-mail addresses to which messages are sent. You can subscribe to a mailing lists typically by sending an email to the contact address with the following in the body of the message: the word subscribe, the name of the list, and your email address. Discussion groups: A particular section within the USENET system typically, though not always, dedicated to a particular subject of interest. Also known as a newsgroup. USENET is a collection of the thousands of bulletin boards residing on the Internet. Each bulletin board contains discussion groups, or newsgroups, dedicated to a myriad of topics. Messages are posted and responded to by readers either as public or private E-mails.

5-42. **E-mail.** E-mail is a means of interpersonal communication that falls somewhere between the immediacy of a phone conversation and the more thoughtful but slower exchange of ideas previously done by writing letters and memos. The specifics of using it vary greatly according to the mail software being used.

5-43. **Commercial On-line Information Services.** Users cannot directly dial up to the Internet using a modem, but must gain access through an internet service provider. Many offices, universities and large businesses provide access for their employees. For access at home, a user can either subscribe to one of the commercial on-line services, which provide Internet access in addition to their other features or to one of the dedicated Internet server companies located in most cities.

Chapter 6

Training

"You must remember that whether you wear one stripe or six, one bar or silver eagles you automatically become an "Army spokesman" when you are approached by the press. Within 24 hours the words of that Army spokesman can be flashed world wide, particularly if they can be construed as criticism....... Everything you say should have the ultimate aim of furthering that effort. Your approach to the questions of the press should emphasize the positive aspects of your activities and avoid gratuitous criticism. Emphasize the feeling of achievement, the hopes for the future, instances of outstanding individual or unit performance and optimism in general. But don't destroy your personal credibility by gilding the lily. As songwriter Johnny Mercer put it, "You've got to accentuate the positive and eliminate the negative."

- U.S. Continental Army Command memo to advisers in Vietnam.
The Military and the Media,
William M. Hammond, 1988

TRAINING OBJECTIVES

6-1. The cost of learning in combat is too high. Therefore, commanders and staff section chiefs must conduct cost- and time-effective staff training during peacetime. In order to conduct effective training, every consideration regarding PA activities should be evaluated.

6-2. Evaluation should include, but is not limited to, analysis, planning and execution of the operation. The PA staff, in conjunction with other members of the organization, must train in the initial planning of the operation. In order to train effectively, the following considerations should be evaluated:

6-3. **Analysis.** Specified and implied tasks are identified at home station for deployment, mission and redeployment requirements.

- Feasibility of all tasks are determined and assessed.
- Consider any PA guidance issued by higher headquarters and determine specialized training requirements.
- Consider all phases of host nation operations and the effect on your operation and organization.

- Determine the availability of required assets and equipment for all phases of the operation.
- Consider the availability and use of internal and external assets -- joint, RC and combined -- to augment known shortfalls.
- Determine any transportation requirements for your organization or attached personnel.

6-4. **Planning**. Planning includes PA annexes, command directives, DOD guidance, regulations and directives stipulated by the commander's intent and mission requirements.

- PA annexes should contain a brief, general description of the situation and intent to conduct operations.
- Planning should also consider enemy forces and host nation personnel and their relationship to the operation.
- PA planning should be considered for all attachments and detachments currently known or listed under the task organization. This includes planning for assets and requirements from home station to redeployment.
- Training should be conducted emphasizing procedures for handling the transmission of all information. Theater-unique requirements may call for special training scenarios.
- All logistical and administrative requirements need to be addressed in the planning and training process.

6-5. **Execution.** The staff, soldiers and media understand the commander's intent in terms of his command position and operational mission. The operational area ground rules are included in the PAG and are widely disseminated throughout the organizations.

- Encourage local media to accompany deployed units.
- Adequate vehicles have been identified and assigned to transport PA personnel and media throughout operational areas. This includes vehicles, aircraft and available transportation assets for transporting all media personnel and products. Consideration of transportation should include all support activities in the forward area of operation.
- Public Affairs personnel in the media center will prepare releases of specific events to disseminate in the area of operations. News media inquiries are anticipated, received, researched and answered as quickly as possible.
- Local security review policies are clear and will not delay the transmission of information.
- Executions of the theater policy for registration policies and local media pool operations have been completed. Non credentialed media are identified and escorted to the rear for registration. Once registered, media personnel will be allowed to cover stories and interact with organizations in the area of operations.

6-6. The staff section must be able to:

- Cope with the unexpected. For example, media arriving at the unit level without PA escort or media not registered with the joint media operations center.

- Separate fact from fiction. For example, media having misinformation which they believe as fact and trying to confirm it.
- Coordinate well amid confusion. For example, be able to access the PA staff at the MOC.
- Think clearly while under great stress. For example, have a unit PA representative (UPAR) who will handle escorted/nonescorted media for the unit.
- PA elements develop training plans from assessments of their performance proficiency on their METL.
- Detailed training plans for Public Affairs Detachments are contained in ARTEP 45-500-10-MTP, for the Mobile Public Affairs Detachment in ARTEP 45-413-30-MTP and for the PAOC, ARTEP 45-326-50-MTP.
- The training plans in those documents can be readily adapted to train other types of PA elements, such as a division PA staff.

PRINCIPLES

6-7. All PA training must be based on the training principles in the FM 25-series:

- **Train as combined arms and services team**. Do not train in a vacuum. Train with the unit you support.
- **Train as you fight**. Conduct realistic training. PA leaders must move soldiers out of the office and into the field to conduct training, including joint and combined operations. Set up a media operations center and conduct operations for a day.
- **Use appropriate doctrine**. All PA leaders should be familiar with FM 3-0 (100-5), *Operations*; FM 3-13 (100-6), *Information Operations*; and the PA manual, FM 3-61 (46-1), *PA Operations.*
- **Use performance-oriented training**. Performing tasks under field conditions with appropriate evaluation enables your soldiers to perform better under actual conditions.
- **Train to sustain proficiency**. PA units must take advantage of all training opportunities to sustain proficiency, participating in all CPXs, FTXs and deployments that their supported units are involved in.
- **Train to challenge**. Training for PA tasks must create the kind of pressure PA soldiers will face in actual situations. Ensure it is challenging but realistic.
- **Train using multi-echelon techniques.** The entire PA chain-of-command must participate in training for it to be realistic and effective.
- **Train to maintain**. Soldiers and leaders must keep equipment in a high state of readiness in support of training and for deployment.
- **Make commanders the primary trainers**. The unit commander is the primary training manager for the unit. The commander assigns primary responsibility to officers for collective training and to NCOs for soldier training.

TRAINING OPPORTUNITIES

6-8. Regardless of type, size or configuration, PA elements/units must train with the units they support. PA leaders must evaluate the ability of their units to support their real-world missions successfully and determine areas of weakness.

6-9. PA leaders must determine what training opportunities they can participate in realistically and coordinate for support from higher headquarters. If support is not provided, then unit leaders must create their own unit training program that simulates the tasks they will be asked to do when deployed.

6-10. For example, PADs should deploy with brigades to NTC or JRTC to practice their unit mission and capabilities and be evaluated accordingly. If the PAD is tasked with garrison duties, these duties must be retasked to allow the unit to train and maintain.

6-11. PA elements and units should take advantage of the excellent training opportunities offered by the JRTC and NTC. Both centers conduct media-on-the-battlefield training for visiting maneuver units. Public affairs elements should accompany the maneuver brigades and establish field media centers to take maximum advantage of these resources. It is imperative that PA elements be able to function effectively in austere environments. They cannot perform to their maximum capability unless they train accordingly. The key is repeated, tough, realistic training.

Historical Perspective

An initial PA team was sent to Somalia to set up a media operations center in December 1992. Mogadishu at the time was a war zone with many fronts. Initially, U.S. forces set up operations at the airport. Under the constant roar of C-5s, the PA team tried to organize a MOC. They had a very difficult time dealing with the lack of power and water. Eventually they were able to set up in a schoolyard. A "JIB-in-a-box" arrived with computer equipment, but it was all 220-volt — power in Somalia is 110-volt.

For the duration of the operation, personnel at the MOC learned to live and work under austere conditions, where food, water, power, and communications were constant problems. They eventually were able to set up a briefing room with benches, an old couch and a bulletin board for posting media opportunities, and conducted media operations under austere conditions.

(After Action Report 1992)

6-12. Home station CPXs and FTXs are fundamental training events that provide PA elements relatively low-cost opportunities to practice the full spectrum of wartime PA tasks on a reduced scale.

6-13. A visit by even a single reporter can be used to exercise a broad range of media support tasks including establishing a field media center; arranging interviews; preparing subject matter experts, individual soldiers and commanders to meet the media; preparing fact sheets; responding to media queries; registering media; arranging escorts and transportation, etc.

PA TRAINING FOR NON-PA AND PA PERSONNEL

6-14. Public affairs training is not limited to training for PA soldiers. PA elements/soldiers also have a responsibility to train non-PA soldiers and family members in military/media relations. Organic PA elements must develop training programs for their supported units. These can take the form of classes as well as pre-deployment briefings for units and family members.

6-15. Training for commanders and subject matter experts. Unit commanders and individual subject matter experts must be prepared to be interviewed. It is a PA responsibility to ensure they are familiar with the ground rules and know to restrict their statements and comments to their area of expertise. Details on how to do this are contained in the Soldier's Manual task, "Prepare a Spokesman to Address the Media," #224-176-4013.

6-16. Training for unit soldiers. Individual soldiers must be advised of the inevitability of media presence during military operations. Classes for units can be part of sergeant's time training and consist of teaching soldiers what they should and should not talk about when meeting the press. All soldiers should be encouraged to represent themselves as soldiers and encouraged to speak about the jobs they perform for the Army.

6-17. Training for family members. PA training for family members consists of educating them on their rights and responsibilities when interacting with the media. Family members often know more about specific operations than should be revealed to the media. They must be advised not to discuss information, which may be used by the enemy against their spouse's unit, such as details about troop movements, destinations, missions etc. They must also be advised that they have the right to refuse to talk to the media. Family member briefings should be a standard element of pre-deployment family support group activities.

6-18. Basic soldier skills. Public affairs soldiers are soldiers first and public affairs practitioners second. It is imperative they are well trained in basic soldiering skills. Public affairs elements and units must make time to train to standards on common soldier tasks that allow them to effectively shoot, move, communicate and survive on the battlefield.

6-19. Readiness. All PA elements and units must have, and exercise, detailed load plans. Soldiers should be aware of the importance of maintaining all equipment and vehicles in a state of readiness for deployment.

6-20. PA NCOs must ensure soldiers participate in preparations for overseas movement (POM) so they are administratively and medically prepared for worldwide deployment. Soldiers should be aware of the importance of wills, shots, powers-of-attorney, personal data and allotments.

6-21. Staff Section and Unit Training. A state of operational readiness to conduct combat or non-combat operations must be attained and maintained. This level of readiness is accomplished by preparing individuals, shifts and staff sections to perform assigned tasks and other duties at the desired standard of proficiency in advance of assigned operations.

6-22. Standards of performance must be set so the section can evaluate its performance. Examples of these standards include:

- Understanding DOD's policy statement -- Commanders will ensure maximum unrestricted disclosures of unclassified information to news media representatives consistent with operational security, guidance from higher headquarters and the privacy of individuals concerned.
- Escorted and unescorted media -- If the media has a PA escort, you may agree to an interview after the escort explains some basic ground rules. If not escorted, ask media personnel to accompany you to the command post, NCOIC or OIC and contact higher headquarters/public affairs officer.
- Conducting an interview -- Military personnel have the right to deny media interviews. If a soldier elects to provide the media with an interview, he should only discuss those things which he has direct responsibility or personal knowledge, and ensure an escort (PA or non-PA) is present. See Appendix N for more details.

THE INTERVIEW PROCESS

6-23. The Interview Process:

- Prepare for the interview. Consider the type of questions the media will ask, and think what your answer will be. When possible, ask for PA assistance (UPAR or PA personnel).
- Relax and be yourself. Imagine that the reporter is someone you know and talk with him in a relaxed manner.
- If cameras are present, ignore them and talk directly to the reporter. Be brief and concise. Remember, a TV news story will use only 10- to 15-second answers.
- If you need time to respond, ask the reporter to restate the question. A simple pause before answering the question is sufficient.
- Use simple language and avoid military jargon (i.e., military abbreviations or acronyms). If you must use military terms, explain what they mean.
- Use appropriate posture and gestures. Relax and be yourself.
- Answer only one question at a time. If asked multiple questions, answer the most important one first, or answer the one you're most comfortable with.
- Always try to end your comments on a positive point. This is your opportunity to tell your unit's story.

6-24. Things not to do:

- Do not allow media to videotape recognizable landmarks nearby, sensitive equipment, interior of tactical operations centers or other sensitive areas. Use OPSEC as guide on this matter.
- Do not answer speculative questions or give opinions concerning real or hypothetical ("what if") situations.
- Do not use the expression, "No comment." A more appropriate comment would be, "We don't comment on future operations." or

"I'm not qualified to respond to your question." or "That information is classified, so I can't discuss it."

- ***Never*** lie to the media.
- Consider everything you say to the media as "on the record." ***Never*** make "off the record" comments.
- Don't lose your temper when media representatives ask questions you consider inappropriate or foolish. Such questions are usually rooted in ignorance rather than in malice.
- Do not discuss operational capabilities, exact numbers or troop strengths, numbers/types of casualties, type of weapons systems or future plans. Use general terms like approximate, light, moderate or heavy.
- Don't repeat a negative phrase in response to a media representative's negative question (e.g., Q: Since your unit is poorly trained, can you really deploy? A: We're well trained and ready to go.).
- Staff section and unit training should be integrated into headquarters FTXs and CPXs in order to perform adequately within the unit's operational staff during real-world missions.

TRAINING EXERCISES

6-25. Training exercises will vary from major FTXs to CPXs and Tactical Exercises Without Troops (TEWTs). Each training opportunity provides the staff element realism, the opportunity to experiment and the ability to face situations. These challenging training exercises enable soldiers and units to tell the Army's success stories.

6-26. Training exercises also enable the commander and staff to:

- Emphasize the tactical SOP.
- Execute plans in a dynamic, hostile environment.
- Exercise bold solutions.
- Exercise contingency plans.
- Experience possible defeat without the penalty of combat loss.
- Work toward goal-oriented performance standards by team building while providing PA support.
- Training exercises also allow PA personnel to become operationally, strategically and culturally aware, and puts them in a unique position to interface with the news media personalities which shape the perceptions of our national and international audiences.

TRAINING FOR MEDIA FACILITATION

6-27. Facilitating media in the military environment includes three phases -- planning, preparation and execution.

6-28. **Planning.** The Media Operations Center receives requests for registered media to visit units in their areas of operation. The MOC will contact a unit and coordinate the time and location for the visit, as well as notifying a PA or nonPA escort for a mission. The escort will need to know the unit's location and whether the unit 's UPAR (if applicable) will travel with the group or meet them at the unit. The unit will need to know the number of media visiting, the duration of the visit and the status of the unit for the last 24 hours (i.e., has the unit been in contact with the enemy and has there been any significant actions).

6-29. The escort will plan the route to the unit. The number of media going will determine the number of escorts and vehicles needed for the trip.

6-30. **Preparation**. The JIB or escort will notify the media and tell them the time and location of departure and estimated time of return. This allows the media to plan for their supplies and equipment for the trip. The escort should meet with the media and update them on the unit's status, find out if the media has any specific requests before departing, cover safety points and OPSEC requirements, and ensure they have everything they need. This will also provide the escort with a direction for the types of questions or issues the media may address. The information concerning the unit's situation will allow the escort to develop a list of Q&As for the commander and individuals whom the media will interview.

6-31. The escort should meet with the drivers and review the route to the unit, cover contingency plans and determine an inspection time before movement. The escort should also contact the unit to reconfirm the visit and their location, as well as providing a list of the names and agencies of media visiting them. The escort should meet the media at the vehicles to insure they have what they need for the trip.

6-32. **Execution.** The escort should leave as scheduled and travel as the tactical or non-tactical situation dictates.

6-33. When arriving at the unit, the escort should have the driver park the vehicle as directed by the unit and have media personnel wait at the vehicles (if possible, with military supervision). The escort needs to request that the media not take photographs of the unit's position for operational reasons. Let the media know you are going to get the commander and will return as soon as possible. If the situation is hostile, get the media inside the perimeter and secure them away from equipment and the command post.

6-34. The escort should meet with the UPAR and let him/her know who is in the media group. Ensure that the unit will be able to provide the interviews the media requested.

6-35. The escort should brief the commander and let him know the ground rules, covering possible Q&As. Answer any questions or concerns he may have about the interview. Let him know that you will be present during the interview to assist him. When possible and if mission requirements permit, the commander should be available for the media. Review the unit's OPSEC requirements to ensure you understand what cannot be photographed.

6-36. The escort should determine where the commander wants to meet the media. If the area is not appropriate for conducting interviews (may be based on PA experience or media request), provide an alternative recommendation to the commander.

6-37. The escort should move the media to the interview point and introduce the commander to the group. After the interview, the escort should facilitate the rest of the coverage with the help of the UPAR. Do not allow the media to linger in the area of operations and become mission detractors. Wrap up the visit, return to your vehicles and depart. Upon returning to the JIB, the escort should be available to back brief the PAO.

6-38. The requirement to maintain proficiency in the full range of public affairs collective and individual capabilities and skills places a high priority on tactical training for both AC and RC PA soldiers. Tactical training participation enhances knowledge of battlefield requirements, increases unit cohesion, and forms the basis of experience needed for operational planning, mobilization, deployment and mission success.

6-39. PA will continue to face expanded missions in the joint arena. Quality PA training provides sufficient numbers of trained PA personnel to conduct joint and combined PA missions. Training must emphasize the joint perspective, enhance interoperability, and contribute to each combatant CINC's PA Mission. Exercise participation is critical to the training of PA personnel.

6-40. Trainers must fully incorporate a broad array of PA activities into all types of exercises and war gaming. These activities can be injected into computerized battle simulation as the training exercise driver. They should also be used in BCTP command post exercises, and JRTC, CTC, and NTC rotations. Seminars, area assessments and TEWTs are all forms of training that also provide relevant, realistic training.

Chapter 7

Community Relations

Public opinion about the Army is greatly influenced by the actions of each command. What the command does for its local community or fails to do affects the perceptions and attitudes of the American people, upon whom the Army depends for its support and existence. This applies not only to official acts but also to unofficial acts, which by their commission or omission affects public opinion. This principle also applies to individual members of the Army, their dependents and Army civilian employees in their personal contacts with the civilian community. Conducting community relations is a vital element to successful public affairs operations. Commanders and public affairs officers (PAO) must seize on key opportunities to gain and maintain links to internal and external publics.

EFFECTIVE COMMUNITY RELATIONS

7-1. PA personnel act in concert with veteran's groups, civic leaders and local populations to increase understanding and build support for Army activities. Army support to and participation in public events is based on the fact that the Army belongs to the American people. Common ownership requires that Army resources be used to support events and activities of common interest and benefit.

7-2. Effective community relations requires:

- Command supervision at all levels.
- Appreciation of public opinion and attitudes toward the Army
- Planning definite actions and positive policies.
- Implementing programs in a competent, professional and responsible manner.
- Constant evaluation of continuing programs to measure their effect upon the public and the command.
- Sharing the results of the program.

7-3. Commanders must maintain continual liaison with persons and organizations in the local community to help resolve common problems and develop cooperation and understanding between the installation and the local community. Community relations develop an effective two-way channel of communication between the Army and the community. PA does this by capitalizing on opportunities for better relations and resolving potential and actual areas of conflict.

7-4. Community relations projects or programs may be supported by use of exhibits, equipment and facilities. Exhibits consist of displays such as mission exhibits models, devices and other information and orientation materials at conventions, conferences, seminars demonstrations, exhibits, fairs or similar events. Also included are exhibits displayed on military installations during open house programs.

COMMUNITY RELATIONS ACTIVITIES

7-5. The goals of community relations is to develop an open, mutually satisfactory, cooperative relationship between the installation and the community. A successful community relations program improves the community's perception of the Army and its appreciation for the installation and the soldiers, family members and civilians who are part of the installation. It is based on openness and honesty. Community relations objectives are community assistance and social improvements for the community in which the military must work and live.

Historical Perspective

Fort Eustis, Va. started a pilot program in 1992 called Operation Self-Enhancement to give high-risk middle school students the opportunity to visit the post and focus on careers, teamwork and self-esteem. The program was so successful that it has become an annual event. Students receive light military training through an array of testable tasks and obstacles presented by members of a cadre team. This training helps students build their self-esteem and self-confidence and affords them the opportunity to interact with positive role models. This also gives the students a better idea of what the Army is about.

7-6. Community Relations activities include:

- **Speakers Bureaus.** Speakers are an effective means of developing understanding of the Army, stimulating patriotic spirit and informing the public about the activities of the installation, its units and its soldiers. Commanders should establish an installation speakers bureau and encourage soldiers of all ranks to participate in the installation program.
- **Community Liaison.** Maintaining liaison through informal community relations councils can establish and maintain open communications with community officials and organizations. Councils can be charged with a variety of responsibilities, such as developing and promoting new ways for members of the command to actively participate in local community activities, capitalizing on opportunities for better relations and resolving potential and actual areas of conflict. Community liaison can also involve recognition of private citizens, local community leaders and citizen groups and organizations for their support of the Army by public service awards. Commands can further community liaison through membership in civic, business and professional organizations when

the goals and objectives of those organizations are beneficial to the Army and their programs and projects are consistent with Army interest.

- **Ceremonial Units.** The band, color guard and other ceremonial units participating in public events are excellent ways to accomplish community relations objectives. These representatives of the Army serve as ambassadors to the civilian community and promote patriotism, interest in the Army, and awareness of the professionalism of our forces.
- **Exhibits.** Exhibits and displays of Army equipment, historical materials, models, devices and other information are other community relations activities that can enhance understanding of the Army and the installation. They provide an excellent opportunity for interaction between our soldiers and members of the local community and can communicate the professionalism, readiness and standards of our forces.
- **Open House.** Open houses may be scheduled to coincide with Armed Forces Day, the Army Birthday, service branch birthdays or anniversaries which mark the history of the installation, a unit or community events or in support of media day. An Open House gives the local community an idea of who we are and what we do. They also have the opportunity to visit us on the installation-- at our job site.
- **Physical Improvements**. Community service physical improvements focus on ensuring that the physical infrastructure is as safe as possible and provides the fullest possible range of support to the population. These activities encompass a wide range of programs that do not compete with the services provided by contractors and businesses in the local civilian community.

7-7. Some examples of physical improvements are:

- Construction projects that enhance the recreational, educational, environmental or cultural facilities of the community, such as building community picnic areas and hiking and biking trails.
- Demolition projects that enhance the safety and appearance of the community, such as the removal of unstable playground equipment.
- Projects that create or enhance a safe, clean environment, such as the removal of debris from a community wildlife area or painting a community recreation center.

TOWN HALL MEETINGS

7-8. Town Hall meetings provide installation commanders with an unfiltered means of communicating ideas to internal and external communities. This tool for conveying important information and ideas about the command cannot be underestimated in its effect and should not be planned haphazardly.

7-9. Commanders, PAOs and staff directorates must work together to produce an effective community relations product.

7-10. Prospective town hall meeting planners must understand, and properly apply, the correct type of town hall meeting. With a focus on the type of meeting and probable audience, the planner can begin the process of planning and conducting the event.

7-11. As part of the plan, the planner must determine the likely audience for the meeting, including attendees from internal and external audiences. He must also evaluate possible attendance by key publics. The planner should develop a standing operating procedure (SOP) to ensure each mechanism of the process is in place for the scheduled event.

7-12. Finally, post-event analysis is imperative to accurately assess its effectiveness.

7-13. The PAO must develop systems to quickly assess the feedback data and activate a follow-up plan that will maintain confidence from the community that town hall meetings are meaningful events.

TYPES OF TOWN HALL MEETINGS

7-14. Installation commanders can stage various types of town hall meetings. The commander must determine which meeting type, or hybrid, is appropriate for disseminating information and gaining useful feedback from internal and external publics.

7-15. The following meeting program structures have inherent strengths and weakness; knowing the potentials for message delivery will assist the commander in making his decision:

- **Commander - Expert Format**: This meeting (Figure 7-1), is characterized by the commander attending with key staff members facing a live audience. Typically the commander and his staff will give presentations and then field questions from the audience. Usually, attendance is open to the public.
- The primary advantages to this format center on the open nature of the meeting.
- This meeting provides the commander with an opportunity to provide detailed presentations with time being a minimal constraint. He also receives instant feedback from the types of questions from the audience and the passion with which questions are asked.
- Meetings in this format are likely to be seen as the most easily accessed by the internal and external publics.
- Among the disadvantages of the format is the ability to reach large audiences and control the conduct of the meeting.
- Unless the meeting is taped for later airing on the commander's cable access channel, the audience is often narrowly focused.
- Although the possibility exists that large audiences will attend, it is also possible small or narrowly focused audience will limit the general effect the commander seeks.
- An open meeting can also become the forum for unruly or disgruntled audience members to incite others or attempt to draw the commander into an open confrontation.
- This factor can be mitigated through the use of question time-limits and use of a moderator (other than the commander), but cannot be

totally eradicated. (As with any other public event, security must be a consideration in the planning process).

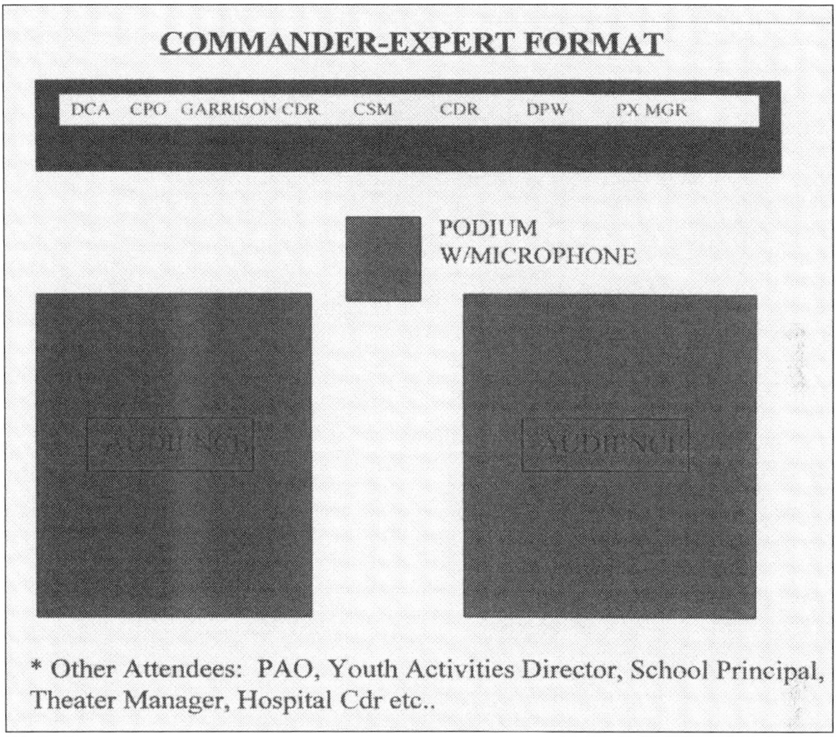

Figure 7-1. Commander Expert Format

7-16. Commander Access TV Channel Format: This meeting format (Figure 7-2) uses the commander's cable access channel to air the event. Normally, the commander, CSM and Garrison Commander (and other staff members as necessary) give a presentation. No live audience is in attendance. The commander provides a set of phone numbers, allowing questions to be called in. The staff operates the phone bank, accepts the questions, directs the questions to the appropriate staff agency, and delivers the answers by 3x5 card to the on-air panel. The panel members read the questions/answers to the viewing audience.

7-17. This format offers advantages focused on control and distribution of the product. The venue and the setting are completely controlled by the commander and his staff. The staff screens questions and, thus, no surprises will occur. Indeed, if questions/answers are given to the commander, he may choose to either not answer or return the card for more information.

7-18. Other advantages include an ability to re-run the meeting as often as desired and provide copies to local cable providers, many of which are interested in using the product for airing.

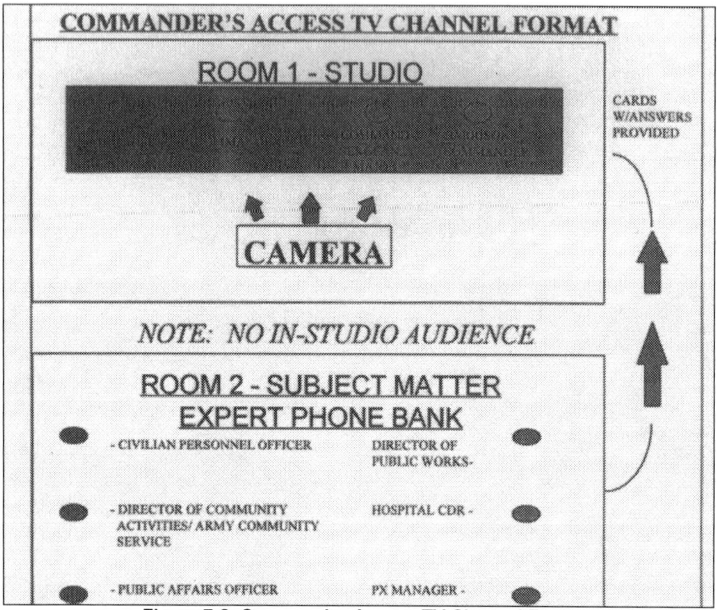

Figure 7-2. Commander Access TV Channel Format

7-19. Disadvantages to this format center on a difficulty in gauging feedback, competing for viewing audiences on television, and providing personal contact with the commander. With no attending audience, collecting feedback data is problematic. Follow-up questionnaires, using statistical probability methods, provide the only means of gaining reliable feedback. The likelihood of viewers to "channel-surf" is high due to the specific nature of questions from callers. Holding the interest of viewers, all of whom have multiple other viewing options available, is difficult. Finally, by appearing on television, in uniform, with a phalanx of staff members, the commander risks appearing to be speaking from "the mount." Audiences may view the commander as speaking down to them and being out-of-touch to their concerns.

7-20. Key members from the internal and external publics form a roundtable discussion (Figure 7-3). The topics are set by published agenda, with some time left for open discussion. The meeting results are published for general distribution. Media are usually invited as participants.

7-21. Roundtable discussions with key publics offer advantages to the commander by providing information to the individuals who represent overlapping and wide constituencies. By setting an agenda, the commander can deliver focused messages with a high likelihood of the messages later reaching targeted audiences. It allows the commander and potentially key staff officers to deliver presentations and provide follow-up information to those in attendance.

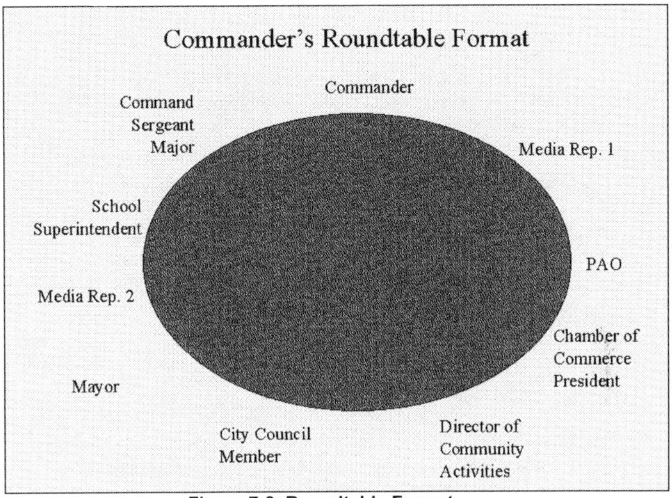

Figure 7-3. Roundtable Format

7-22. The principal disadvantage is the distance between the commander and the publics. The publics will receive information about the meeting second-hand and in a potentially filtered state. Gaining feedback may also prove challenging (but not necessarily impossible). Other disadvantages may be in the limits of the audience makeup; key publics, which are not invited, may resent the slight.

7-23. Commander-in-the-round format. A room is set-up that will allow the commander to be in the middle of all the attendees. Normally, he will stand and walk around in the circle formed by attendees. The meeting is usually open to the public.

7-24. The primary advantage to this format is the close contact the commander shares with his audience. Audience members may feel that barriers are lowered because the commander is close in proximity and no staff members are buffering their access. The commander also can realize feedback very quickly and can gain some appreciation for the resonance of his ideas with the assembled audience.

7-25. Such closeness with the audience can also be a significant disadvantage. Limiting surprises and controlling potentially unruly audiences is extremely difficult in this format. Further, because of the commander's reliance on his personal notes and memory, his ability to provide detailed information and multi-media presentations may be restricted.

7-26. Mitigation against such limits is dependent on the site design. Other disadvantages include problems associated with other "open" formats, including a possibly non-representative makeup of the audience.

7-27. Before deciding on which format is appropriate for the information, the commander must have a plan for what must be achieved.

7-28. Characteristics of an effective plan:

7-29. Answers the "why?" questions. The planner must understand the purpose of the meeting to correctly advise the commander on the format and substance to begin coordination of the plan. As described above, the various formats each have strengths and weaknesses that will assist the planner to shape the meeting.

7-30. Routinely Scheduled. Normally, meetings can be scheduled quarterly or monthly. Use the installation planning cycle to ensure proper coordination and notification of public meetings.

7-31. Site Plan. Checking and securing a site for the town hall meeting is dependent upon the type of meeting selected. Plan for the site early in the process and establish the layout of the site in detail.

7-32. Calendar Check. Before scheduling events, check local and regional calendars for possible conflicts. For example, scheduling a meeting on the same night as the local high school homecoming football game may prove disastrous.

7-33. Presentation Submission Deadline. Coordinate with appropriate staff agencies responsible for preparing presentations. Ensure presentations are properly staffed and approved. As PAO, establish a firm submission cut-off date.

7-34. Focused Presentation. Inform the staff of the commander's intent for the meeting. If the commander wants a particular theme addressed, ensure staff agencies adhere to the parameters of the intent.

7-35. Media Invitation List. Invite local and regional media including print and electronic outlets. Develop relationships with individual reporters and provide background material as necessary. (Local newspapers, radio stations and television stations can often assist in publicizing meetings as well).

7-36. Publicity Plan. Ensure all available avenues are used to publicize meetings. Included in this process are the post newspaper and radio station (where available), normal distribution, staff meetings, E-mail delivery and chain of command communications. Take special care to invite key publics by individual invitation and phone call follow-up.

EVALUATING FEEDBACK

7-37. Assessing the effectiveness of town hall meetings is essential to developing community-related policies and courses of action for the command. PAOs can use standard statistical measurements using survey techniques to gauge the level and intensity of views of the various publics. Other analyses can be derived from follow-up media content analysis, letters to the commander and post newspaper, and reactions at the meetings. None of the methods described here will render a perfect picture. PAOs must exercise good judgment and personal insight when advising the commander of analytical results.

7-38. **Response Follow-up**. Investigating and responding to issues raised at town hall meetings are critical to public perceptions of the level of care the commander applies to community operations. Just as a maneuver commander sees battle damage assessment (BDA) as crucial to determining the efficacy of fires, commanders and PAOs must determine the true productiveness of community relations programs. Commanders will often feel compelled to promise action, such as investigation or immediate problem resolution, during a meeting. Staff agencies are normally the conduit for actions (only occasionally will the commander personally provide the requisite service).

7-39. PAOs should provide the staff oversight of the response mechanism. PAOs can devise a recurring memorandum that provides the commander (usually through the chief of staff), details about the status of actions. All staff agencies should receive updates routinely.

7-40. Along with the response mechanism, PAOs can use command information and media relations channels to inform the publics of problem resolutions. For example, if an issue raised at a town hall meeting indicated that the local recreation center was routinely opening two hours late each day, describe the measures taken to alleviate the problem. The PAO can use the post newspaper or radio station to provide lists of problems and resolutions.

CONCLUSION

7-41. Community relations and activities are vital to instilling and maintaining the confidence of internal and external publics in our great Army. Commanders and PAOs cannot leave the prospect of successful relations to chance. Too much is at stake. Careful selection of the type of town hall meetings to be used must be taken. The event must be planned with the attention to detail required for all military operations. Each part of the plan has unique importance and cannot be overlooked. To ensure the effectiveness of the operation, PAOs are compelled to build a clear mechanism for evaluating outcomes. Finally, the command must provide conspicuous follow-up responses to issues raised to complete the process. Town hall meetings provide an excellent opportunity for commanders and the various publics to interact and improve community institutions.

COMMUNITY ASSISTANCE

7-42. Community assistance applies the skills, capabilities and resources of the needs and interests of the local community. Providing support for and participating in events and activities which are beneficial to both the Army and the community, builds on a long tradition of "America's Army" helping American communities. Identifying opportunities, which advance the interests of both the Army and the community, is an important objective for every commander.

7-43. Community assistance projects and operations must impact positively on the unit or individual soldier, enhance unit or individual readiness and contribute toward the common good of the community. Army commanders must ensure that their initiatives are not competitive with local resources or services and do not benefit any particular interest group and will not result in monetary or service remuneration in any form for unit members or the unit as a whole.

7-44. Increasing public awareness and understanding of the Army, inspiring patriotism and enhancing the Army's reputation as a good neighbor is a goal of community assistance. Community assistance activities can help build unit morale and esprit de corps. These activities also provide an excellent opportunity for soldiers to serve as role models, which not only enhances recruiting efforts, but also serves to motivate soldiers by promoting their self-esteem and furthering their sense of service to the nation.

READINESS ENHANCEMENT

7-45. Certain community assistance activities enable a commander to train soldiers, enhance individual and unit readiness, maximize use of assets and foster a positive training environment where soldiers can become involved in realistic, "hands-on" training opportunities. Projects should be selected which exercise individual soldier skills, encourage teamwork, challenge leader planning and coordination skills, and result in measurable, positive accomplishment. Finally they should enable a unit to exercise use of its equipment, resulting in training opportunities that can generate greater operator efficiency for future missions.

7-46. Examples are:

- Medical Readiness Program. The Medical Readiness Program is an activity in which Army medical unit personnel, together with state medical emergency officials, plan and provide support in the form of diagnosis, treatment, and preventive medical, dental, and veterinary care to citizens in remote areas of the U.S. or its territories. The program is designed to enhance the unit's medical readiness, provide unit training opportunities and serve the public in locations where medical care is not otherwise available. The program may not compete with local private medical care that may be available.
- Air Ambulance Participation. The Military Assistance to Safety and Traffic Program. (MAST) is a proven example of Army support to civil authorities. This program permits the utilization of Army aviation assets to conduct emergency air evacuation and recovery efforts.
- These projects contribute to the health and welfare of the community, making the Army an integral partner in community progress and development. They can enhance the ability of the local community to support itself and its people, to provide the best possible services to the citizens and to promote a positive, healthy safe environment.
- Community service activities are those which focus on improving the community, its infrastructure and its ability to serve the local population.

7-47. **Social Improvements.** Community service social improvements, which focus on making the social environment as healthy as possible, provide the widest range of support to the population. They encompass many projects including:

7-48. Support to youth programs, such as scouting and programs that provide assistance to special need audiences such as the Special Olympics.

7-49. Examples are:

- The Civilian Youth Opportunities Program (Challenge). This is a youth program directed at attaining a high school diploma, providing job training and placement, improving personal and social skills and providing health and hygiene education and physical training. Soldiers work with civilian leaders to provide a comprehensive support package, ranging from choosing appropriate clothing to attending residential training facilities.
- Youth Physical Fitness Clinic Program. The National Guard encourages fitness and combines academic and athletic achievement by helping schools conduct competitions in selected athletic events. This program also establishes a separate scholar-athlete category for those students with a 3.5 or higher grade point average.

7-50. Involvement in ventures and projects that enhance the educational or cultural climate of the community, such as adult literacy, reading or community theater programs.

7-51. Examples are:

- Civilian Community Corps. This program provides managerial, organizational and technical skills for disadvantaged Americans seeking the skills for success. Through this program, the Army helps participants become productive citizens. In exchange, participants perform a wide range of community service activities that improve the foundation of American society. This program encourages intra-governmental cooperation on the federal level. It also encourages partnerships with industry, education, state, federal and local governments.
- Science and Technology Academies Reinforcing Basic Aviation and Space Exploration Program. (STARBASE) This program is an innovative partnership of professional educators, military personnel and corporate sponsors. It promotes science, mathematics and technology basics for primary through secondary schools. Using National Guard resources to spark student interest, the program develops strong self-esteem, provides excellent role models, promotes positive attitudes and develops goal-setting skills.
- The guiding principle behind community relations efforts is that the installation and the community have a common interest in providing the best possible support for each other. A cooperative relationship exists, because soldiers stationed at the installation receive life support from the community while many of the civilians who make up the community receive life support from the installation. The interdependence of the military installation and the civilian community can involve economics, education, health care, basic services, quality-of life issues and many others.
- The community relations goal of local commanders is to develop an open, mutually satisfactory, cooperative relationship between the installation and the community. These efforts improve the community's perception of the installation and the soldiers, family members and civilians who are part of the installation.
- Participation in community relations activities is an effective method for projecting a positive Army image, making the best use

of assets, providing alternative training opportunities and enhancing the relationship between the Army and the American public. Activities vary widely, ranging from individual soldier involvement to full Army participation. They are characterized by detailed coordination between the military command and community authorities. They fulfill community needs that would not otherwise be met, enhance soldier and unit morale, skills and readiness and improve the mutual support between the military and civilian communities.

Chapter 8
PA Organizations

THE BRIGADE

8-1. The Brigade PAO is the lowest level to which the Army has assigned organic Public Affairs Assets.

8-2. Working as both a special staff officer and as a member of the Brigade's planning team, the BDE PAO acts as the spokesperson for the unit, advisor to the Commander, and provides Public Affairs guidance and planning to commanders at all levels.

8-3. To support and conduct Public Affairs Operations within the Brigade, the BDE PAO has the following functions, organization and equipment:

8-4. Specific functions of the Brigade PAO, as outlined within the core competencies are:

- **Public Affairs Planning**
  - Advise the commander and staff on PA implications of plans and actions.
  - Write Brigade PA annex and matrices using operational tools, terms, graphics, and concepts
  - Execute the plan
  - Plan for future operations.
  - Command and control attached Public Affairs assets within the brigade AO.
  - Tactically communicate with PA units and supported combat units in the Brigade's Battlespace.
- **PA Training**
  - Train and supervise stringers to assist their commanders in conducting their internal information programs.
  - Train soldiers, family members, DACs and everyone habitually associated with the unit to comfortably and successfully communicate with the media.
  - Utilize Train the Trainer.
  - Evaluate public affairs training programs of subordinate commands.
  - Train subordinate commands to facilitate the media.

  - Train leaders and soldiers to protect information products and information systems from compromise and intrusion by practicing security at the source.
- **Media Facilitation**
  - Assist media to gain access to units, soldiers and commanders.
  - Evaluate subordinate command's media facilitation plan.
  - Assist subordinate commanders to coordinate transportation.
  - Conduct situation briefs as needed.
  - Monitor media operations within the Area of Operations. (Maintain accountability of the media to preclude fratricide. Accomplished through reports from escorts or through electronic tagging and monitoring.)
  - Respond to and mediate potential media problems; react quickly to coordinate and facilitate information issues in Brigade's Battlespace.
- Information Strategies
  - Monitor and analyze the local Military Information Environment.
  - Provide public affairs support to the G5 / S5 for the development and implementation of civil affairs programs.
  - Support higher echelon Public Affairs requirements for information.
  - Gather Open Source Information to help build relevant information.
  - Assist Commanders to conduct internal and external information to include Hometown News Release Program.
  - Monitor local news media products (Visual, print and audio) and analyze for PA implications.
  - Monitor and Analyze the local Military Information Environment.
  - Protect digital images, information products and PA and non-PA information systems from compromise and intrusion.

8-5. Organizational Structure:

- Captain , 46A.
- Sergeant, 46Q

8-6. Equipment:

- Computer with compatible software and communications hardware
- Access to the Tactical Internet and tactical battlefield radio communications
- Access to FM, Satellite, and video and cellular communications
- Access to Army Battle Command System, (MCS\P or CSS\CS)

8-7. Transportation:

- HMMWV

8-8. Addititional Support Requirements. PAO requires linguistic support from the Civil Affairs soldiers or contract civilians attached to the Brigade.

SAMPLE PA EXECUTION MATRIX BRIGADE PUBLIC AFFAIRS OFFICER

	PHASE I	PHASE II CROSS LD	PHASE III DEFEAT	PHASE IV SEIZE OBJ. LION	PHASE V RECONSIT.
BRIGADE	PAD CIVILIAN MEDIA				PAD AND CIVILIAN MEDIA
1-66 AR	INTERNAL	EMBEDDED MEDIA-- PRIORITY OF EFFORT			EXTRACT MEDIA
1-4 AV	AVIATION WEEKLY	SAME	SAME	SAME	AVIATION WEEKLY DEPARTS
1-5 IN (L)	OWN THE NIGHT MAG.				
SCOUTS		MEDIA REPS LINK PRIOR TO DEPLOY			
BSA	FOOD PREP QTRLY	SAME	SAME	SAME	SAME
MEDICAL	SAFETY STORIES				MEDIA - INJURED SOLDIERS

THE DIVISION

8-9. The Division PAO is the next level to which the Army has assigned organic Public Affairs assets. The division is largely self-sustaining and capable of independent operations. The division is a unit of maneuver, organized with varying numbers and types of combat, combat support (CS) and combat service support (CSS) units.

8-10. The division may be armored, mechanized, medium, light infantry, airborne or air assault; each can conduct operations over a wide range of environments.

8-11. Working as both a special staff officer and as a member of the division planning team, the DIV PAO acts as the spokesperson for the division, advisor to the Commander, and provides Public Affairs guidance and planning.

8-12. To support and conduct Public Affairs Operations within the division, the DIV PAO has the following functions, organization and equipment:

8-13. Functions: Assumption #1: DMAIN is not in country

- Gather, analyze and disseminate open source information, focusing on global media, to increase the situational awareness throuhout the command.
- Gather and disseminate multimedia products throughout the command, and to external and home station audiences.
- Conduct Public Affairs research and write the PA Estimate of the situation.
- Write the PA plan for OPORDERS, OPLANS, and TACSOPS.
- Monitor and analyze American and foreign public sentiment of current operations from available media sources for PA implications and advice.
- Monitor and analyze battlespace communications (visual, audio, FM, satellite) for PA implications.
- Monitor news media products (Visual, print and audio) and analyze for PA implications.
- Facilitate Media Operations.
- Subfunctions:
  - Assist media to gain access to units, soldiers and commanders.
  - Coordinate air and ground transportation on a non-interference basis.
  - Conduct daily situation briefs as needed.
  - Respond to media queries.
  - Validate media's credibility, expertise, knowledge, purpose and legitimacy.
  - Train and supervise unit-level PA representatives (Command Information NCOs) to assist their commanders command information programs.
  - Train soldiers, family members, DACs and everyone habitually associated with the unit to comfortably and successfully communicate with the media.
  - Survey soldiers, family members, DACs and other members of the internal audiences to measure effectiveness of the command information program.

8-14. Assumption #2: DMAIN is in country

- Advise the commander and staff on PA implications of plans and actions
- Gather, analyze, and disseminate Open Source Information, focusing on global media to increase the commander's situational awareness.
- Gather and disseminate multimedia products throughout the command, and to external and home station audiences.

- Conduct Public Affairs research and write the PA Estimate of the situation.
- Write the PA plan for OPORDERS, OPLANS and TACSOPS
- Monitor and analyze American and foreign public sentiment of current operations from available media sources for PA implications and advise
- Monitor and analyze battlespace communications (visual, audio, FM, satellite) for PA implications
- Monitor news media products (Visual, print and audio) and analyze for PA implications
- Coordinate and integrate all information-related functions (PSYOPS, CA, VI, Joint, Combined and Interagency PA) into the PA plan
- Conduct Information Operations
- Act as a conduit for CI products from the field, sanctuary and commercial sources for input into the commander's information program. Provide command information to soldiers, family members and Department of the Army Civilians.
- Facilitate Media Operations
- Subfunctions:
    - Coordinate air and ground transportation on a non-interference basis
    - Assist filing stories, video and photographs on a non-interference, reimbursable basis
    - Conduct daily situation briefs as needed
    - Respond to media queries
    - Validate media's credibility, expertise, knowledge, purpose and legitimacy
    - Conduct Primary and Secondary accreditation
    - Primary--Full accreditation of non-accredited media
    - Secondary--Process media previously accredited at Corps and higher
    - Survey soldiers, DACs and other members of the internal audiences to measure effectiveness of the command information program

8-15. Organizational Structure:

- At a minimum an embedded PA division section has:
    - Major 46A and two Captains 46A
    - Master Sergeant 46Z
    - Specialist 46Q
    - Specialist, 46R
    - PFC 46Q

8-16. Equipment:

- Computer with compatible software and communications hardware and tactical fax machine.

- Access to the Tactical Internet and tactical battlefield radio communications (SINGARS, MSRT etc.)
- Access to FM, Satellite, video and cellular communications
- Access to Army Battle Command System, (MCS\P)

8-17. Transportation:

- Light Division--HMMWV
- Armor and Mech Division-- HMMWV and Trailer

8-18. TAC 1 AND 2 Missions and Functions

TACTICAL COMMAND POST PA SECTION

8-19. The TAC CP, Public Affairs Section is task-organized based on METT-TC. It is the center of gravity for immediate internal and external communication, resolution of Public Affairs issues and violations of ground rules by media representatives. It acts as a conduit to the Information Operations Cell in the DMAIN adding immediacy to the Open Source Information process.

8-20. Public Affairs personnel deploy with the lead elements during any contingency and remain with the TAC CP to assist the commander, provide internal command information to deployed soldiers and limited external information to home station. They conduct media facilitation to expedite the flow of information to the America public while freeing the commander and his soldiers to conduct their mission.

8-21. The TAC CP, Public Affairs Section maintains connectivity with commanders and global information sources. It contributes to and monitors the common relevant picture, and synchronizes collection and dissemination efforts of soldiers far forward to internal and external audiences.

8-22. Particularly during split-based operations, the PAO task organizes his section to best serve the commander, his soldiers and the American public's need for information.

8-23. Assumption #1: DMAIN is not in country

- Act as the division spokesman.
- Advise the commander and staff on PA implications of plans and actions.
- Execute the plan.
- Assist DMAIN to gather Open Source Information to build the common relevant picture.
- Conduct Information Operations.
- Gather and disseminate multimedia products throughout the command and to external and home station audiences. Act as a conduit for CI products from the field, sanctuary and commercial sources for input into the commander's information program.
- Monitor and analyze battlespace communications (visual, audio, FM and satellite) for PA implications.
- Monitor local news media products (Visual, print and audio) and analyze for PA implications.

- Coordinate, integrate, and synchronize all information-related functions (PSYOPS, CA, VI, Joint, Combined, and Interagency PA).
- Synchronize Public Affairs assets (internal and external) within AO.
- Facilitate Media Operations:
  - Assist media to gain access to units, soldiers and commanders.
  - Coordinate transportation on a non-interference basis.
  - Assist filing stories, video and photographs on a non-interference, reimbursable basis.
  - Conduct daily situation briefs as needed.
  - Respond to media queries.
  - Monitor media operations within the AO.
  - Respond to and mediate potential media problems; react quickly to coordinate and facilitate information issues throughout the AO.
  - Validate media's credibility, expertise, knowledge, purpose and legitimacy.
  - Provide seamless connectivity for media accountability throughout the AO.
  - Conduct Primary and Secondary accreditation.
  - Primary--Full accreditation of non-accredited media.
  - Secondary--Process media previously accredited at Corps and higher.

8-24. Assumption #2: DMAIN is in country

- Act as the division spokesman.
- Advise the commander and staff on PA implications of plans and actions.
- Execute the plan.
- Assist DMAIN to gather Open Source Information to build the common relevant picture.
- Monitor battlespace communications (visual, audio, FM and satellite) for PA implications.
- Facilitate Media Operations:
  - Assist media to gain access to units, soldiers and commanders.
  - Conduct daily situation briefs as needed.
  - Monitor media operations within the AO.
  - Respond to and mediate potential media problems; react quickly to coordinate and facilitate information issues throughout the AO.
  - Provide seamless connectivity for media accountability throughout the AO.
  - Conduct Secondary accreditation.
  - Secondary--Process media previously accredited at Corps and higher.

8-25. The division public affairs section provides public affairs support to the division commander and to divisional units deployed in support of combined or joint operations. The division PAO has operational and tactical control over all PA TOE organizations assigned or attached to the division and coordinates closely with embedded PA sections within brigades or divisions to carry out PA operations.

8-26. The division PA staff, when deployed, is augmented by one PAD and one MPAD per three combat brigades. The division public affairs section, if augmented only by a Public Affairs Detachment, operates the division media operations center.

8-27. Traditionally, divisions have operated as part of a corps. In corps operations, divisions normally comprise 9 to 12 maneuver battalions, organic artillery battalions and supporting CS and CSS units. Divisions perform a wide range of tactical missions and for limited periods are self-sustaining. Corps augment divisions as the mission requires.

Chapter 9

Information Operations

INTEGRATING PUBLIC AFFAIRS

9-1. Information Operations involve a variety of disciplines and activities that range from electronic warfare and physical destruction through cyberwar and information campaigns. Public affairs is a related activity of IO, and contributes to overall operational success, both real and perceived.

9-2. Successful integrated IO requires coordination of themes, messages and activities in order to leverage the massing of information effects. When synchronized with other military operations, IO is a combat multiplier.

9-3. Information campaign objectives cannot be neatly divided by discipline, such as PA, CA and PSYOP. The responsible organization cannot be easily determined solely by looking at the medium, the message or the audience. For example, information about weapons turn-in policy and collection sites may be disseminated through a variety of means. This could include direct contact by Civil Affairs personnel with local populations and their leaders; PSYOP print and broadcast products; news releases, press conferences and other media facilitation coordinated by PA.

9-4. In accordance with joint doctrine (Joint Pub 3-61, Doctrine for Public Affairs in Joint Operations), public affairs are an operational function designed to contribute to the overall success of joint operations. For Public Affairs, the audience may be internal or external, but the objective is constant: Soldiers, participants and the public must understand objectives, motives and the nature, scope and duration of friendly actions. The relevant audiences important to the commander are not limited to soldiers and the American public, but are also international as well as local to the operation.

9-5. Synchronized information operations contribute to military campaigns in a variety of ways. These contributions may:

- Gain or sustain support for the U.S. or coalition position
- Reduce the need for combat forces
- Influence events with non-lethal means
- Counter propaganda and disinformatio
- Discourage adversary offensive operations
- Deter hostile actions

- Undermine illegitimate regimes
- Support the maintenance of coalitions

9-6. Information Operations during peacekeeping operations:

- Promote peaceful cooperation
- Lower friendly force requirements
- Counter propaganda and disinformation
- Reduce friction leading to hostilities
- Gain and maintain the initiative
- Shape opponent plans and operations

9-7. These goals may not be achieved solely by tactical level information operations, but rather, may be theater and national-level issues that are reinforced by tactical-level message dissemination. This requires horizontal and vertical integration of themes and messages to achieve a massing of information effects.

IO STAFF ORGANIZATION

9-8. Composition of the Information Operations battle staff/coordination council or other such element is flexible and tailored to the operation and desires of the commander. (See figure 9-1). Notional IO staff structures are included in FM 3-13 (100-6), Information Operations.

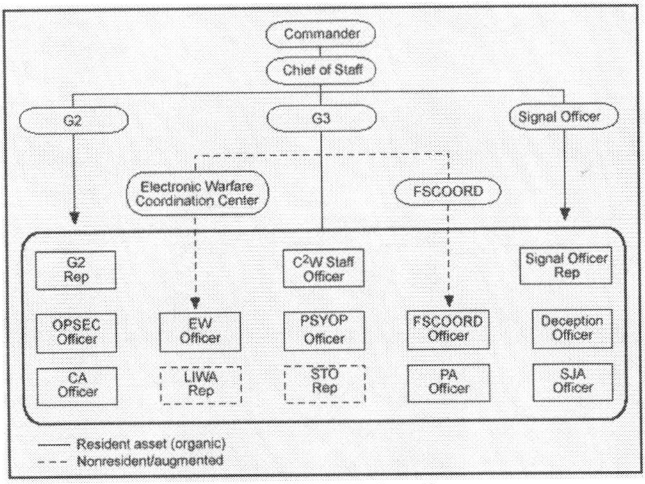

Fig 9-1 Notional IO Battlestaff

9-9. The IO cell is often headed by the G-3 or his designated representative, and includes representatives of a variety of organizations. The staff may include, but not necessarily be limited to the G2, G3, G5, G6, PAO, PMO, JAG, PSYOP, Electronic Warfare Officer, Political Advisor, Joint Military Commission representative, Fire Support Coordinator and Targeting Board representative.

9-10. The successful accomplishment of a specific mission may require the close coordination and synchronization of the range of information activities as well as maneuver elements. While the IO cell is lead by the G3 or his designate, IO coordination is the responsibility of the Land Information Warfare Activity field support team. While providing a public affairs representative to the IO Cell, PAOs must maintain a clear and direct link with the commander.

LIWA

9-11. The Land Information Warfare Activity, an INSCOM element, provides commanders with field support teams (FST) that serve as IO advisors in addition to effecting the synchronization and coordination of the range of activities that support IO. LIWA field support teams do not serve as functional area specialists, but rather, coordinate the activities of those elements. For example, the LIWA FST members may be from the military intelligence and PSYOP branches, but do not serve as the commander's intelligence analyst or PSYOP planner, or for that matter, Public Affairs advisor. They do, however, coordinate the actions and products of these and other activities in support of the IO plan.

PA SUPPORT TO INFORMATION OPERATIONS

9-12. PA participation in IO involves no completely new tasks but does require a broadened scope of operations. PA support to IO requires analysis of the Global Information Environment (GIE) and the operational environment, as well as synchronization of efforts with other organizations and agencies to ensure themes and messages are consistent and deconflicted.

9-13. PA in IO requires PA staffs to be fundamentally proactive rather than reactive. Often, actions may be taken and products developed to assist command achievement of a desired end state. This is more than merely reacting to events with a press release or conference

9-14. PA actions and events that support IO include print and electronic products, news releases, press conferences and media facilitation. PA advises the commander on how the operation is being perceived and portrayed and also provides guidance to unit commanders and soldiers. This includes regular talking points and themes for commanders and preparing soldiers to interact with the press. It's a means of emphasizing selected issues and positions--speaking with one voice.

PA ESTIMATE AND IO PLANNING

9-15. The starting point for PA contributions to Information Operations is the Public Affairs Estimate. (See Appendix C). The PA estimate consolidates information on the audiences, media presence, public opinion, personnel available and PA guidance.

9-16. This is not a static document created at the beginning of an operation, but must be continually updated to reflect changes in the operational situation and environment. Issues to consider include:

9-17. **Audience analysis.** Who are the audiences, both internal and external? What are their information needs? How do they get their information: television, radio, newspapers or word of mouth? Is the media state-run or independent? Does the audience population have telephones, cell phones, fax machines or Internet connections? These devices are frequently found even in developing countries and must be considered during the analysis of information channels.

9-18. **Media presence.** What media representatives and organizations are in the area of operation? Are they radio, television or print? Are they state-run or independent? What is their political slant? Are they pro- or anti-coalition? Are they receptive to coalition information products such as news releases or other print or electronic products? Is the local media interested in live interviews with U.S. commanders and soldiers?

9-19. **Public Opinion.** What are the opinions/beliefs of the local populations; of the international community; of the U.S. national population?

9-20. **Personnel available.** What is the available Public Affairs force structure (PADs, MPADs, BODs, CPIC/JIB staff, unit organic PA staff and individual additions), translators, Combat Camera and administrative staff.

9-21. **PA guidance.** What guidance has been received from higher levels? Official positions on theater issues are naturally not developed at the tactical level. What is the theater strategic/national command authority position? This is often coordinated and deconflicted at all levels via conference calls and other communication means.

IO CAMPAIGN CYCLE

9-22. There are four stages to an IO campaign cycle: capability development, assessment, planning and execution. The execution stage is accompanied by evaluation-- during and after the mission -- in order to adjust operations as needed and after the operation to gather lessons learned.

9-23. Capability development:

- Identify local resources and available external support. Theme and message delivery can take many forms, including radio/television, handbills, leaflets, loudspeakers, soldiers, displays, the Internet, internal information products, USIA, Voice of America, print and electronic news releases, press conferences, direct contact with parties, leaders, officials and citizens. Direct contact may include military liaisons, Civil Affairs personnel, diplomatic contact, or any form of personal interaction.
- Establish processes and procedures
- Collect, organize and store relevant information

9-24. Assessment

- Perform mission analysis
- Obtain commander's guidance
- Define IO goals and objectives
- Conduct risk assessment
- The assessment phase includes a mission analysis, clarification of the commander's guidance, initial identification of IO goals and objectives and a risk assessment. Goals and objectives include a

determination of the desired end state and what must be done to achieve it. This may mean inducing others to take or not take certain actions, or have the information to make certain decisions that will support the goals of the operation. Public Affairs is not in the business of shaping beliefs and attitudes of populations, but can provide factual information that enables people to make informed decisions.

- Risk Assessment
  - Consequences for command if information operation fails?
  - Potential unintended effects?
  - Operational success too reliant on IO?
  - Can IO campaign be used against U.S. or coalition?
  - Force protection issues?
  - Compromise or loss of impartiality?

9-25. Planning

- Develop and coordinate themes
- Determine the best implementation means
- Delineate tasks and responsibilities
- Identify feedback and measures of effectiveness channels
- Prepare implementation order
- During the planning phase, specific themes are developed and coordinated with all members of the IO cell. Message/theme delivery methods are determined and specific tasks and responsibilities are assigned. The IO cell may use a synchronization matrix to effectively manage IO events. For example, this matrix will indicate specific actions, events or products each member organization of the IO cell will execute or produce to support the plan. For example, a specific event may require PSYOP leaflets and broadcasts, PA press releases, news conferences and interviews with soldiers and Civil Affairs meetings with local officials and community leaders. These activities are coordinated on the matrix, ensuring deconfliction of resources, messages and products.
- Measures of effectiveness and feedback indicators vary widely and should be identified in the planning process. They may include questions raised by the media, editorials and commentaries, statements by public officials, postings to internet newsgroups and forums, demonstrations and protests, statements during meetings, responses given to public opinion surveys, behaviors during specific events, as well as other SIGINT and HUMINT collection and analysis.
- The product of the planning phase is a synchronization matrix and execution schedule. The matrix is then coordinated with the overall synchronization matrix, ensuring that IO is coordinated across the BOSs.

9-26. Execution

- Conduct the mission.

- IO monitoring must be conducted throughout the execution of the event and during follow-up review, feedback and evaluation.
- If necessary or possible, alter mission if evaluation determines it is not successful or unexpected responses occur.

9-27. **Evaluate**

- Assess the effectiveness of the operation
- Determine preventive methods, document lessons learned and apply to next operation

Appendix A

DoD PRINCIPLES OF INFORMATION

The DoD Principles of Information are contained in DoD Directive 5122.5, Change 1. They chart the course for all DoD Public Affairs activities, and apply to the full continuum of day-to-day activities and operations. It is the commander's responsibility to ensure that all planning for military activities and operations efficiently and effectively achieve the goals set by these principles.

DOD PRINCIPLES OF INFORMATION

A. Timely and accurate information will be made available so that the public, Congress, and the news media may assess and understand the facts about national security, defense strategy, and on-going joint and unilateral operations.

B. Requests for information from organizations and private citizens will be answered in a timely manner. In carrying out this policy, the following principles of information apply:

(1) Information will be made fully available, consistent with statutory requirements, unless its release is precluded by current and valid security classification. The provisions of the Freedom of Information Act and the Privacy Act will be complied with in both letter and spirit.
(2) A free flow of general and military information will be made available, without censorship or propaganda, to the men and women of the Armed Forces and their family members.
(3) Information will not be classified or otherwise withheld to protect the government from criticism or embarrassment.
(4) Information will be withheld only when disclosure would adversely affect national and operations security or threaten the safety or privacy of the men and women of the Armed Forces.
(5) The Department's obligation to provide the public with information on its major programs and operations may require detailed public affairs planning and coordination within the Department and with other government agencies. The sole purpose of such activity is to expedite the flow of information to the public; propaganda or publicity designed to sway or direct public opinion will not be included in Department of Defense public affairs programs.

Appendix B

THE GUIDELINES FOR COVERAGE OF DOD COMBAT OPERATIONS

DOD MEDIA GUIDELINES

The DoD Media Guidelines, issued as Change 3 to DoD Directive 5122.5, provide the following guidelines for coverage of DoD combat operations:

A. Open and independent reporting will be the principal means of coverage of U.S. military operations.

B. Pools are not to serve as the standard means of covering U.S. military operations. But pools may sometimes provide the only feasible means of early access to a military operation. Pools should be as large as possible and disbanded at the earliest opportunity--within 24 to 36 hours when possible. The arrival of early access pools will not cancel the principle of independent coverage for journalists already in the area.

C. Even under conditions of open coverage, pools may be appropriate for specific events, such as those at extremely remote locations or where space is limited.

D. Journalists in a combat zone will be credentialed by the U.S. military and will be required to abide by a clear set of military security ground rules that protect U.S. forces and their operations. Violation of the ground rules can result in suspension of credentials and expulsion from the combat zone of the journalists involved. News organizations will make their best efforts to assign experienced journalists to combat operations and then make them familiar with U.S. military operations.

E. Journalists will be provided access to all major military units. Special operations restriction may limit access in some cases.

F. Military public affairs officers should act as liaisons but should not interfere with the reporting process.

G. Under conditions of open coverage, field commanders will permit journalists to ride on military vehicles and aircraft whenever feasible. The military will be responsible for the transportation of pools.

H. Consistent with its capabilities, the military will supply PAOs with facilities to enable timely, secure compatible transmission of pool material and will make these facilities available whenever possible for filing independent coverage. In cases when government facilities are unavailable, journalists will, as always, file by any other means available. The military will not ban communications systems operated by news media organizations, but electromagnetic operational security in battlefield situations may require limited restrictions on the use of such systems.

I. These principles will apply as well to the operations of the standing DoD National Media Pool system.

Appendix C

PUBLIC AFFAIRS ESTIMATE

SAMPLE PA ESTIMATE

FM 6-99 (101-5)

____

Classification

Headquarters
Place of issue
Date, time, and time zone
Message Ref. no.

PUBLIC AFFAIRS OPERATIONS ESTIMATE NO. __________

References: Maps, charts, or other documents.

Time zone used throughout the estimate:

1. MISSION

This paragraph lists the command's restated mission from a public affairs perspective.

2. THE SITUATION AND CONSIDERATIONS

The paragraph describes the strategic and operational media environment in which the operation is being conducted and identifies the critical factors that might impact on the command's mission -- the "action and reaction" within global media channels. It identifies the media environment across the operational continuum, describing it from "austere" for low media interest and capability in a limited AOR communication infrastructure to "dynamic" for high media interest and capability in a high-tech AOR infrastructure. At a minimum, this paragraph must include:

a. Information environment. This paragraph describes the characteristics of the operation and the information environment in the area of operations. It identifies any activities or issues affecting the over-all mission and the command's public affairs objectives.

b. Media presence. An assessment of the news media presence in the theater of operations prior to deployment and the likely presence of additional news media during the conduct of operations. This assessment should address the authority under which media representatives are operating and the degree of control that can be imposed on their efforts.

c. Media capabilities. An assessment of the media's information collection and communication technology, specifically identifying their level of visual information acquisition and satellite communication capabilities. It includes an analysis of the logistics support, transportation assets, and host-nation communications infrastructure available to them.

d. Media content. An assessment of the global media's presentation of information and their agendas, and an analysis and prioritization of the potential strategic and operational issues confronting the command in the news media. This media content analysis will provide an evaluation of the quantity of coverage and the nature of that coverage.

e. Public opinion. Assessment of national and international attitude about the operation and command, leaders, and soldiers conducting it. This paragraph should include both the perceptions held by major audience groups and the relative solidity or strength of those attitudes. A public opinion analysis should include as a minimum an analysis of the following groups:

- American public
- Civilian political leaders
- Coalition and allied forces and their publics
- International audience
- Internal command audience
- Home station public

f. Information channel availability. An assessment of the information channels available for the communication of information in and out of the AOR. It identifies the means available to the commander for receipt, transmission, and dissemination of voice, data, text, graphics, and digital visual imaging. It describes command, coalition, and local national facilities and equipment available, to include an analysis of available telephone lines for voice and data transmission, the accessibility of audio and video channels, the prevalence of private communications devices such as soldier-owned cellular telephones, facsimile machines, computers, portable radios and televisions, still and video cameras, and the nature and flow of the information possible through these channels.

g. Information needs. This is an assessment of the information needs of the previously identified key publics. It analyzes key internal and external audiences and assesses theirs news and information expectations. It identifies the types of information made available to these key audiences.

h. Personnel situation. Describes the present dispositions of public affairs personnel and units that affect the PA mission, and the assets needed and available. State known personnel problems, if any, that may affect the PA situation. Consult the personnel estimate for details. (Examples of personnel include shortages of PA NCOs and skilled operators for equipment.)

i. Public Affairs situation. This summarizes current PA objectives and identifies specific courses of action for each objective. At high levels of command, detail information in a summary with a reference to an annex to the estimate. Subparagraphs will include all current (PAG) from OSD (PA).

j. Logistical situation. State known logistic problems, if any that may affect the PA situation. See logistic estimate for details. (Examples of logistic problems include the lack of transportation and adequate facilities.)

k. Assumptions. Until specific planning guidance from the commander becomes available, you may need assumptions for initiating planning or preparing the estimate. Modify these assumptions as factual data or planning guidance becomes available.

3. ANALYSIS OF COURSES OF ACTION

Analyze each course of action based on the public affairs objectives in paragraph 2i. Indicate problems and deficiencies. At a minimum, subparagraphs should include media facilitation and support, news and information provision, and force training and support.

Analyze each COA from a PA point of view to determine its advantages and disadvantages for conducting PA. The detail in which the analysis is made is determined by the level of command, scope of operations, and of urgency of need.

4. COMPARISON OF COURSES OF ACTION

Compare each course of action. List advantages and disadvantages of each course of action under consideration. Include methods of overcoming deficiencies or modification required for each course of action.

5. CONCLUSIONS

a. Indicate whether you can support the command mission (restated in paragraph from the public affairs viewpoint.

b. Indicate which COAs you can best support.

c. List major public affairs deficiencies, which commander must consider. Include specific recommendations about methods of eliminating or reducing their effects.

____

(Public Affairs Officer)
ANNEXES: (As required)

DISTRIBUTION:

(NOTE: The headings listed in this assessment are for example only. Use headings appropriate to your command's operations.)

____

Classification

PA Operations Estimate

Appendix D

PUBLIC AFFAIRS ANNEX

SAMPLE PA ANNEX

FM 6-99 (101-5)

____

(Classification)

(Change from oral orders, if any)

Copy ___ of ___ copies
Issuing headquarters
Place of issue (may be in code)
Date-time group of signature
Message reference no. _____

ANNEX__ (PUBLIC AFFAIRS) TO OPERATIONAL ORDER ______

References: Maps, charts, and other relevant documents

Time zone used throughout order:

1. SITUATION

A brief general description of the situation, information affecting public affairs support, which paragraph 1 of the OPORD does not cover, and intended purpose of this annex.

a. Friendly forces. Outline the higher headquarters' plan (and PA annex) and adjacent unit PA plans. Provide information on friendly coalition forces, which may impact the PA mission. Note PA resources supporting the unit. (Who, where, when), (higher, allied and adjacent headquarters).

b. Attachments and detachments. Identify all augmenting PA units supporting this command and all attached/assigned subordinate units. Include effective dates, if applicable.

c. Enemy forces. List information not included in the OPLAN/OPORD, which may impact the PA mission. (Who, where, when, disinformation, rumors, propaganda and OPSEC).

d. Media. Identify media in the area. (who, where, pools, US. international, local-host country).

e. Assumptions. List any additional assumptions or information not included in the general situation, which will impact the PA mission.

2. MISSION

Clearly, concisely state the public affairs mission. (Internal information for deployed and non-deployed forces, media facilitation and staff operations).

3. EXECUTION

a. Concept of operation. Briefly summarize the public affairs operation plan. Include PA priorities. (Intent --access, information, welfare, morale, will to win) (Concept--who, where, what, why, when) (Specifics--task to a subordinate; who is to do what, where, when, covers non-PAs too, actions with media: credential, train, transport)

b. Outline of PA tasks. Identify and assign supporting PA tasks to each element of subordinate and supporting units. Assign specific tasks to elements of the command charged with public affairs tasks, such as requirements for PA augmentation.

c. Coordinating instructions. Give details on coordination, task organization and groupings. List instructions, which apply to two or more subordinate elements or units. Refer to supporting appendixes (PA assessment) not referenced elsewhere. (Public Affairs Guidance, media in country, media enroute with US forces, media contact report, handover checklist, and task organization).

4. SERVICE SUPPORT

a. A statement of the administrative arrangements applicable to this operation. If they are lengthy or are not ready for inclusion in the OPORD, these arrangements may be issued separately and referenced here.

b. A statement of the logistical arrangements applicable to this operation. Specific coordination should be included if possible, but arrangements may be issued separately and referenced here, if they are too lengthy. (Class I-IX and water), (Services: billets, medical, laundry and mortuary), (Transport: ground, air, TOE, tasked rented/leased, contracted).

5. COMMAND AND SIGNAL

List signal, visual imaging and satellite communications policies, headquarters and media center locations or movements, code words, code names, and liaison elements. (PAO location, media center, JIBs, sub-JIBs, phones, faxes, e-mail and web page).

ACKNOWLEDGE:

NAME (Commander's last name)
RANK (Commander's rank)

OFFICIAL:
APPENDIXES: (List PA assessment appendix)
DISTRIBUTION:

Classification

Appendix E

PUBLIC AFFAIRS GUIDANCE

SAMPLE PA GUIDANCE

Public Affairs Guidance (PAG) is the operational tool that guides commanders and their public affairs officers in the application of doctrine and policy during major military operations, exercises, and contingencies. The information below is tended to assist local commanders in preparing and obtaining approved guidance.

DoD policy requires that PPAG be provided to the Assistant Secretary of Defense-Public Affairs (ASD-PA) by the Unified and Specified commands and others, as required for all major operations.

This requirement includes major training exercises that could attract national and/or international attention. PPAG may not be used without ASD-PA approval.

Upon receipt of the warning order, the commander, through his PAO, should request PAG from high headquarters. PAG may be included in alert notification or operational orders (see Part Three: Operational Planning). Commanders of major units/commands will direct their PAOs to prepare PPAG to forward the proposal through MACOM and Unified/Specified command PA channels to ASD-PA.

Commanders of Unified/Specified commands should ensure that the PPAG has been coordinated with appropriate organizations within the theater of operations whenever possible (e.g., embassies, country teams, host governments, subordinate commands).

Upon receipt of the PPAG, the ASD-PA coordinates and staffs the PPAG within the DoD and Department of State.

The ASD-PA then issues a message either approving, modifying, or disapproving the PPAG. PPAG is broken down into subject, references and then eight paragraphs (Information/explanation, Purpose and coordination for PPAG, PA Approach, Public Statement, Q&As, Contingency Statement, Miscellaneous Information, Point of Contact). The format for PPAG follows:

SUBJECT

The subject line of the PPAG should state "PROPOSED PUBLIC AFFAIRS GUIDANCE - followed by the exercise and/or event name (U)." For coordination, it is best if the subject is unclassified.

If an exercise or event is so sensitive that the actual name cannot be used, an unclassified short title should be used; e.g., "PROPOSED PUBLIC AFFAIRS GUIDANCE - CC-(U)."

REFERENCES

Pertinent messages or other documents shall be cited in the reference section. If the PPAG is based on PA policy in the Significant Military

Exercise Brief, then the SMEB message Date-Time-Group (DTG) shall be listed.

EXERCISE INFORMATION

The **first paragraph** of the PPAG shall explain the references, the exercise, and any significant existing or anticipated problems associated with the exercise. The information in this paragraph is not for release so may remain classified after the PAG is approved for release. This paragraph may restate some PA information from the SMEB.

COORDINATION INFORMATION

The **second paragraph** shall explain the purpose of the message; identify it as being fully coordinated and theater-approved; request ASD-PA approval and specify the date it is required for use. If the PAG is transmitted to the ASD-PA before it is fully coordinated, it is the responsibility of the submitting command to ensure that the ASD-PA is promptly informed of the results of the remaining coordination. The submitting command should always follow-up a PPAG message with a phone call to ensure that the primary addressee(s) is aware that the message is en route. When the submitting command is a supporting CINC from outside the supported CINC's AOR, the supported CINC is responsible for theater coordination.

PA APPROACH

The **third paragraph** shall discuss the public affairs approach for the exercise; i.e., active or passive. This may be a restatement of the PA policy indicated in the SMEB.

- Active Approach
  - For this discussion, an "active approach" involves efforts made to stimulate public or press interest such as distributing press releases and inviting the press to observe the exercise. If an exercise or event is to be publicly announced, this paragraph shall state who will make the announcement, the method of announcement, and preferred time, and date for the announcement. If unusual circumstances prevail, the rationale for the recommendation should also be included. Part I and II exercises shall normally be announced by the ASD(PA) by issuing a news release (blue top). Other lesser exercises or training deployments, if announced by the ASD(PA), normally shall be made by release of a memorandum for correspondents (MFC). The preferred release time and/or date of exercise announcements is 1200 Eastern Standard Time (E.S.T.) on either Tuesdays or Thursdays in conjunction with the normally scheduled DoD press briefing. If a combined announcement is desired with a host country, complete details of the methods, time, and procedure shall be included in this paragraph. The active approach is recommended whenever possible to ensure appropriate media coverage of specific commands and/or units.
- Passive Approach
  - A "passive approach" is where no action is taken to generate media and/or public interest in an issue or activity beyond answering specific inquiries. If a passive approach is desired,

the PPAG shall so indicate and specify that the PAG is for response to query (RTQ) only. It shall also specify who is authorized to respond; e.g., "Only OASD(PA) may RTQ," or "All of the following addressees may use this PAG for RTQ only." To de-emphasize an event, it is best to authorize release or RTQ at the lowest possible level.

PUBLIC STATEMENT

The **fourth paragraph** shall contain a statement that explains the exercise and/or event. The statement shall be for public release in an active PA approach or for RTQ in a passive PA approach. For ease of coordination, each paragraph of the statement shall be identified as a sub-paragraph of the message; for example: The following statement is for initial public release: (TEXT FOLLOWS): QUOTE.

EXERCISE (NAME).....

THE EXERCISE WILL.....

PREVIOUS MILITARY.....

FOR ADDITIONAL INFORMATION,

CONTACT.....(UNQUOTE).

As indicated above, the last paragraph of the statement shall identify points of contact where additional information may be obtained.

QUESTIONS AND ANSWERS (Q&AS)

The **fifth paragraph** shall contain a list of proposed Q&As to enable the user to respond to the majority of anticipated questions. They should all be contained in one paragraph and should be numbered sequentially; e.g., Q1, A1; Q2, A2; Q3, A3, etc. Q&As are for use in both active and passive PA approaches, but are strictly for RTQ only and shall not be given to media as handouts.

CONTINGENCY STATEMENT

The **sixth paragraph** of the PPAG shall contain a contingency statement to be used efore release of the final PAG. Usually, the contingency statement should be that we don't discuss exercises before they have been formally announced. However, this approach can be modified, as appropriate, depending on the circumstances of the exercise. If a contingency statement is not required, so state in Paragraph 6 of the PAG.

MISCELLANEOUS INFORMATION

The **seventh paragraph** shall contain other pertinent information to include the following items (when a certain sub-paragraph is not applicable, so state):Media Information Centers (e.g., Joint Information Bureaus (JIBs), Press Information Centers (PICs), exercise PA elements, etc.) Discuss whether centers are joint or combined; delineate who is responsible for the establishment; give generic description of its composition (e.g., U.S. Army desk (O-4 and E-6/E-7), U.S. Navy/Marine desk (USN O-4/O-5, and USMC E-5/ E-6), etc.); establish the center's functions (coordination of all exercise media and/or PA activities, clearance of U.S. military-generated news material before release, production of news material for release, escort of accredited news media representatives); etc.

- Command Relationships
  - Designation of sole approving authority for all exercise-related news materials; procedures for the release and/or clearance of information (to include list of addressees for notification in case of accident and/or incident); request for participating commands and/or units to ensure that the media center is action and/or information addressee on all messages with potential PA impact (to include incident and/or accident reports); hometown news release requirements and/or instructions (passive PA approach may make hometown releases inappropriate); etc.
- Media Coverage
  - State whether media coverage is encouraged or solicited, giving rationale; news media transportation instructions; point of contact (POC) and procedures for handling such requests; requirements for news media representatives (valid passport, working media visa, local accreditation requirements, funds for food, lodging, return travel (if military air is not available), etc.); instructions regarding assistance to continental United States (CONUS)-based units for handling request from news media for accompanying travel before and following public announcement of the exercise; etc.
- DoD National Media Pool
  - Each exercise is a potential opportunity for activation and deployment of the DoD National Media Pool to cover exercise activities. As a minimum, planning should include arrangements for local ground and/or air transportation, special clothing or equipment to be provided, messing, billeting, protection of media equipment and gear, local escort requirements, and communications support for filing of pool products. Sponsoring commands shall indicate whether the exercise should be considered for a pool deployment. Identify the primary POC should the pool be activated.
- Internal Media and Audiovisual Coverage
  - Provide instructions on assistance that will be provided to this effort; degree of freedom of movement (to include whether escorts are necessary); screening of visual information (VI) materials upon completion of exercise; sponsoring command POC for handling internal information matters; etc. Also include guidelines for Armed Forces audiovisual teams documenting the exercise.
- Media Opportunities
  - If known well enough in advance, provide chronology of potential exercise events that would be of interest to media.
- Miscellaneous PA Considerations

- Indicate any other proposed PA activities or considerations; if there are none, then so state.

POINTS OF CONTACT

The **eighth paragraph** shall state the originating POC's name and phone number.

DECLASSIFICATION INSTRUCTIONS

Declassification instructions shall be the last part of the message and in accordance with subsection 4-207 of DoD 5200.1-R (reference (d)).

Appendix F

PUBLIC AFFAIRS OPERATIONS BRIEFINGS

BRIEFINGS

The purpose of a briefing and the desired response or result determines the briefing technique. Basically, there are four briefing types: the information briefing, the decision briefing, the mission briefing, and the staff briefing.

The Information Briefing. The information briefing informs the listener and deals primarily with facts and background information. The Information Briefing contains an introduction to the subject and the scope of the subject area. It then presents the high-priority information requiring immediate attention and complex information involving complicated plans, systems, statistics, or charts. It may also explain controversial situations or information, which require elaboration.

The Decision Briefing. The Decision Briefing includes many of the elements of the information, but goes further by seeking a decision from the decision-maker. At the beginning, the briefer clearly states that he is seeking a decision; at the end, he requests the decision.

The Mission Briefing. The Mission Briefing gives special instructions, amplifies the mission, elaborates on new orders, or assigns taskings to subordinate elements. This briefing usually follows the five-paragraph operations order format. But the briefer may also choose the information briefing format.

The Staff Briefing. The Staff Briefing informs the commander and staff of the current operational situation. Its purpose is to generate a coordinated or unified effort and in a tactical environment. It serves to keep the entire staff aware of each section's activities, thus aiding coordinated action. While there is no specific briefing format, commanders usually tailor this type of briefing to fit their information needs. PAOs address the major PA activities and the PA implications of the operational situation and other staff sections activities.

In a commander's staff briefing, the PAO is responsible for providing a summary of the "global information environment (GIE)." The PAO should present this information at the beginning of the briefing, following the intelligence summary. The PAO's GIE summary, combined with the G2/S2's intelligence preparation of the battlefield (IPB) briefing, helps to complete the picture of the operational situation, which the other elements of the command must react.

Preparation. After conducting the initial analysis of the situation, the briefer gathers and organizes information, prepares visual aids, and then rehearses, briefing only what is essential. Before developing the briefing, the briefing officer or NCO must know and understand:

- the higher headquarters mission
- the higher commander's intent
- the commander's personal guidance and intent
- all aspects of the PA estimate of the situation
- the command's objectives

While much of the information required for a GIE presentation can be found in previously published PA Area Studies, a significant amount of situational information should pulled directly from the PA Estimate of the Situation and updated to present the latest information available.

BRIEFING FORMAT

The PA briefing format should follow an outline similar to the PA Estimate of the Situation:

- Mission - Command mission including PA objectives
- Situation and Conditions
- Information environment
- PA situation
- Media presence
- Media capabilities
- Media content
- Public opinion
- Information channel availability
- Information needs
- Current Operations/Actions
- Future Operations/Actions

Appendix G

PUBLIC AFFAIRS AREA STUDIES

FORMAT FOR PA AREA STUDIES

Public Affairs area studies provide the PA specialist and the PA planner a starting point to begin their campaign planning process. These studies are produced either to address long-term general background information or to address immediate short-term needs.

BASIC STUDY

A PA study is a document that succinctly describes the most PA-pertinent characteristics of a country, geographical area, or region. It serves as an immediate reference for the planning and conduct of PA operations.

TITLE PAGE

The title page of a Media Environment Study (MES) must show the country, area, group, or other subject of the study. It should also show the classification, the copy number, the date-time group, and command post location. (The originating unit should keep a record of the names, grades, and duty assignments of the authors.)

EXECUTIVE SUMMARY

The executive summary should address the strategic significance of the area under study. (The authors should write the executive summary last, in a clear, brief, accurate, and coherent form.)

NOTE: The commander reading the summary should us its information to decide how to employ Public Affairs in that area.

TABLE OF CONTENTS

Executive Summary

Introduction

Chapters

1. History and Ideology
2. Government and Politics
3. Foreign Relations and Policy

4. Society and Culture
5. Economy
6. Military Establishment
7. Communication Process and Effects

Endnotes

APPENDIX

A. Country Summary

B. Government Structure

C. Communications Facilities

D. Glossary

Bibliography

Distribution

INTRODUCTION

The introduction outlines the study's intent, cites the directive requiring the study, and explains the study's format.

The study's intent is to provide a summary of aspects of subject country significant to GIE. The study should identify psychological vulnerabilities, characteristics, insights, and opportunities that exist in subject country.

It is prepared as a basic source document for further development of estimates, plans, and annexes.

Although this study can help develop contingency plans, it is not tailored to any particular plan. Rather, its neutral data and insights can be used to analyze possible political and military developments in the region.

Insert here a paragraph referencing the authority directing the study and stating the research cutoff date and provisions for updating the study.

Focus the PA study on the GIE aspects of the many topics addressed. Do not view this document as a comprehensive and self-contained area study.

Instead, use it as a complement to such other standard references as the Department of the Army Area Handbook Series. In addition, the PA study should include:

Results from standard open source products.

Tries to be more analytical than descriptive in nature, making it subject to varying individual perspectives.

Works well when PA specialists compare it with studies on other countries in the region or area.

Is not a U.S. foreign policy statement or comprehensive analysis of subject country, except in areas with direct GIE relevance.

Insert here a statement of U.S. policy goals toward the country in question. This information comes from the proper USA Country Plan, Department of State Policy Memorandum, or similar document, in the priority order.

Because certain gaps exist in getting PA study material (classification level, availability of complete and timely information, or time limits on research), listing these gaps here to aid future research and guide PA study users to further inquiries.

Insert here issues or GIE-relevant material (such as an area map) not included, addressed, or completely answered elsewhere.

CHAPTER 1

HISTORY AND IDEOLOGY

In this chapter, review the evolution of the state and its people, focusing on aspects having GIE and Public Affairs significance. Do not detail chronology of the country's development. Keep in mind, however, the country's history has an important relationship to the country's historical perspective, attitudes, and current world position.

Because of its special importance to PA, in this chapter, cover historical issues thoroughly. An historical analysis of current political, economic, and military policies gives PA personnel a solid base for the study.

CHAPTER 2

GOVERNMENT AND POLITICS

In this chapter, summarize the country's political system. Give a description of its political power sources, policy making process, and the political complexities of the government.

When discussing the political system, pay close attention to the role of individuals, special interest groups, and political parties. Include the population's political attitudes, values, and view of the political system. Also discuss the government's function in society.

CHAPTER 3

FOREIGN RELATIONS AND POLICY

In this chapter, summarize the country's foreign relations. Describe its political alignment in world affairs and its relationship with the United States. Describe the foreign policy of the country. Also analyze and interpret why the country acts as it does in international affairs.

CHAPTER 4

SOCIETY AND CULTURE

In this chapter, analyze the subject country's social setting. Provide the PA personnel with the knowledge needed to understand potential audiences.

Cover the country's social organization and cultural and behavioral patterns and characteristics. Place special emphasis on the society's social values and the role of the family.

Address culture, social organization, education, customs, ethnic composition, and the interrelated effects of religion, language, and history.

CHAPTER 5

ECONOMY

In this chapter, present a brief description of the characteristics, structure, and dynamics of the national economy. Cover the subject country's economic strengths and weaknesses, current economic and labor problems, and economic potential

Describe the country's economic base and the importance of agriculture, industry, and trade. This information helps determine if the present economic structure meets the people's needs.

These economic considerations explain many of the sociological conditions that impact public opinion. Address society's perceptions of the wisdom of government economic policies.

Also describe individual or group perceptions of how members of society stand to gain or lose from those policies.

CHAPTER 6

MILITARY ESTABLISHMENT

In most countries, the military establishment involves itself in internal politics as well as external defense. Even when the military establishment does not directly compete for political power, its actions influence social and political development. Analyze the following topic areas:

- Emergency of the modern military establishment.
- Military roles in the political, economic, and social spheres, and the effects of those rules.
- Issues creating cohesion or conflict within the armed forces.

Leadership.

Conflicts within the military establishment.

Extent, quality, and influence of foreign military aid.

CHAPTER 7

COMMUNICATION PROCESS AND EFFECTS

List essential information about communication patterns for the implementation of a PA program. Include the following information:

Manner and social means of communication (not technical data on communications facilities).

Languages and language groups, nonverbal communication, and nonverbal symbols specific to the country's culture or cultures.

Distinctive styles in rhetoric or visual arts, including dramatic, poetic, and musical forms. All these forms are significant to PA.

Data on the society's formal and informal leadership positions where the incumbents are key communicators and opinion leaders.

Analysis of the reading and listening habits of the society.

Analysis of printed formats.

Analysis of media effectiveness. Also address freedom of the press issues, if any.

ENDNOTES

LIST PUBLICATIONS, DOCUMENTS, AND OTHER SOURCES USED IN PRODUCING THIS BPS. NUMBER TEXT REFERENCES SEQUENTIALLY WITHIN THE STUDY.

APPENDIX A

COUNTRY SUMMARY

Give the reader a brief overview of the subject country, its geography, and its people. These background data and statistics should include the following items:

Country. Identify the country, tell when formed, and show previous control.

Government. State briefly the type of government, method of appointing or electing leaders, and length of terms. List current leader or leaders and political power in country.

Size, location, and geographical subdivision. List the size of the area in square miles or kilometers, and give the general location. Show any geographical subdivision, such as coasts, mountains, and flatlands.

Population. State the number of people and the area density. Show the heavily populated areas.

Languages and dialects. List the official language, languages spoken by the population, percentage of population speaking each language, and areas of the concentrations.

Labor. Outline the total work force, the area of endeavor, and the percentages.

Religions and sects. List the religions of the area and the percentage of the population that practices each.

Education. List the types of systems and the primary emphasis of each.

Literacy rates. Latest statistics.

Health. List the general conditions of the populace. Describe the medical care system.

Justice. Describe the justice and court systems.

Administration. Outline the breakdown of the governmental and judicial districts, counties, or precincts.

Transportation. List the methods of transportation available and include the total capabilities. This information may include the number of airlines, airfields, kilometers and kinds of highways, and kilometers of waterways and depth.

Armed forces. List organization and strengths.

Police. List the types and areas of responsibilities.

APPENDIX B

GOVERNMENT STRUCTURE

List the formal government structure, key positions, and organizations of the country. Outline the chain of government control, including political parties, if applicable. A schematic diagram may be helpful.

Include only branches of government and their key positions, not names.

APPENDIX C

COMMUNICATION FACILITIES

Give a brief overview of the subject country's media facilities. Include the facilities' locations and levels of technical sophistication. Cover printing, publishing, and the distribution of radio and television receivers, studios, transmitters, and relay facilities. Include news service facilities. Write this appendix as if the U.S. PA units will use this equipment or contract for its services. Since PA personnel may get operating supplies or repair parts from in-country sources, provide the following information:

- Make.
- Model.
- Type.
- Series.
- Name of manufacturer.

Any other technical information on the repair or operation of this equipment.

GLOSSARY

Prepare a glossary that lists in alphabetical order all acronyms and foreign words used in the study. List also all words and terms that have special meaning and need to be defined.

BIBLIOGRAPHY

List the source material used. Include the name of the author, the title of the publication, the publisher, and the date of publication.

DISTRIBUTION

Dissemination is accomplished by the originating agency for the recipients within PA. Include in the distribution list the identification of recipient agency (by code), the number of copies furnished, and the office symbol of the recipient.

Appendix H

BRIEFINGS AND PRESS CONFERENCE FORMATS

NEWS BRIEFING AND PRESS CONFERENCE FORMAT

Before the Presentation

Know your publics
Anticipate interests, concerns and questions
Consider the latter in preparation

Prepare your presentation
Develop a strong introduction
Develop a maximum of three key messages
Assemble your supporting data
Prepare audiovisual aids
Practice

Prepare for answering questions
Anticipate what questions will arise
Prepare answers to those questions
Practice questioning and responding

The Opening Statement

A strong opening statement sets the tone for the press conference or news briefing and is crucial in attempting to establish trust and build credibility. The elements of a strong opening are:

Introduction
A statement of personal concern
A statement of organizational commitment and intent
A statement of purpose and plan for the meeting

Key messages and supporting data
A maximum of three "take-home points"
Information to support the key messages

Conclusion
A summarizing statement

Total time for all presenters should be 15 minutes or less. Do not have too many presenters. Three is usually sufficient.

Introduction

Remember that perceived empathy is a vital factor in establishing trust and building credibility and your publics assess it in the first 30 seconds.

Examples are:

Statement of personal concern: "As a resident of this community I'm interested in the safety and well-being of our families and neighborhoods."

Statement of organizational commitment and intent: " I'm here to share with you the knowledge and confidence I have in the military's ability to assist the citizens of our community. They have been trained in their occupational skill to assist with the task at hand."

Statement of purpose and plan for the presentation: Today I would like to share with you the most current information regarding the (incident.) I will also be available to answer additional questions or to continue the discussion.

Key messages and supporting data

The key messages are points you want your publics to have in their minds after the presentation. They should:

Address central issues.
Be short and concise.
Examples are:
"We have trained personnel and emergency response plans in place to aid in protecting the health, safety and welfare of the public. We are working with local and state officials to handle the incident.

"We are actively responding to the emergency...."

To develop your key messages:

Brainstorm
Think freely and jot down all pieces of information you wish to communicate.

Select key messages
Identify the most important ideas. Repeat the process until you list is down to three items.

Identify supporting data
Other information you listed probably provides support to your key messages. Organize it to reflect this.

Conclusion

Restate verbatim your key messages.

Add a future action statement --- What is your organization going to do about this problem in the short and long term?

Appendix I

WAIVER OF LIABILITY STATEMENT

Whereby, I __ . NAME __________________________ passport no:____________ am about to travel with __________forces, and whereas I am doing so entirely upon my own initiative, risk and responsibility; now, therefore, in consideration of the permission extended to me; I do hereby for myself, my heirs, executors and administrators, remiss, release and forever discharge ________ and its member officers, agents and employees acting officially or otherwise, from any and all claims, demands actions or causes of action, on account of my death or on account of any injury to me or my property which may occur from any cause during my stay, travel as well as all ground, flight or sea operations incidents thereto.

I also agree to withhold any classified information, which may be accidentally disclosed to me, and to respect embargo restrictions, which may be imposed on information which, if disclosed, may jeopardize operational security. During my stay with _______ forces, I will not interfere with operations. I understand that failure to comply with these security restrictions will result in the loss of authorization to accompany ______and may result in cancellation of my press registration.

____________________________ ____________________________

Signature **Witness**

____________________________ ____________________________

Printed Name **Nationality**

Address:____

Please provide the following information for a person to be notified in an emergency (preferably next immediate relative):

____

____

Appendix J

SAMPLE PRE-DEPLOYMENT CHECKLIST

INTRODUCTION

A multitude of factors make each deployment uniquely different from other deployments. Each factor must be carefully examined to determine its impact on the mission, actions before deployment and what equipment is taken.

(Unit Basic Load). In all other instances the chances of drawing the UBL are remote. Certainly if the unit is deploying for war it would draw the unit's basic load of ammunition Exercise or deployment duration will affect the quantity of expendable supplies.

This checklist, like all others, is based on what has occurred before and what we have come to expect in the future. As equipment and missions change, so too must the checklist. Bottom line, expect the unexpected and react accordingly; use the checklist as a guide to help you through deployment preparation. Add to it whenever the need arises.

CHECKLIST ITEMS

1. Each HQ element and unit should have a 45-day supply of expendables identified and set aside for contingency operations. This supply should be inventoried quarterly and stockage rotated accordingly.
2. Maintain and update a master list of all areas and topics that can affect individual readiness. This includes all shots, dental status, family support plan, check to bank, etc.
3. The PAO or detachment commander (or deploying team leader) should:
   a. Review and update PA estimates annexes and plans.
   b. Obtain and comply with applicable published Public Affairs Guidance.
   c. Inventory and inspect TOE equipment for accountability and serviceability before deployment.
   d. Figure total weight and cube of equipment before moving to assembly area.
   e. Check dependency clause in TOE document or OPLAN to determine who will be supporting and who is supporting us. Ensure responsible parties know the relationship. Establish liaison with units OPCON; attached or any new parent organization.
   f. Ensure accompanying equipment not on TOE is listed on interim authorization document (such as facsimile machines or cellular phones).
   g. Prepare an internal OPORD for your element detachment or team.
   h. Review the essential elements of friendly information (EEFI) contained in the base OPLAN/OPORD and ensure each soldier understands them.
4. All leaders must consider the need for the following:
   a. Will flak vests be required?
   b. Should each soldier take a footlocker instead of a dufflebag?
   c. Will weapons' carrying/security cases be needed/available? (Will weapons and protective masks be required?)
   d. Will desert or jungle uniforms be required, and if so, what fund can be used to pay for alterations, sewing and patches?
   e. Is there an opportunity for a service contract to be initiated at the deployed location? Cameras, video, computers, etc.? if not, and cameras go down and must be swapped out, what is the plan?
   f. Will the deployment be considered TDY? If so, who prepares orders? What fund cite will be used? Are rental vehicles available? (Are we TDY under field conditions and required to carry meal card?)

g. Will a fund cite be made available after redeployment to pay for repairs? (Make this request soon after being tasked - don't wait until redeployment to find out you've got to use unit funds to repair equipment.)
h. If departing from location other than current station, what type of transportation to that location is to be used for personnel and equipment?
i. Will personnel and equipment travel together at all times? (Insist that they do whenever possible!)
j. Does each team have a credit card holder for authorized payments or purchases?
k. Has an express mail system been established to ensure timely transit of products to home station? With what frequency? (If you are the ARFOR or sub-JIB, have the division PAO's/PAD's establish a plan to get products to you for review, release or use).
l. Under what conditions will the MPAD or team(s) work? Fixed site, field conditions, etc. Will we need to deploy our own tents for sleep and work?
m. Is the heat a condition that will affect computers, cameras, batteries, etc? If so, is air conditioning available? Refrigerators for film? What about humidity problems? Air conditioning may help but can dehumidifiers do better...consider charcoal bags placed in shipping cases.
n. From what unit(s) do we draw support...rations, billets, fuel, etc.?
o. If one team is deployed for a lengthy duration, is there an opportunity for the teams to be rotated?
p. If deploying a risograph or other commercial printing machine, what quantity of copies and frequency (daily, weekly) is desired? What is the plan for distribution of copies? What is the plan for paper replenishment?
q. Has every effort been made to ensure deployed asset has commercial phones available to assist transmitting digital photographs, responding to query, accidents and incidents, fax capability, etc.
r. Has unit/team packing list been carefully checked to ensure the easy-to-forget yet must- have items are not forgotten. Glue sticks, chalk, scotch tape, blank overheads, etc.

Appendix K

PAO SOP OUTLINE

Chapter 1, Alert Notification Procedures
- Annex A, Notification Procedures
- Annex B, Section Telephone Contact Roster

Chapter 2, Individual Preparation
- Annex A, Individual Admin Checklist
- Annex B, POV Storage Plan, Procedures for Completing Post POV Storage Forms.
- Annex C, Personal Property Storage Plan
  - Section 1, Power of Attorney
  - Section 2, State of Obligations
  - Section 3, DD Form 1299, Application for Shipment/Storage of Personal Property.
  - Section 4, DD Form 1701, Household Goods Inventory
- Annex D, Recommended Personal Readiness Equipment
- Annex E, Family Member Pre-deployment Checklist

Chapter 3, Unit Preparation
- Annex A, Unit Equipment List

Chapter 4, Tactical Vehicle Preparation
- Annex A, Vehicle Preparation Standards
- Annex B, Vehicle Load Card
- Annex C, Vehicle Inspection Checklist
- Annex D, Joint Airlift Inspection Record
- Annex E, HMMWV Configuration and Load Plan

Chapter 5, Rear Detachment Operations
- Annex A, Communications
- Annex B, Logistics Coordination

Chapter 6, Public Affairs Checklists
- Annex A, Pre-Deployment Checklists
- Annex B, Guide for Media Interviews
- Annex C, METL and Tasks, Conditions, Standards
- Annex D, DoD Principles of Information
- Annex E, Policy on the News Coverage of U.S. Military in Combat
- Annex F, PA Guidance on Terrorism Counteraction
- Annex G, PA Guidance on Counter-Drug Operations
- Annex H, Standard PA Ground Rules
- Annex I, Spokesperson Guidelines
- Annex J, Command Unique Media Operations Center Guidelines

Chapter 7, Field Operations
- Annex A, Tactical Uniform
- Annex B, Installation Security
- Annex C, Tactical communication
- Annex D, Personal Hygiene
- Annex E, Morale, Welfare and Recreation

Appendix L

EXAMPLE OF MEDIA OPERATIONS CENTER

Operations of a media center will need the following support:

- Communications
- Vehicle support (day-to-day operations and media transport -- tactical or non-tactical as needed)
- Billeting and rations for media center personnel
- Admin support personnel for 24-hour operations
- Office space (hard site if possible) and power as needed

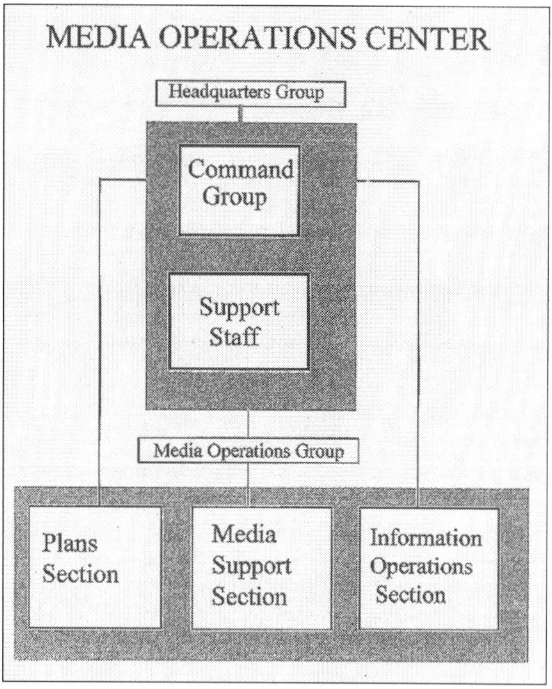

Figure L-1. Media Operations Center

Appendix M

NEW MEDIA INQUIRY FORMAT (SAMPLE)

This format is intended as an example only and should be adapted to local needs and SOPs. In addition to query forms, PAOs should maintain a separate log of all inquiries.

MEDIA QUERY SHEET

Query Number:____________

Deadline:________________

CALLER'S NAME:____

CALLER'S NEWS ORGANIZATION:________________________________

CALL TAKEN BY:_______________TIME:____________DATE:____________

QUESTION (use reporter's precise wording):_____________________________
____
____
____

RESPONSE (if written release is made, attach a copy):___________________
____
____
____
____

SOURCE OF INFORMATION/COORDINATION (e.g., G-3, surgeon): ___________
____

RELEASED TO:____

TIME/DATE:____

RELEASE METHOD: In Person_______ Phone____________ News Release_______

RELEASED BY:_______________ RELEASE NUMBER:________________

Appendix N

GUIDE FOR MEDIA INTERVIEWS

GUIDELINES

1. Considerations

a. When human safety or other serious concerns are involved, deal with those considerations first.

b. Communicate only information that is approved for external distribution. Always tell the truth.

c. Know to whom you are speaking. Get the person's name and telephone number, if necessary.

d. Do not be intimidated. You may tell a reporter that you need to clarify an important matter before you can answer questions.

e. Talk from the public's viewpoint. Avoid jargon. Speak within the audience's frame or reference.

f. If the questions do not lie within the framework of approved statements or within your area of expertise, find the appropriate technical advisor or spokesperson.

g. State the most important fact at the beginning. Place your own headline on the answer.

h. Attack problems in your answers, not people.

i. Do not repeat offensive or negative language. Do not let other people put words in your mouth.

j. Direct questions deserve equally direct and forthright answers.

k. Do not exaggerate the facts. Listen to how your answer "sounds" when spoken.

l. Ignore cameras and microphones. Talk to the reporter.

m. During videotaped interviews, it is all right to stop your statement and start over.

n. Do not say "no comment." Explain why you do not have an immediate answer.

o. Keep your composure, even if a news reporter gets snappy.

p. Be prepared to provide sufficient evidence for statements you make.

q. Be especially alert about photos. You have little control over photos taken off military reservation property, but you have every right to control photos taken on the military reservation.

r. Be aware of your surroundings and follow local OPSEC rules when determining interview location

2. What will be asked?

a. What happened and where? When did this occur

b. Are there injuries or deaths as a result? How many and to whom?

c. What actions is the unit taking to control the situation?

d. Have chemicals or other hazardous substances been released into the environment? What kinds? How much?

e. What types of hazards are presented to people off-site?

f. Have off-site emergency response personnel been notified? Which ones?

g. Are unit operations shut down?

h. Has the site or facility been evacuated?

i. How many people are employed at this site?

j. What do you do at this site?

k. How old is the facility? Does it meet current regulations?

l. Why did this situation occur? (DO NOT SPECULATE.)

m. Are there safety rules covering the situation? Were they violated?

n. Has a Site Emergency Response Plan been activated? What does that involve?

o. Tell me about your organization?

p. Will this situation have national ramifications, or will its effect likely be limited to a single site or region?

q. How much money is this going to cost the taxpayers?

r. Is there insurance coverage for the loss or damage? How much?

s. Are commanders handling the situation locally or is a higher headquarters taking control?

t. Has this occurred anywhere within the unit before? Why weren't you ready?

u. What do your soldiers think about this situation?

v. For accidents and incidents, don't speculate causes. Use "ongoing" investigation statements.

Appendix O

MEDIA RESEARCH AND ANALYSIS

AN INTRODUCTION

Since the early part of the 20th century, when there was almost no interest in the size of audiences or in types of people that comprised various audiences, business leaders and their organizational communicators have increasingly come to rely on research for nearly every major decision they make. This expanded demand for information has created a particular demand for media communication research capabilities, specifically -- the development of a scientific basis for media analysis and media effects.

The importance of mass communications research and media analysis is partly due to the realization that gut feelings or reactions are not entirely reliable or credible bases for decisions. Although common sense is often accurate, Army commanders and other decision-makers need additional, more objective information to evaluate problems, especially when lives are at stake. Thus, the past 50 years have witnessed continuing evolution of media analysis, combining research and intuition to create a higher probability of success.

This evolution has resulted in a "scientific" approach to media research known as media content analysis.

SCIENTIFIC RESEARCH: A DEFINITION

Scientific Research is defined as a **"systematic, controlled, empirical, and critical investigation of hypothetical propositions about the presumed relationships among observed phenomena."** This definition contains the basic terms that are necessary in defining the method of scientific research and describes a procedure that has been accepted for centuries. In the 16th century, for example, scientist Tycho Brahe conducted years of systematic and controlled observation to prove wrong many of Aristotle's theories of the universe. By gaining an understanding of the phenomena, he challenged the accepted beliefs and knowledge of the time with his own hypotheses. Thus, scientific research was begun.

Whether we realize it or not, we all conduct research as a matter of course in our day-to-day life whenever we speculate about the possibility of something -- we start with an idea or concept and test it.

All research begins with a basic question or proposition about a specific phenomenon -- for example, Why do Americans usually support the soldiers within the Army when they may not support the operation the soldiers are involved in? What factors determine why Americans will support the political justification for military involvement? What types of messages are most effective in garnering support for American forces?

The answers to these questions can be forecast to some degree with well-designed research studies. There are some difficulties, however. The Army media analyst faces the problems of determining which data collection methods can most accurately provide answers to the questions at issue, and in gaining adequate access to information prior to and during military operations. In the pages that follow, we will describe the methods and procedures PA professionals may use in overcoming these difficulties.

RESEARCH APPROACHES

There are several research approaches or "methods of knowing" which have been used to conduct studies: intuition, authority, and science.

In the **intuition** approach, one assumes that something is true because it is "self-evident" or "stands to reason." An example of this type of thinking would be if some Public Affairs leaders resist efforts to perform area studies because they believe they already "know" their AO.

The **authority** method seeks to promote a belief in something because a trusted or credible source says it is true. Here, the emphasis is on the source, not the methods the source may have used to gather his information.

The **scientific method** approaches learning as a series of small steps, with each step identifying more specific information and leading to a more clearly identifiable conclusion.

For example, one study or source provides only an *indication* of what may or may not be true; the "truth" is found only through a series of objective analyses.

This means that the scientific method is self-correcting in that changes in thought or theory must be continually reviewed, that issues and situations require constant monitoring.

The scientific method has become a valuable tool to produce accurate and useful data in mass media research. This annex focuses solely on the scientific approach and forms the fundamental basis for media content analysis in Army Public Affairs.

MEDIA RESEARCH: THE SCIENTIFIC METHOD

The goal of Public Affairs media research is to provide the methodology to support situational assessment, planning and decision-making that is fast, inexpensive, reliable and valid. The application of scientific methodologies to media research by Public Affairs personnel accomplishes this goal.

Five basic characteristics, or tenets, distinguish the scientific method from the other methods of research. A research approach that does not follow these tenets cannot be considered a scientific approach:

- **Scientific research is objective.** Science tries to rule out eccentricities of judgment by researchers. When a study is undertaken, explicit rules and procedures are constructed and the researcher is bound by them, letting the chips fall where they may. Objectivity also requires that scientific research deal with facts rather than interpretations of facts.

- **Science is empirical.** Researchers are concerned with a world that is knowable and potentially measurable. (Empiricism derives from the Greek word for "experience.") Analysts must be able to perceive, understand, and classify what they study and reject nonsensical explanations of events. For example, a newspaper editor's claim that declining readership rates are "God's will" would be rejected by scientific researchers because such statements cannot be perceived, classified or measured. Experience shows that there are usually easily identifiable reasons for declining readership.
- **Scientific research is systematic and cumulative.** No single research study stands alone, nor does it rise or fall by itself. Astute research analysts always use previous studies as building blocks for their own work. One of the first steps taken in conducting research is the review of all available literature on the topic so that the current study will draw on the heritage of past research.
- **Scientist attempt to find order and consistency in their findings.** In its basic form, scientific research begins with a single, carefully observed event and progresses ultimately to the formulation of theories and laws. A theory is a set of related propositions that present a systematic view of phenomena by specifying relationships among concepts. Researchers develop theories by searching for patterns of uniformity to explain and describe the information collected.
- **Scientific research is predictive.** Science is concerned with relating the present to the future. In fact, scientific researchers strive to develop theories because they are useful in predicting behavior. The importance of theories lies in their ability to predict an outcome or an event successfully. If a theory generates predictions that are supported by data, and the results are always the same, the theory can be used to make predictions in other similar situations.

MEDIA ANALYSIS STEPS

Evaluation of a problem must follow a standard sequence of steps to increase the chances of producing relevant data. Analysts who do not follow a prescribed set of steps increase the amount of error possible in the study.

These steps are:

- Select a topic (issue, situation, perception, or belief).
- Review existing research and other available information on the topic.
- Develop hypotheses and research questions.
- Determine an appropriate methodology, format or design.
- Collect relevant data.
- Analyze and interpret the results.
- Present the results in appropriate form (Information Paper, PA Study or PA Estimate)
- Validate and replicate the study when necessary.

The use of the scientific method of research is intended to provide an objective, unbiased evaluation of data pertaining to an issue or event. To investigate hypotheses systematically, media analysts must follow these steps. However, merely following the eight steps does guarantee that the research is good, valid, reliable or useful.

A countless number of intervening variables (influences) can destroy even the most well-planned research effort. Unanticipated events occurring during the research period may impact the results and they must be accounted for during the process. However, PA analysts must remain focused on the purpose of the research effort and not lose sight of the original objectives.

STEP 1 – SELECTING A TOPIC, DETERMINING RELEVANCE

Selecting a research topic is usually not a concern for Public Affairs analysts -- planning guidance, current situations, and most importantly, the operational issues confronting our commands, will guide the application of media content analysis. In most instances, the Public Affairs analyst will receive planning guidance well in advance of an operation, which will help determine the issues to be addressed.

Once the basic subject has been chosen, the next step is to ensure that it is relevant to the operation or situation at hand. This can be accomplished by answering six basic questions.

- What is the goal of this research effort?
- Is the subject too broad?
- Can the subject really be studied?
- Is the subject significant?
- Can the results of the research be generalized, communicated and understood?
- Does the issue lend itself to analysis?

Underlying all eight steps of the Media Analysis process is the necessity for validity. In other words, are all eight steps (from topic selection to data analysis to presentation and interpretation) the correct ones to follow in trying to answer these questions?

The answers to these questions will help focus the research you must do, make information gathering easier, and ensure the results are valid.

STEP 2 -- REVIEW OTHER RESEARCH AND INFORMATION

Media analysts should always begin studies by consulting all literature, research, and other information available on the topic. The review provides information about what work has been done, how it was done, and what the results were as they apply to a given subject. It not only allows analysts to learn from (and eventually add to) previous media research, but also saves time, effort and money.

The review also helps to identify the facts pertaining to the situation being studied.

Completed media content analysis also provides a starting point for PA leaders who will follow in your position after you move on.

STEP 3 – DEVELOP HYPOTHESES AND RESEARCH QUESTIONS

After the general research are has been identified and the existing information reviewed, the analysts must state the problem or issue as a workable **hypothesis** or **research question.**

(Example: "The American public is losing confidence in the Army's ability to protect its soldiers, resulting from the media's portrayal of Army leaders as negligent and soldiers as lacking competence in avoiding fratricide.")

A hypothesis is a formal statement regarding the relationship between variables and is tested directly. In the example cited above, those variables are the news media, the messages they send, and the perception and understanding of those messages by the American public. With a hypothesis, the predicted relationship between the variables is either true or false. Identifying the degree of "trueness" or "falseness" and their implications is essential to the development of information campaign strategies.

On the other hand, a research question is a formally stated question intended to provide indications about something, and is not limited to the relationships between variables. Research questions are generally used in situations where an analyst is unsure about the nature of the problem under investigation. The intent is merely to gather preliminary information. Research questions are generally used to identify the focus and scope of a research project.

STEP 4 – DETERMINE AN APPROPRIATE METHODOLOGY/RESEARCH PROGRAM DESIGN

Given the variety of situations facing Public Affairs personnel, different approaches to media research are required. Some issues lend themselves more toward survey methodology via telephone, E-mail, or standard mail; others are best attacked through in-person interviews. Still other problems necessitate a controlled evaluation situation designed to eliminate extraneous variables by targeting analysis to specific media types. (An example of this approach would be a study of how a newspaper covered a specific story over a six-month period.)

The approach selected by the analyst depends on the goals and purpose of each particular study. Regardless of whether the problems or issue being addressed is a local one, affecting only a fraction of a community audience, or a national issue affecting us all, all research requires a design of some type. All procedures, including all variables, samples, and measurement instruments, must be pre-designed with hypothesis and research questions in mind.

There are four characteristics of research design that should be noted if a research study is to produce reliable and valid results:

- **Accurate setting**. For a study to have external validity, the study must be conducted as an historical account of the situation during the time frame studied. The analyst must have a clear understanding of the events unfolding around him and attempt to document as much related information as possible.
- **Clear cause-and-effect relationships**. The analyst must make every attempt to identify spurious dependent relationships and

weed them out. The results of a study can be interpreted with confidence if and only if all confounding effects are identified.

STEP 5 – COLLECT RELEVANT DATA

STEP 6 – ANALYZE AND INTERPRET THE RESULTS

STEP 7 – PRESENT THE RESULTS IN APPROPRIATE FORM (PA STUDY OR PA ESTIMATE)

STEP 8 – VALIDATE AND REPLICATE THE STUDY WHEN NECESSARY

Appendix P

INFORMATION PROGRAM EVALUATION

METHODS

How our audiences perceive the Army is critical to the success of all operations we are involved in. Internally, the Army's people require certain information to function effectively. The more they know and understand, the better they perform. Information about the operation, the unit's particular mission, how the commander feels about the situation, and a host of other subjects are of interest to both soldiers and civilians. Externally, the general public has specific needs for information about what their Army is doing and how they are doing it. This appendix explains methods for measuring success in the conduct of PA information programs.

Command information is communication between the commanders and those commanded. Command Information is different from the Public Affairs function of Information Provision in that it is the commander's responsibility to inform his people. Commanders must communicate their intentions and the troops, community, indeed the general public, must know his concerns and intentions. It is especially important to note that PA Information Provision techniques and procedures are just one channel that the commander may use in communicating to his audiences.

A poorly recognized fact is that the communication links between the commander and his audiences occur on various levels and assorted channels. This type of communication no longer fits the "top-down" communications model of the cold war Army. The explosion of today's digital technology has provided individual soldiers, civilians, family members, and the general public the ability to bypass rigid, controlled, vertical communication systems in favor of the common user, multidirectional, reciprocal, simultaneous, real-time transactive communications systems. Americans have the power to bypass the gatekeepers and ignore canned, shoddily produced, dated industrial age information products, in favor of accessing on-line information services or the Internet directly. Information is passed in all directions, continually. These audiences will have access to many more sources of information, which makes evaluating the effects of Army PA Information Programs all the more difficult.

This explosion of information technology has also highlighted how critical it is for PA elements to stay up to date on communications technologies, information services, and socio-economic trends of these forces at work.

Despite the rapid change in the information environment, the general steps for evaluating information program effects has remained the same:

- Determine the command's mission and the commander's method for accomplishing that mission.
- Identify all the various audiences interested in information related to the command, its members, and the mission.

- Identify a public opinion baseline -- the template against which new public opinion information will be compared.
- Gather identify all the messages communicated and identify which audiences' received such information.
- Gather information on the information program impacts. This is done reviewing unit newspapers, letters to editors, responses to information programs fact sheets, formations, surveys, and interviews. Check related bulletin boards on all on-line information services. Check related Newsgroups on world-wide-web nodes, which commonly carry related information. Monitor discussion groups on on-line services.
- Attend commander's calls, staff meetings, formations, briefings, and other gatherings where audience reaction, troop morale and like information will be discussed.
- Evaluate the knowledge of the targeted organizations. This is accomplished through in-person question/answer surveys and interviews, E-mail surveys, and electronic town hall meetings, etc.
- Coordinate with other staff elements addressing similar information issues (SJA, Chaplain, PMO, IG, etc.).
- Produce a summary of information gathered in an impact assessment.

Appendix Q

PRINCIPLES OF PA SERVICE AND STANDARDS

INTRODUCTION

1. This checklist identifies the primary tasks associated with the functions of Public Affairs, and establishes standards for successful accomplishment of those tasks. Standards of service equate to minimum exceptions of an operational commander and will be used to judge unit readiness, leader effectiveness, and individual soldier performance.

2. The following definitions apply to this checklist:

Austere: No existing PA units, assets or Army signal information infrastructure available in area of operations upon deployment. Commercial communications infrastructure is not available. PA elements must perform all missions and provide all PA support using organic personnel and equipment.

Existing: PA units, assets and Army signal information infrastructure in place before deployment. Commercial communications infrastructure is available. Deploying PA element assumes duties of or augments organic PA elements. Existing personnel and equipment augmented by additional PA elements to accomplish PA mission and provide PA support.

FUNCTIONS/TASK	CONDITION	STANDARD
Planning		
1. Perform Global Information Environment analysis	Existing	Austere
2. Develop PA Estimate (COAs)	Existing	Austere
3. Develop PA Strategy (plan)	Existing	Austere
4. Develop PA Guidance	Existing	Austere
5. Coordinate PA Annex	Existing	Austere
Media Facilitation		
1. ID media support requirements	Existing	Austere
2. Register news media	Existing	Austere
3. Coordinate media support	Existing	Austere
4. Provide media orientation	Existing	Austere
5. Coordinate news media interviews	Existing	Austere
6. Plan/coordinate news briefings	Existing	Austere
7. Establish Media OPS Center	Existing	Austere
8. Operate MOC	Existing	Austere

9. Provide media assistance/escort	Existing	Austere
Information strategy		
1. Identify target audiences	Existing	Austere
2. Identify information needs	Existing	Austere
3. Develop info themes/CMD messages	Existing	Austere
4. Gather info/develop products	Existing	Austere
5. Disseminate information to deployed forces	Existing	Austere
6. Disseminate information to families/home station audiences	Existing	Austere
7. Disseminate information to national/local news media	Existing	Austere
8. Disseminate information to general public	Existing	Austere
Public Affairs Training		
1. Identify training needs	Existing	Austere
2. Develop unit PA Training plan	Existing	Austere
3. Develop Family Spt Training Plan	Existing	Austere
4. Develop Senior LDR Training Plan	Existing	Austere
5. Conduct unit PA training	Existing	Austere
6. Conduct Family SPT Training	Existing	Austere
7. Conduct Senior LDR Training	Existing	Austere
8. Prepare SMEs for Media interviews	Existing	Astere
9. Evaluate training efforts	Existing	Austere

Appendix R

PRIVACY ACT/FREEDOM OF INFORMATION ACT

INTRODUCTION

This checklist addresses personal information about soldiers and any living persons that can or cannot be released under the provisions of the privacy act.

CHECKLIST ITEMS

- AGE (date of birth) = releasable. This information is public record.
- HOME OF RECORD/PRESENT ADDRESS = In most cases, home of record can be released if no street address is given. There is no general rule for disclosure of this information. Widely different circumstances surround each incident, and judgment is made on a case-by-case basis. In most cases, the person's present geographical location may be provided (city, state), but not the street address. In each case, the desires of the actual person or next of kin should be considered.
- MARITAL STATUS/DEPENDENTS = RELEASABLE. This information is public record, including names, ages and sex of dependents.
- AWARDS/DECORATIONS/CITATIONS = releasable.
- EDUCATION/SCHOOLING/SPECIALTY = releasable. Major area of study, school, year of graduation, degree and specialty designator is releasable.
- RACE = In most cases, NOT releasable. However, where the fact of an individual's race is relevant in providing essential facts to the Press, it may be released (such as in a racially oriented protest or altercation.)
- CHARACTER OF DISCHARGE:
- ADMINISTRATIVE = NOT releasable, unless the individual provides his written consent.
- PUNITIVE = releasable. This includes discharges resulting from courts martial.
- DUTY STATUS = releasable.
- PERSONNEL BOARDS:
- Results of promotion boards and augmentation boards are releasable.
- Results of administrative discharge boards and aviator flight boards are NOT releasable.
- PHOTOGRAPHS IN THE CUSTODY OF THE DEPARTMENT OF DEFENSE = releasable, unless they warrant an invasion of anyone's personal privacy.

Appendix S

PA GUIDANCE ON TERRORISM COUNTERACTION

In view of the continuing media interest in the subject of terrorism, the public affairs guidance contained in this message is provided to assist PAOs in responding to media queries and in developing local contingency plans.

TERRORISM IN GENERAL

a. **U.S. POLICY**. All terrorist acts are criminal. The U.S. Government will make no concessions to terrorists. Ransom will not be paid and nations fostering terrorism will be identified and isolated.

b. **RESPONSIBILITY**. Department of State is the lead agency for response to international terrorist incidents that involve U.S. military personnel and facilities outside the U.S. The administrator of the Federal Aviation Administration is responsible for terrorist incidents that affect the safety of DoD personnel or property aboard an aircraft in flight. When terrorist incidents occur at military installations within CONUS or its possessions (Puerto Rico, Guam, Virgin Islands, American Samoa, and trust territories), the FBI will be the lead agency. If the FBI declines to exercise its authority, military authorities may take appropriate action within the limits of their responsibility to resolve the incident.

c. **NOTIFICATION.** All terrorist incidents should be reported immediately through channels, to OASD/PA. No public release of information regarding a terrorist incident should be made without OASD/PA approval except for cases involving public safety.

COMBATING TERRORISM

Combating terrorism can be divided into two major areas: counter-terrorism (offensive measures) and anti-terrorism (defense measures).

- Counter-terrorism. The following statement may be used in response to queries regarding counter-terrorist forces:
- "The U.S. Government has trained forces and equipment from all four services to cope with terrorist incidents. We have also said that command and control elements for these forces exist and have been exercised. These elements report to the Joint Chiefs of Staff, as do other command and control elements for military operations. We do not comment on any details concerning the circumstances under which these forces may be deployed, their identity, or tactics."
- Requests for interviews or filming of counter-terrorism training will not, repeat will not, be approved.
- Requests for photos of counter-terrorist forces personnel or their training will not be approved.
- Because of the clear association/confusion surrounding the relationship between special operations counter-terrorism forces,

all requests for interviewing or filming special operations forces and or training will be approved by OASD/PA.

- Questions beyond the scope of the above guidance should be referred to OASD/PA..
- Anti-terrorism. The following guidance is applicable to media request for information pertaining to anti-terrorism.
- DoD officials, senior leaders, commands and knowledgeable individuals may discuss the subject of anti-terrorism as it pertains to those areas/installations for which they are responsible.(Anti-terrorism measures and procedures should be discussed in a general manner without going into a checklist of specific details.)
- Media requests to film anti-terrorist training will be approved on by OASD/PA.
- Photos of anti-terrorist training should be forwarded through channels for approval.

QUERIES

In response to queries regarding a possible or real terrorist threat at a particular base/installation/activity the PAO may acknowledge, if appropriate, that increased security measures have been/will be taken without going into specific details of all the measures taken. PAOs may, when appropriate, acknowledge the obvious.

For example, increased security measures such as increased guards at the gate or additional patrols, if they are obvious to the public may, in some cases, be acknowledged. PAOs should, however, exercise care and prudent judgment in any discussion of these or other security measures, which have been/will be implemented.

Appendix T

BROADCAST OPERATIONS

This section focuses on the radio and television services provided by Armed Forces Radio and Television Service at the unified command level and the coordination necessary by Army component commanders to ensure operational area support. It identifies and explains the AFRTS mission and its capabilities. It also discusses planning considerations and theater broadcast information requirements.

AFRTS CONTINGENCY BROADCAST OPERATIONS

The scope of the ABS mission of providing AFRTS radio and television news, information and entertainment programming to DoD personnel stationed overseas greatly expands during wartime to include support of global contingency requirements. As new contingency plans are developed based on emerging joint and Army doctrine, ABS must consider how the additional broadcasting personnel and equipment resources needed to support a rapid deployment broadcasting mission can be obtained while simultaneously meeting increased requirements in existing unified command theaters. The immediate response necessary to meet contingency requires the development of AFRTS appendices to Unified Command Operations Plans (OPLAN).

Army Broadcasting Service (ABS) is the Unified Command AFRTS Planner (UCAP) for the U.S. Southern Command and the U.S. European Command. ABS also has Geographic Area Planner (GAP) responsibilities for U.S. Forces Korea (USFK). ABS and AFRTS networks within these unified commands are responsible for updating and maintaining appendix content under provisions of the American Forces Information Service (AFIS) Concept Plan for Peacetime and Wartime Operations for the Armed Forces Radio and Television Service (AFRTS CONPLAN 98-1) and DoD Directive 5120.20-R

Contingency and/or Wartime Plans define the Mobilization/Contingency mission. Force structure to support these plans may be packaged as blocks, deployable units or detachments for ease of planning.

According to CONPLAN 98-1 and DoD Directive 5120.20-R, Unified Commanders and Subordinate Commanders, with the advice of ABS, determine the type of AFRTS Flexible Response Option (FRO) necessary. Unified and Subordinate Command support is required for the current levels of support contract and for any mission changes which affect AFRTS service in the CINC's area of operation.

AFRTS Mission

Provide live news, information and entertainment programming, free of censorship, to all DoD elements in place or deployed worldwide.

Provide U.S. military theater commanders with sufficient electronic media resources to effectively communicate DoD, Army, theater and AO command information.

Concept of Operations

Existing facilities and services will be the first AFRTS assets used to respond to AFRTS taskings.

Upon implementation of a Contingency Plan or OPLAN, AFRTS assets in the AO come under the direct operational control of the Unified Command for the period of the operation. When the operation is terminated, normal command relationships will be resumed.

The AFRTS Commander reports directly to the Unified Commander responsible for the theater of operations.

The AFRTS Commander retains direct command authority over AFRTS personnel and equipment.

The AFRTS Commander is responsible for all AFRTS matters concerning the operation, accomplishing direct coordination for the Unified Command with AFIS/AFRTS for all AFRTS issues requiring AFIS/AFRTS action, and managing all AFRTS assets involved in the operation in accordance with the AFRTS CONPLAN 98-1.

The AFRTS Commander will be a designated member of all public affairs meetings as a member of the staff.

The AFRTS Commander is authorized direct coordination with other members of the unified command's staff to work specific AFRTS support requirements.

All requests for command internal information or emergency announcements from AO organizations or personnel will be forwarded to the director of public affairs for approval.

All public information released by public affairs officials will be available for AFRTS use.

AFRTS radio and television electronic news gathering (ENG) will be dedicated for the AFRTS "on-air" mission and in direct support of the Unified Commander's internal information program.

AFRTS will provide full, factual and timely internal information and news to military audiences in the AO, consistent with national and operational security, and host country sensitivities.

AFRTS will follow Operations security (OPSEC) and communications security (COMSEC) rules.

AFRTS OPERATING PARAMETERS

AFIS is responsible for all AFRTS satellite programming services and overall policy and guidance for their use.

The AFRTS Broadcast Center (AFRTS-BC) is responsible for providing non-local and non-theater radio and television programming material to AFRTS facilities in the AO, except as outlined in Flexible Response Options (FROs).

The Television-Audio Support Activity (T-ASA) in Sacramento, California, is responsible for providing technical and logistical support to AFRTS.

The AFRTS facility in the AO will provide service based upon initial Flexible Response Options (FROs) and continue operations until directed to modify its services by AFRTS, the Unified Command or as the mission requirement dictates.

AFRTS will provide AO-wide announcements as required on both radio and television in order to facilitate unified command needs.

At unmanned repeaters and cable distribution systems, local officials may have AFRTS personnel make local announcements, if possible, in coordination with the AFRTS commander or on-site command representative, if approved by the director of public affairs for the operation.

AFRTS NETWORK COMMANDER OR DESIGNATED REPRESENTATIVE IN THE AREA OF OPERATIONS

The AFRTS network commander or designated representative will be co-located with the unified command director of public affairs.

The AFRTS network commander or designated representative will have a command function with direct operational command authority over all resources assigned to support the AFRTS mission in the AO.

The AFRTS commander or designated representative will ensure a logistics and engineering function responsible for providing advice and assistance to maintenance personnel assigned to AFRTS outlets, and maintaining unmanned equipment. This function will assist the AFRTS chief engineer in developing new equipment support requirements as changes occur in the AO.

The AFRTS network commander or designated representative will ensure internal information ENG coverage of unified command activities of interest to the members assigned in the AO. In joint service situations, Army AFRTS representatives may also be responsible for the production and duplication of radio and television internal information products for use at AFRTS outlets and television programming for DoD, or satellite cabled sites in the AO.

UNIFIED COMMAND RESPONSIBILITIES

The unified command provides logistic support for AFRTS. This includes vehicles, POL, and supply requirements in the AO as noted in the OPLAN. Also included is vehicle maintenance and POL for all AFRTS contingency vehicles. If additional forces are deployed to support AFRTS, the unified command assigns additional vehicles to support the expanded maintenance and production requirements.

If security and/or intelligence forces determine that AFRTS facilities have been identified as a potential target by hostile forces, the unified command will notify the AFRTS facility and provide security to targeted facilities.

If contract communications support is terminated during the implementation of the OPLAN, the unified command provides communications support for AFRTS use in the AO.

This includes existing long-wire or microwave systems for distributing the AFRTS signal and support for telephones, facsimile transmission and computer equipment for AFRTS.

If support cannot be obtained from existing assets, the unified command should be prepared to augment the AFRTS mission as outlined in the OPLAN.

The unified command provides personnel, administrative, vehicle and other logistic support for all AFRTS personnel assigned in the AO and those deployed to supplement that force. This includes unit line numbers and entry clearances required for all deployed personnel supporting the AFRTS mission.

The unified command obtains country clearance for construction of any temporary transmitter towers required due to expanded AFRTS service, which may occur as the operation unfolds.

The unified command provides electrical backup power for AFRTS facilities if contract services are terminated during the implementation of the OPLAN.

The unified command is responsible for obtaining necessary broadcast frequencies in consultation with the host-nation government to meet AFRTS broadcast requirements.

Flexible Response Options

Although each operation will differ, the following are general concepts of AFRTS Flexible Response Options (FROs) available for peacetime engagements, wartime operations and stability and support operations in an area where little or no AFRTS service exists or where crisis situations require a modification to existing AFRTS services. The unified command AFRTS planner (UCAP) is responsible for developing specific equipment, support and manning requirements to implement the AFRTS FROs that best support the specific operation.

The Unified Commander for the area of operations must request AFRTS radio and/or television services or for a change in present level of service before deployment. The Unified Commander requests AFRTS assistance through the unified command AFRTS planner responsible for the area of operations. The request will then be forwarded to ABS and AFIS for final approval.

FRO One: Direct to Ship (DTS) Service Support System. DTS is an U.S. Navy peacetime capability that provides news, sports, information and entertainment to audiences on ships at sea. A wartime adaptation of the service can provide immediate access to three radio and two television channels for land-based audiences including geographically separated units down to the lowest level. A deployable AFRTS kit containing an individual receiver decoder (IRD) provides service. This service provides a single-source 24-hour capability of receiving all services, but only one channel can be accessed at a time and no local or theater command information would be available.

FRO Two: Satellite Direct Radio and Television (SDRTV). SDRTV is an unmanned AFRTS satellite service that can be provided to virtually any land based audience on the globe with up to 10 stereo radio channels and six television news, sports, information and entertainment channels. In the European theater it will include a regionally generated signal. Service is provided at a single location using a deployable AFRTS kit containing a simultaneous receiver decoder (SRD) that provides a capability of receiving all services and accessing up to six radio and television programming sources at a time. As a public affairs option,

SDRTV provides an internal information data stream that can be accessed with the addition of a computer, printer and proprietary software to the SDRTV equipment package for use by public affairs activities in providing support to deployed populations. Unified command public affairs offices and the UCAP should consider coordinating the use of the additional capability whenever a manned public affairs activity is deployed.

FRO Three: Manned Radio Systems. Signal Distribution Systems. These deployable systems include audio and video transmission and cable systems that provide a capability to distribute, DTS, SDRTV or manned radio service to an expanded autonomous geographic area such as a base camp or Air Force base in an AO.

FRO Four: Manned Radio Systems. These deployable systems provide a capability for local, live internal information and radio news. Various types and sizes of local radio systems exist that can be used to establish a range of services from simple local break away "radio-in-a-box" to a full service facility with local production capability. Some of these systems will include radio transmitter that can provide limited signal distribution without deployment of FRO Three. This system can provide a limited single-source radio service to outlying populations that are not served by FRO One or Two, are not available or would not be appropriate programming sources.

FRO Five: MOOTW Management, Local TV and Network Live Radio. These deployable personnel and systems support the development of an AFRTS management function to oversee dispersed AFRTS operations and will add local television and network-wide live radio capability using organic distribution systems. The capability will establish a network to support operations in an AO comprising a large peacetime engagement of medium to long duration encompassing a large geographic area where the operations commander requires near real-time internal information capability. The system provides network administrative, computer, maintenance, engineering and operations support functions. These functions provide the unified command director of public affairs with AFRTS management expertise not normally available on the public affairs staff. The television service system is designed to produce AO information that can also be fed to the AFRTS Broadcast Center for rebroadcast to DTS/STRTV audiences worldwide.

FRO Six: Theater Satellite Radio and Television Operations (TSRTO). In a major regional conflict where large force deployments are planned the AFRTS Broadcast Center will dedicate one channel of radio and one television channel for use by the UCAP to broadcast directly to the theater of operations. Programming will include time shifting "prime time" so that each 12-hour shift receives prime time programming in the first four and-one-half hours of off time. In conjunction with FRO One or Two, this will provide a virtual network capability to the AO commander and the PAO. It will appear to the audience as if the broadcast was occurring in the AO when in fact it is originating from the Broadcast Center. Initial spots can be unsophisticated radio readers and character-generated (CG) messages on television. If there is a manned radio facility providing theater and operation-specific internal information in place, they will forward copies of all AO specific spots via computer to the Broadcast Center for use initially as television CG messages within the dedicated TV channel. As with all internal information, they will be developed in coordination with the PAO in the AO. If there is insufficient AFRTS manning in place which would be the case if there were numerous geographically-separated operating locations, the internal information will be supplied, via computer, by the AO public affairs office. Base and component command PAOs will be advised of the scheduling of theater programming and encouraged to provide service-unique spots for use in theater. Service will continue until a significant drawdown occurs, the operation stabilizes to the point where local TV spots are not needed and the presumption of pre-conflict programming will serve the majority of the deployed population or an AO based network begins serving the operation.

Appendix U

COMMUNITY SURVEY

COMMUNITY SURVEY EXAMPLE

I. The area

A. Geographical description
1. Of areas surveyed--size of cities, counties.
2. Of surrounding area, if pertinent.
3. Climate, topography, annual and seasonal temperatures, rainfall, etc. (one sentence will suffice for each.)
4. Are the industries dispersed or centralized? Attach a map of the Area indicating the location of the principal plants. The map should show the names and numbers of principal streets and highways furnishing access to these plants.

B. Population
1. Of city
2. Of area.
3. Of labor market area, if different from above.
4. Breakdown by sexes, color, native or foreign born, educational level, percentage of homeowners, etc.

C. Industrial data
1. Types of industries and number of each, labor force of each, key products of the area, and additional data as considered applicable.
2. Does one type of industry dominate the area? If so, give pertinent information regarding the industry.

II. Manpower

A. Labor market rating
1. Is department of labor market classification, a,b,c, or d?
2. Include supporting statistical data.

B. Unemployment
1. Totals and percentages of skilled, semiskilled, and unskilled.
2. What types of skills are most commonly available?

C. Source of labor supply

1. Compared to the World War II years, what are the reserves of women, handicapped, older-age groups, part-time workers, and school graduates?

2. Has there been much intermigration to total population?

D. Occupational classification of area workers. What are the most common occupations of the area? The less common?

E. Skills in shortage category. List, with numbers of each, if available. Make a comparison of this list with the national shortage list.

F. Area wage schedules
   1. List the wage schedules of major occupations and industries.
   2. How do they compare with national averages?
   3. How do they compare with neighboring areas? With competing areas?

G. Requirement of defense industry in area
   1. Is manpower available for present production schedules? Current planned production?
   2. What skills are lacking for production schedules, both present and future?
   3. Do employers ordinarily use training programs? If so, give some Examples.

H. Other pertinent information
   1. Include current work stoppages, if any; record of work stoppages During last 10 years.
   2. Are workers highly organized? Principal unions?

III.Industrial facilities

A. Facilities suited or adaptable for defense production.
B. List facilities with current and World War II products.
C. Give current and capacity employment.
D. Give types of machinery.
E. What defense contracts are held or sought?
F. Vacant factory space. Describe space and indicate production potential.

IV.Housing

A. Housing regulations.
   1. Is it a critical defense housing area under public law 96? P. L. 139?
   2. Local rent control?

B. Housing units available.
   1. Number for sale, including 1, 2, and 3 bedrooms. Price ranges. Are the prices reasonable?
   2. Number for rent, including above information. Apartments available. Number, size, price ranges. Are the rents reasonable?

3. Sleeping rooms available. Number, price ranges. Are the prices reasonable?
4. Building permits issued (in past 12 months)
5. Number for houses--1, 2, 3, or more bedrooms.
6. Number for apartments.
7. If houses, number for rent and contemplated rental prices.
8. Housing units contemplated .
9. Number and sizes. How many bedrooms?
10. Number of these for rent.
11. Estimated rental rates.

C. Builders

1. Adequate number of experienced builders?
2. Do they have trained skeleton force?
3. Can other necessary housing construction workers be secured?
4. Is land available? Under option?
5. Can materials be obtained?
6. What bottlenecks?

D. Building capital.

1. Is capital for the building of housing and rental units readily available? If so, on long- or short-term loans?
2. What are the sources of this capital?
3. Does the community object to construction of more housing units now?

V.Adequacy of housing

For present work force? For expanded production? (quote a figure or percentage, such As peak load in world war ii or 50 percent above present.) This figure should be adequate to cover planned defense expansion known to you at time of Survey.

VI.Other community facilities and services

Discuss each of the following items as to adequacy for the present work force and for an expanded work force; give specific facts for each, as pertinent (yes and no answers are not adequate.)

A. Water
B. Electric power
C. Gas
D. Sewerage

*Note: for items 1 through 4 above, describe sources of supply, capacity, reserve storage, current use, reserves on hand, plans for expansion--whether on hand or projected.

E. Transportation: types and numbers
F. Highway and road systems: are the roads serving the area adequate and in good? What is the present traffic load and the peak capacity of these roads? Describe any unsatisfactory factors. What action is contemplated or considered necessary to assure free traffic movement within the area? (survey requests will furnish, whenever possible, specified information on any industrial and

defense manpower requirement changes under consideration for the area.) Contact with local, state, and federal highway authorities should be made, if necessary, to explore fully this phase of the survey.

1. Schools: number of each type of school, crowding, shifts, new construction, etc.
2. Hospitals: number, number of beds, population per bed.
3. Doctors: number, population per doctor and per dentist.
4. Fire protection: size, ratios, and ratings.
5. Police protection: size, ratios, and ratings.
6. Shopping centers and shopping hours.
7. Recreational facilities: number of each type.
8. Churches: all denominations.
9. Sanitation service (garbage collection).
10. Laundries, dry-cleaning businesses, barber shops, beauty shops, etc.
11. Banking facilities (include arrangements for shift workers).
12. Hotels: number and number of rooms, scale of rates, etc.
13. Restaurants and other eating places.
14. Newspapers: number (morning, evening).
15. Municipal government (form, etc.).
16. Tax rate: local, county, state.
17. Cost of living index: get whatever information is available Compare local figures to national index.

*Note: include reference material, maps, booklets, etc., If possible.

Appendix V

AUDIENCE SURVEY

Audience surveys systematically gather information about the effectiveness of CI programs and products as they relate to a particular group of people. The commander and the PAO to make decisions about management and direction of an internal information program or product use the results.

PUBLIC AFFAIRS REQUIREMENT

The PAO will conduct a readership survey at least every two years (every three years for the Reserve Components). Coordination with the DOIM for possible computer and analysis support is recommended. Additionally, AR 600-46 can provide information on conducting surveys. The survey will provide data on distribution effectiveness, reader awareness and acceptance, readership and perceived usefulness of standing features and topics covered, and opinions of the value and effectiveness of the publication. Repeated surveys will provide trend data.

The survey may include any or all of the 20 questions listed in the Readership Survey (RCS: SAOSA-223) (app H) in AR 360-81. However, surveys not using these tested questions must be pretested to ensure validity before being used in a survey. Survey respondents will be selected using probability-sampling techniques.

Informal surveys, such as those included in a newspaper or conducted randomly/haphazardly with a few people, are not substitutes for readership surveys. This does not preclude an editor from periodically publishing a coupon or set of questions to solicit informal feedback that is not statistically projectable.

Before administering the survey, the survey managers must coordinate with the agency that will provide response analysis to be sure questionnaires; answer sheets, data entry program, or any other materials are appropriate and usable. Survey conduct may be included in the command's CE publication contract and may also be contracted by the command for Army Funded newspapers, providing funds are available.

When civilian employees are surveyed, PAOs should also coordinate with the civilian personnel officer for local union notification requirements. Completed questionnaires may be analyzed by the local Director of Information Management (DOIM) to provide percentages of responses to survey questions. Where computer support is available (from the local DOIM or DRM), responses will be analyzed using a program such as the Statistical Program for the Social Sciences package. Questionnaires must be constructed using the parameters of available software.

A written discussion of findings and conclusions drawn from the survey will be forwarded within 60 days after the survey is completed through the local commander, appropriate major command, to HQDA (SAPA-CI-PMN), Room 2E625, The Pentagon, WASH DC 20310-1510.

As a minimum, the report will contain the survey statistics, an analysis of the data, identification of strengths and problem areas (e.g., distribution, more sports, etc.), recommended improvements and changes to editorial policy, and an indication that the commander has reviewed the results.

Surveys may be conducted any time. However, no newspaper's survey report on file at HQDA should be older than 3 years (4 years for the Reserve Components). This allows for the time to conduct a survey.

A copy of the most recent survey will also be submitted with the annual CI Program Assessment Report (DA Form 510-R), unless the survey was previously submitted to the MACOM and OCPA-HQDA.

The PAO will conduct electronic media surveys at least every two years (every three years for the Reserve Components).

SURVEY ADMINISTRATION

Among the more common methods of conducting surveys are the mail survey, face-to-face interview, and telephone interview. The mail survey is the preferred method for purposes of this requirement, although other methods, managed properly, may be used.

Survey respondents will be selected in a totally random manner (e.g., simple random, stratified, or systematic, using probability sampling procedures). Survey managers will select samples, which achieve at least a minimum of a +/- 5 percent reliability (error margin) at the 95 percent confidence level.

Sample sizes shown for the various reliability levels (e.g., +/- 5 percent error margin) are the number of usable responses received, not the number of questionnaires to be sent out. For a population of 5,000, 357 usable responses will accurately reflect, to within +/- 5 percent, what the entire 5,000 member audience would have said, had it completed the survey.

Experience with mail audience surveys shows that they realize an approximate 30 to 35 percent response rate. Therefore, send out at least three times as many questionnaires as are needed for analysis. Remember that incentives encourage responses.

Maybe the local MWR office or similar staff agency could provide bumper stickers, discount coupons, or other incentive for completed responses.

FOCUS GROUPS

Focus groups. One of the most effective ways to learn how CI products are being received is to conduct focus group interviews. Focus group interviews are structured group discussions in which representative members of the audience are brought together to discuss one or more CI products or issues. These interviews or sessions can examine the effectiveness of products or programs, gain suggestions for improving existing products or programs, and determine the need for new products or programs.

The key to effective focus group interviews is proper planning. Focus group organizers must determine who will participate, and what are the specific objectives of the session (e.g., what topics or issues will be discussed, what specific questions will be asked, what is to be done with the results, etc.). The method of selecting participants should be determined and the location for the meeting secured. While there is no optimal size for focus groups, generally groups of six to 10 individuals are manageable. Group makeup (officer/NCO/enlisted, men/women, military/civilian, active/reserve, retired/family member) depends on the objectives of the session. Generally, homogeneous groups are preferable. Often it will be necessary to hold more than one focus group session to obtain information needed to evaluate a particular CI program or product.

The moderator or group leader should be someone skilled in interview techniques and knowledge about the product or program being evaluated. It is often best not to have a high-ranking individual as the moderator with a group of junior enlisted or young family members, as free flow of information and opinions may be inhibited. The group leader must facilitate the discussion, not serve as an interrogator.

Focus group sessions should be informal. Participants should be encouraged to speak whenever they wish; the moderator should focus the discussion on the topics without being overbearing. If participants agree, it will be useful to videotape their comments for use in evaluating the session.

It is important that all group members understand that their honest opinions are being sought, and that the session is intended as a positive method of improving CI within the organization. No punitive actions should occur as a result of these sessions.

Appendix W

PA Lessons Learned

"What this century's history teaches us is that the Army's real strength is its ability to change and adapt to the period's requirements. Our ability to change was the key to victory in two world wars and a cold war, and it will be the foundation for our future success."

--General Dennis J. Reimer

INTRODUCTION

Explosive developments in information age technology have made the prospect of sharing lessons and ideas across a wide audience a reality today. With ready and easy access to E-mail and the Internet, soldiers can distribute documents, graphics, and photographs with lightning speed.

This appendix is based on an article published in the Center for Army Lessons Learned (CALL) ***News from the Front!***

This section provides public affairs officers (PAOs) with a tool for capturing observations and an outlet for rapid analysis and dissemination of tactics, techniques and procedures (TTPs) to the force. The initial focus is on defining and narrowing the scope of TTPs. It is important for officers and soldiers in the field to clearly understand the process before proceeding to methodologies of collection. The next section provides a structure for developing a narrative product for publication by CALL. Such a product will provide a coherent article of information, which can be quickly used by the force. The final section describes the observation-gathering process. Understanding the process for collecting data will prove invaluable to the operational planner and for producing effective training tools for the future.

DEFINITION AND LIMITS

Focusing the collection effort is central to capturing meaningful observations. Although CALL regularly sends combined arms assessment teams (CAATs) to major exercises and actual operations to gather observations, units from the field, including any public affairs section or detachment, can provide great insight by planning for the collection of information. In fact, only the Army as a whole can make CALL a continuing conduit of information for use by soldiers.

By using the structure and tools described below, units can provide useful TTP by establishing a collection effort as part of the originating operation order (OPORD), with almost no interference with normal operations. Indeed, the tools will enhance planning for future (and remedial) training by incorporating the capturing of TTP into the plan.

TTP are often limited to the specific operation or exercise. The function and use of TTPs are analogous to legal precedents. In law, if given circumstances of a case are generally similar to a prior case, it is assumed that a judgmental decision for the present case should be the same. However, circumstances in law often have aspects that are unique and must be considered before rendering a new decision. When applying TTP, study prior situations in context and use the lessons prudently.

STRUCTURING THE PAPER FOR USE BY CALL

PAOs at all levels can build upon the after-action review (AAR) process in the plan by producing a publishable document. In almost all exercises, units learn and consequently implement improvement measures. By employing the structure below, units can effectively share information throughout the force -- not only from Combat Training Center (CTC) rotations but also from home-station training and exercises away from the training centers.

Do not view the structure below as a rigid construct. Rather, it should serve as a point of departure for unit writers. Although quantitative material is useful for commanders and researchers, make this document narrative in format. Use graphics to support the narrative, if possible. Bring together data into a cohesive product that other units can readily use without resorting to sifting through large amounts of charts, lists, and disjointed bullets.

- Type of unit. Describe the type of unit the PAO supported (mechanized infantry division, separate brigade).
- Context of event. Summarize the general setting for the exercise or operation. (See Exercises and Actual Operations below. More operational context information is provided in this section.)
- Commander's comments. If possible, the commander can provide a brief (one or more paragraphs) commentary on public affairs operations. Work closely with the unit's executive officer or chief of staff for such input.
- Interaction with PSYOP, Civil Affairs, Signal. As information operations continues to grow and doctrine is further developed, interaction between various agencies will also continue to expand. While ensuring coordination with PSYOP and civil affairs operations, PAOs will continue to recognize the separation in functions of the organizations required by law. Discuss the coordination measures used.
- Media Relations.
- Summary of events. Provide a summary of events. Were press conferences and interviews scheduled and executed? What was the pace of daily operations? What was the routine daily schedule?
- Command messages. In developing this section, answer the following questions in detail: What were the command messages? More importantly, did the command messages come through to print or broadcast? Were any command messages distorted or misinterpreted? How can clarity be improved for the next operation?
- Summary of higher headquarters' public affairs guidance (PAG). Write a one or two paragraph summary of the initial and follow-up PAG received from higher headquarters. (Provide a complete copy as an appendix.) Provide answers to the following questions

following the summary: How did PAG influence operations? Was the PAG clear and meaningful? Were excerpts used to create lower level command messages?

- Media contacts. Describe the types and numbers of media contacts. Did the unit encounter numerous print-journalist requests? Electronic requests? Were there patterns in the requests? What could a future media preparation package contain to answer some questions in advance? How did the PAO prioritize media access? Were major outlets afforded more opportunities?
- Summary of Media Releases. Summarize media releases in one or two paragraphs. What were the major themes? What media received the releases? Were releases used in stories? Were there any comments from members of the media about the releases?
- Media Content Analysis. During and following an event, gather press clippings and, if possible, record electronic media stories about the event. What was the nature of the coverage? What was the tenor of editorial comments? Did command messages get exposure? Was the content of articles generally accurate? What could PAOs do in the future to improve the accuracy of content?
- Command Information Products. The command information program in the field is fundamental in the minds of American soldiers. And, this phenomenon is nothing new to this culture.

CIVIL WAR

"When the (Civil) war entered Pennsylvania, the Philadelphia Inquirer often sold up to 25,000 copies of a single issue to the men in the field. During a lull in the Battle of Cedar Creek in October 1864, observers later remarked that the first thing the men did along the line was to sit down, boil coffee, and pull out their newspapers."

- Because of this intense internal interest in events surrounding the operation and events back home, it is imperative that PAOs adequately address methods and practices used to inform the soldiers. Answer the following questions in the narrative: Did soldiers in the field receive consistent and timely information about operations and world events? What was the distribution method? How was it evaluated for effectiveness? Were reproduction resources available and used adequately?
- Changes Incorporated in the Tactical Standing Operating Procedures (SOPs). From the TTPs gathered, what changes will the unit make to the SOP? Briefly describe why the unit is making the change and what mechanisms to put in place to test the effectiveness of the change.

EXERCISES AND ACTUAL OPERATIONS

Operational information is important for the reader to understand the context in which the PAO operated. Provide the following information for the CALL document to contribute to the reader's full understanding of the event(s):

- Mission - Summarize the unit's mission. (The focus here is not on the PAO's mission, but on the mission of combat headquarters.)
- Commander's Intent - This information is available on the OPORD. By incorporating this information, the reader will have an appreciation of the context in which the PAO operated. Emphasize the components of the intent:
  - Purpose
  - Method
  - Endstate
- New Equipment Used - Was new equipment available and used in the operation? Describe the equipment. Was it useful? What were the additional training requirements for using the equipment?
- New Techniques Used - Did the PAO incorporate techniques which are not described in doctrinal manuals? Describe the techniques employed.
- Structure - What was the structure of the PAO unit or shop? What manning -- required versus on-hand? (Note: Do not provide information that is classified under provisions of unit strength reporting (USR) regulations. Seek to provide a document that is free of classified material.)
- Operational Developments - How did the headquarters change its plan during the operation? How did the change(s) affect PAO operations? The descriptions here will bridge the gap between the original plan and its actual implementation.
- TTPs applied during the mission and for future operations - In bullet narrative, describe TTPs gathered in the operation. The bullets must contain sufficient detail for the reader to understand the situation and application possibilities for future operations. Support the bullets by providing individual observations (see TTP-Gathering Process below) as a combined appendix. The narrative in the base document must stand alone, with the appendix of individual observations providing additional detail.

TTP-GATHERING PROCESS

Units can contact CALL when developing plans for collecting TTP. CALL analysts can provide observer guidance, assist in delineating responsibilities of observers, identify documents or reference for use in developing a collection plan, and describe collection methodology.

- Observations. Individual observations assist in providing the basis for the narrative document described above. Use the form below to capture observations and develop a database for use in narrative development. Provide a copy of each observation to CALL as an appendix to the narrative. (Note: Any document published by CALL will not list units nor individuals by name. Refer to units by level ("the division" instead of "the 101st Airborne Division") and personnel by position ("a brigade chaplain" instead of "Chaplain Jones"). The purpose of CALL publications is to share ideas - not to point fingers.

- Observation Forms. The observation form (Appendix A) can be used for individual observations. A Microsoft Word version is available. Contact CALL via E-mail at call@leavenworth.army.mil or DSN 552-9571 (commercial 913-684-571) to receive a copy of the document. The document contains key components which aid the researcher in preparing analyses:
- Observer Name - The observer's name is used administratively only. No observer's name will appear in a CALL product.
- Administrative Information - Like the observer's name, unit information is used administratively only. Unit names do not appear in CALL products.
- Observation Indicators - Check all the appropriate blocks.
- Interoperability Indicators - Check all the appropriate blocks.
- Environmental Indicators - Check the appropriate block.
- File Name - Employ a system that differentiates each observation. One method is for observers to use name initials combined with sequence number and date (John Smith's first observation of May 5th would read, jsmay0105). Other systems are acceptable if plainly explained.
- Observation Title - Give the observation a brief, distinct title.
- Observation - In one sentence, summarize the observation.
- Discussion - Provide as much detail as necessary to provide a clear picture to the analyst or future reader. The length of the discussion will vary.
- Lesson Learned - In the context of your observation, provide a TTP.
- DTLOMS Implications - Describe how the observation impacts one or more areas in DTLOMS:
    - Doctrine
    - Training
    - Leadership Development
    - Organization
    - Materiel
    - Soldier Support

Include other media support, such as photos, sketches, or slide presentations in support of the narrative text.

Provide the narrative text, appendices and other material to CALL at the following locations: call@leavenworth.army.mil, or Department of the Army, Center for Army Lessons Learned, 10 Meade Avenue, Fort Leavenworth, Kansas 66027-1350.

In addition to maintaining and expanding a database of information, CALL publishes ***News From the Front!***, and a host of other publications for easy use by the force. Importantly, much of what is available has originated from the field -- from soldiers just like you. ***News From the Front!*** is published six times per year and provides a forum for a wide variety of topics of interest to the field. Other publications, including newsletters, CTC Bulletins, special editions, Handbooks, and more, focus on specific topics. Much of the published holdings of CALL can be found on the CALL website, http://call.army.mil, post libraries, or by contacting

CALL at the E-mail address listed above. Various search engines are available on the website to assist researchers.

CONCLUSION

Sharing information is possible with rapid and potentially colossal results. Leaders and soldiers who understand the TTP-gathering process can build plans for the future into every OPORD. By incorporating a plan to collect data and produce a clean narrative product for use by the force, soldiers throughout the Army gain maximum benefit from existing and future advancements in information technology. PAOs can focus on critical elements for successful media relations operations and command information programs in the field. Planners can easily adapt collection plans to exercises or actual operations -- anywhere in the world.

Learning is crucial for continued success on the battlefield. PAO planners must inculcate a practice of gaining a full understanding of the process and incorporating it into future exercises and actual operations.

Appendix X

MEDIA GROUND RULES

OPERATION DESERT STORM GROUND RULES

The following information should not be reported because its publication or broadcast could jeopardize operations and endanger lives:

(1) For U.S. or coalition units, specific numerical information on troop strength, aircraft, weapons systems, on-hand equipment or supplies (e.g. artillery, tanks, radars, missiles, trucks, water), including amounts of ammunition or fuel moved by support units or on hand in combat units. Unit size may be described in general terms such as "company-size, multi-battalion, multi-division, naval task force and carrier battle group." Number and amount of equipment and supplies may be described in general terms such as "large, small, or many."

(2) Any information that reveals details of future plans, operations or strikes, including postponed or cancelled operations.

(3) Information, photography and imagery that would reveal the specific location of military forces or show the level of security at military installations or encampments. Locations may be described as follows: all Navy embark stories can identify the ship upon which embarked as a dateline and will state that this report is coming "from the Persian Gulf, Red Sea or North Arabian Sea." Stories written in Saudi Arabia may be datelined "Eastern Saudi Arabia, near the Kuwaiti border, " etc. For specific countries outside Saudi Arabia, stories will state that the report is coming from the Persian Gulf region unless that country has acknowledged its participation.

(4) Rules of engagement details.

(5) Information on intelligence collection activities, including targets, methods and results.

(6) During an operation, specific information on friendly force troop movements, tactical deployments and dispositions that would jeopardize operational security and lives. This would include unit designations, names or operations and size of friendly forces involved until released by CENTCOM.

(7) Identification of mission aircraft points of origin, other than as land or carrier based.

(8) Information on the effectiveness or ineffectiveness of enemy camouflage, cover, deception, targeting, direct and indirect fire, intelligence collection or security measures.

(9) Specific identifying information on missing or downed aircraft or ships while search and rescue operations are planned or underway.

(10) Special operations forces methods, unique equipment or tactics.

(11) Specific operating methods and tactics, (e.g. air ops angles of attack or speeds, naval tactics and evasive maneuvers). General terms such as "low" or "fast" may be used.

(12) Information on operational or support vulnerabilities that could be used against U.S. forces, such as details of major battle damage or major personnel losses of specific U.S. or coalition units, until that information no longer provides tactical advantage to the enemy and is, therefore, released by CENTCOM.

*** Damage and casualties may be described as "light," "moderate," or "heavy."**

Glossary

AFRTS— Armed Forces Radio and Television Services

BOD—Broadcast Operations Detachment

C2 protect—command and control-protect—see **command and control warfare**

C2W—**command and control warfare**

command and control warfare—The integrated use of operations security (OPSEC), military deception, psychological operations (PSYOP), electronic warfare (EW), and physical destruction, mutually supported by intelligence, to deny information to, influence, degrade, or destroy adversary command and control capabilities, while protecting friendly command and control capabilities against such actions. Command and control warfare applies across the operational continuum and all levels of conflict. Also called C2W. C2W is both offensive and defensive:

C2-protection—To maintain effective command and control of own forces by turning to friendly advantage or negating adversary efforts to deny information to, influence, degrade, or destroy the friendly C2 system.

civil affairs—the activities of a commander that establish, maintain, influence, or exploit relations between military forces and civil authorities, both governmental and nongovernmental, and the civilian populace in a friendly, neutral, or hostile area of operations in order to facilitate military operations and consolidate operational objectives. Civil affairs activities (1) embrace the relationship between military forces and civil authorities and population in areas where military forces are present; and (2) involve application of civil affairs functional specialty skills, in areas normally the responsibility of civilian government, which enhance conduct of civil-military operations. These activities may occur prior to, during, or subsequent to other military actions. They may also occur, if directed, in the absence of other military operations.

combined operation—an operation conducted by forces of two or more allied nations acting together for the accomplishment of a single mission.

command information—see **internal information**

community relations—establishing and maintaining effective relationships between military and civilian communities through planning and active participation in events and processes which provide benefits to both communities.

community relations program—that command function which evaluates public attitudes, identifies the mission of a military organization with the public interest, and executes a program of action to earn public understanding and acceptance. Community relations programs are conducted at all levels of command, both in the United States and overseas, by military organizations having a community relations area of responsibility. Community relations programs include, but are not limited to, such activities as liaison and cooperation with associations and organizations and their local affiliates at all levels; armed forces participation in international, national, regional, state, and local public events; installation open houses and tours; embarkation in naval ships; orientation tours for distinguished civilians; people-to-people and humanitarian acts; cooperation with government officials and community leaders; and encouragement of armed forces personnel and their dependents to participate in activities of local schools, churches, fraternal, social, and civic organizations, sports, and recreation programs, and other aspects of community life to the extent feasible and appropriate, regardless of where they are located.

GIE—**global information environment**

global information environment—all individuals, organizations, or systems, most of which are outside the control of the military or National Command Authorities, that collect, process, and disseminate information to national and international audiences.

ground rules—conditions established by a military command to govern the conduct of news gathering and the release and/or use of specified information during an operation or during a specific period of time.

information age—the future time period when social, cultural, and economic patterns will reflect the decentralized, nonhierarchical flow of information.

information architecture—Description and specifications of information systems to include identification of communicators, information transmitted, equipment specifications and network designs. Includes operational, system and technical architectures.

information operations—continuous military operations within the military information environment that enable, enhance, and protect the friendly force's ability to collect, process, and act on information to achieve an advantage across the full range of military operations; information operations include interacting with the global information environment and exploiting or denying an adversary's information and decision capabilities.

information strategy—a synchronized plan for using all available and appropriate methods of communication to achieve specific goals of informing target audiences.

internal information—communication by a military organization with service members, civilian employees and family members of the organization that creates an awareness of the organization's goals, informs them of significant developments affecting them and the organization, increases their effectiveness as ambassadors of the organization, and satisfies their desire to be kept informed about what is going on in the organization and operation (also known as command information).

JIB—joint information bureau

joint information bureau—facilities established by the joint force commander to serve as the focal point for the interface between the military and the media during the conduct of joint operations. When operated in support of multinational operations, a joint information bureau is called a Combined Information Bureau or an Allied Press Information Center.

joint force—a general term applied to a force composed of significant elements, assigned or attached, of two or more Military Departments, operating under a single joint force commander.

joint operations—a general term to describe military actions conducted by joint forces, or by Service forces in relationships (e.g., support, coordinating authority), which, of themselves, do not create joint forces.

JOPES—Joint Operations Planning and Execution System

joint task force—a joint force that is constituted and so designated by the Secretary of Defense, a combatant commander, a subunified commander, or an existing joint task force commander. Also called JTF.

media facilitation—the range of activities such as providing access and interviews that assist news media representatives covering military operations.

media operations center—facility that serves as the focal point for the interface between the military and the media covering an event or operation.

media pool—a limited number of news media who represent a larger number of news media organizations for news gathering and sharing of material during a specified activity. Pooling is typically used when news media support resources cannot accommodate a large number of journalists. The DoD National Media Pool is available for coverage of the earliest stages of a contingency. Additionally, the combatant commanders may also find it necessary to form limited local pools to report on specific missions.

METT-TC—mission, enemy, terrain, troops, time available and civilians

MIE—military information environment

military information environment—the environment contained within the global information environment, consisting of the information systems and organizations—friendly and adversary, military and nonmilitary—that support, enable, or significantly influence a specific military operation.

MPAD--Mobile Public Affairs Detachment

multinational operations—a collective term to describe military actions conducted by forces of two or more nations, typically organized within the structure of a coalition or alliance.

news media representative—an individual employed by a civilian radio or television station, newspaper, newsmagazine, periodical, or news agency to gather information and report on a newsworthy event.

NGO—nongovernmental organization

nongovernmental organizations—transnational organizations of private citizens that maintain a consultative status with the Economic and Social Council of the United Nations. Nongovernmental organizations may be professional associations, foundations, multinational businesses, or simply groups with a common interest in humanitarian assistance activities (development and relief). "Nongovernmental organizations" is a term normally used by non-United States organizations. Also called NGO. See also private voluntary organizations.

PAD—Public Affairs Detachment. The smallest of the PA units.

PAG—public affairs guidance

PAOC—Public Affairs Operations Center.

private voluntary organizations—private, nonprofit humanitarian assistance organizations involved in development and relief activities. Private voluntary organizations are normally United States-based. "Private voluntary organization" is often used synonymously with the term "nongovernmental organization." Also called PVO. See also nongovernmental organizations.

psychological operations—operations to convey selected information and indicators to foreign audiences to influence their emotions, motives, objective reasoning, and ultimately the behavior of foreign governments, organizations, groups, and individuals. The purpose of psychological operations is to induce or reinforce foreign attitudes and behavior favorable to the originator's objectives. Also called PSYOP.

PSYOP—psychological operations

public affairs assessment—an analysis of the news media and public environments to evaluate the degree of understanding about strategic and operations objectives and military activities and to identify levels of public support. Includes judgments about the public affairs impact of pending decisions and recommendations about the structure of public affairs support for the assigned mission.

public affairs estimate—as assessment of a specific mission from a public affairs perspective.

public affairs guidance—normally, a package of information to support the public discussion of defense issues and operations. Such guidance can range from a telephonic response to a specific question to a more comprehensive package. Included could be an approved public affairs policy, news statements, answers to anticipated media questions, and community relations guidance. Public affairs guidance also addresses the method(s), timing, location and other details governing the release of information to the public.

public information—A general term describing processes used to provide information to external audiences through public media.

sustaining base—the home station or permanent location of active duty units and Reserve Component units (e.g., location of armory or reserve center) that provides personnel, logistic and other support required to maintain and prolong operations or combat.

Bibliography

Joint Pub 3-13. Joint Doctrine for Information Operations. 9 October 1998.

Joint Pub 3-61. Doctrine for Public Affairs in Joint Operations.

Joint Pub 1-02. DOD Dictionary of Military and Associated Terms. 24 March 1994.

DOD Directive 5122.5 Public Affairs Program. 12 February 1993.

DOD Directive 5120.20-R. Management and Operation of Armed Forces Radio and Television Service. 9 November 1998.

FM 3-61 (46-1). Public Affairs Operations. 10 March 1997.

FM 3-100.71 (71-100). Division Operations. 28 August 1996.

FM 3-07.29 (90-29). Noncombatant Evacuation Operations. 17 October 1994.

FM 3-0 (100-5). Operations. 14 June 1993.

FM 3-13 (100-6). Information Operations. 27 August 1996.

FM 3-100.1 (100-15). Corps Operations. 29 October 1996.

FM 3-07.7 (100-19). Domestic Support Operations. 1 July 1993.

FM 3-07.6 (100-23). Peace Operations. 30 December 1994.

TRADOC PAM 525-5. Force XXI Operations. 1 August 1994.

AR 200-1. Environmental Protection and Enhancement. 23 April 1990.

AR 360-5. Public Information. 31 May 1989.

AR 360-61. Community Relations. 15 January 1987.

AR 360-81. Command Information Program. 20 October 1989.

Public Affairs Guidance on National Guard Bureau Environmental Programs, National Guard Bureau Office of Public Affairs, 1994.

Commander's Guide to Environmental Management, U.S. Army Environmental Center, 1995.

Index

A

annex, 3-11, 3-12, D-1, D-2
acquisition, 5-6
area study, G-1
Armed Forces Radio and Television Service, T-1
Army Broadcasting Service, T-1
audience survey, V-1

B

briefings, 4-12, H-1, H-2
brigade public affairs, 8-1, 8-2, 8-3
Broadcast Operations, T-1
Broadcast Operations Detachment, 1-9, 1-10

C

Chief of Public Affairs, 2-2
Civil Affairs, 9-1
civilians, Army, 1-3
commanders and public affairs, 1-1, 1-2, 1-3, 2-1, 2-3, 5-3
community assistance, 7-10
community readiness enhancement, 7-11
community relations, 7-1
community social improvements, 7-12
community surveys, U-1
Corps PA Section, 2-3
crisis planning, 3-2

D

Defense Data Network, 5-10
deliberate planning, 3-2, 3-4, 3-5
Division PA Section, 8-4, 8-5, 8-6, 8-7
DOD Media Pool, 4-10, 4-11
DOD Media Guidelines, B-1, B-2
DOD Principles of information, 4-1, A-1

E

electronic newsgathering, 5-6
estimate, public affairs, 3-6, 3-7, 3-8, 3-9, 3-10, 9-3, 9-4, C-1, C-2, C-3

F

Freedom of Information Act, R-1, R-2

G

global information environment, 1-2
ground rules, media, 4-8
guidance, PA, 3-10, E-1, E-2, E-3

I

information infrastructure, 5-9
information needs, 5-5
Information Operations, 2-4, 2-5, 9-1
IO Battlestaff, 9-2
IO campaign cycle, 9-4, 9-5
Information Program Evaluation, P-1, P-2
Information Strategies, 5-1
Information Strategy Process, 5-4, 5-5, 5-6, 5-7, 5-8
internal audience, 5-1
internet, 5-10
interviews, 6-7, 6-8, 6-9, N-1, N-2,

J

Joint Information Bureau, 4-2

L

Land Information Warfare Activity, 9-3
logistical support, 2-5

M

media faciliation, 4-1, 6-10, 6-11,
media analysis, O-1
media operations center, 4-2, 4-6, L-1
media pools, 4-10, 4-11
media support, 4-6
METT-T, 3-15
Mobile PA Detachment Also MPAD, 1-6, 1-7

O

operational level public affairs, 5-4
operational security, 5-7
organic PA sections , 1-4

P

planning, 3-1
pre-deployment checklist, J-1, J-2, J-3
Principles of Information, 4-1, A-1
Principles of PA Standards and Service, Q-1, Q-2
Privacy Act, R-1, R-2
Psychological Operations (PSYOPS), 9-1
Public Affairs Detachment, 1-4, 1-5
Public affairs elements, 1-3
Public Affairs Operations Center (PAOC), 1-7, 1-8, 1-9
public affairs civilian, 1-3
public affairs NCO, 1-3
public affairs officer, 1-2
public affairs specialist, 1-3

R

registration of media, 4-7
releasable information, 4-9

S

strategic level public affairs, 5-4
SOP, 3-16, 3-17, K-1, K-2
speakers' bureau, 7-2

T

tactical level public affairs, 5-4
tactical command post PA section, 8-8
telecommunication systems, 5-9
Town Hall Meetings, 7-4, 7-5
Training, 6-1
  non-PA personnel, 6-6; unit soldiers, 6-6; family members, 6-6; commanders, SMEs, 6-6; exercises, 6-9, 6-10
terrorism guidance, S-1, S-2

W

Waiver of Liability, I-1

***FM 3-19.15 (FM 19-15)**

Field Manual
No. 3-19.15

Headquarters
Department of the Army
Washington, DC, 18 April 2005

Contents

Page

PREFACE ... iv

Chapter 1 OPERATIONAL THREATS OF THE CIVIL DISTURBANCE ENVIRONMENT ... 1-1
General Causes for Civil Unrest ... 1-1
Crowd Development ... 1-2
Gatherings ... 1-2
Crowd Building ... 1-4
Dispersal Process of a Gathering ... 1-4
Crowd Dynamics ... 1-5
Crowd Types ... 1-6
Crowd Tactics ... 1-7
Tactics Used to Defeat Authorities ... 1-9
Weapons ... 1-9

Chapter 2 CONTROL FORCE OPERATIONS ... 2-1
Civil Disturbance Planning Considerations ... 2-1
Legal Considerations ... 2-9
Multinational Operations ... 2-12
Scalable Effects ... 2-12

Chapter 3 APPREHENSION, SEARCH, AND DETENTION ... 3-1
Legal Considerations ... 3-1
Types of Subjects ... 3-2
Apprehension ... 3-3

Chapter 4 RIOT SHIELD AND RIOT BATON TECHNIQUES ... 4-1
Overview ... 4-1
Riot Shield ... 4-1

Distribution Restriction: Approved for public release; distribution is unlimited.

*This publication supersedes FM 19-15, 25 November 1985.

		Page
	Riot Baton	4-3
	Working as a Team	4-21
	Nonlethal Munitions	4-22
Chapter 5	**NONLETHAL CAPABILITIES SET AND EMPLOYMENT CONSIDERATIONS**	5-1
	Personnel Protectors	5-1
	Personnel Effectors	5-4
	Mission Enhancers	5-12
	Core Capabilities	5-20
	Additional Capabilities	5-22
	Employment Considerations	5-22
	Nonlethal Weapons Capabilities in Formations	5-22
Chapter 6	**CIVIL DISTURBANCE FORMATIONS**	6-1
	Overview	6-1
	Actions Before Movement	6-2
	Rally Point Actions	6-3
	Control Force Formations	6-3
	Extraction Team Operations	6-10
	Lethal Overwatch Teams	6-11
	Reserve Forces	6-12
	Squad Formations	6-13
Chapter 7	**CIVIL DISTURBANCE OPERATIONS IN CONFINEMENT FACILITIES**	7-1
	Crowd Dynamics Within the Confinement Facility	7-1
	Use of Chemical Irritants	7-3
	Record of Events	7-7
	Training	7-7
	Equipment	7-8
	Riot Control Formations	7-9
	Apprehension Teams	7-9
	Forced Cell Move Teams	7-12
Chapter 8	**CIVIL DISTURBANCE TRAINING**	8-1
	Unit Training Strategy Development	8-1
	Training Plan Development	8-1
	Range Setup	8-2
Appendix A	**METRIC CONVERSION CHART**	A-1
Appendix B	**OPERATIONS AND LEGAL CONSIDERATIONS IN THE CONTINENTAL UNITED STATES**	B-1
	Federal Intervention and Aid	B-1
	Roles and Responsibilities of Various Agencies	B-2
	State and Local Government Responsibilities	B-3

Page

Legal Considerations and Constraints B-5
Support to Civil Law Enforcement B-9
Lead Agency Concept and Role of Military B-12

Appendix C **URBAN TERRAIN ANALYSIS** C-1
Significant Factors C-1
Urban Terrain Types C-2
Urban Form and Function C-5
Urban Society C-6
Urban Infrastructure C-6

Appendix D **PRACTICAL APPLICATION** D-1
Overview D-1
Procedures D-1

GLOSSARY Glossary-1

BIBLIOGRAPHY Bibliography-1

INDEX Index-1

Preface

Field Manual (FM) 3-19.15 addresses continental United States (CONUS) and outside continental United States (OCONUS) civil disturbance operations. Today, United States (US) forces are deployed on peacekeeping, peace enforcement, and humanitarian assistance operations worldwide. During these operations, US forces are often faced with unruly and violent crowds intent on disrupting peace and the ability of US forces to maintain peace. Worldwide instability coupled with increasing US military participation in peacekeeping and related operations requires that US forces have access to the most current doctrine and tactics, techniques, and procedures (TTP) necessary to quell riots and restore public order.

In addition to covering civil unrest doctrine for OCONUS operations, FM 3-19.15 addresses domestic unrest and the military role in providing assistance to civil authorities requesting it for civil disturbance operations. It provides the commander and his staff guidance for preparing and planning for such operations. The principles of civil disturbance operations, planning and training for such operations, and the TTP employed to control civil disturbances and neutralize special threats are discussed in this manual. It also addresses special planning and preparation that are needed to quell riots in confinement facilities are also discussed.

In the past, commanders were limited to the type of force they could apply to quell a riot. Riot batons, riot control agents, or lethal force were often used. Today, there is a wide array of nonlethal weapons (NLW) available to the commander that extends his use of force along the force continuum. This manual addresses the use of nonlethal (NL) and lethal forces when quelling a riot.

Appendix A complies with current Army directives, which state that the metric system will be incorporated into all new publications.

The proponent for this publication is HQ, TRADOC. Send comments and recommendations on Department of the Army (DA) Form 2028 (*Recommended Changes to Publications and Blank Forms*) directly to Commandant, US Army Military Police School (USAMPS), ATTN: ATSJ-DD, 401 MANSCEN Loop, Fort Leonard Wood, MO 65473-8929.

Unless this publication states otherwise, masculine nouns or pronouns do not refer exclusively to men.

Chapter 1

Operational Threats of the Civil Disturbance Environment

As the Cold War ended with the fall of the Berlin Wall and the collapse of Soviet communism in 1989, the Army was no longer concerned with containing the aggressive Soviet Union. The US Army began to transform because the battlefield on which it would fight had changed. Operations Just Cause, Joint Endeavor, Desert Shield, and Desert Storm have all been recognized as containing nontraditional battlefields. Following these operations, the Army was called upon to conduct peacekeeping, peace enforcement, and humanitarian assistance operations. These operations required US forces to protect refugees and humanitarian workers and keep hostile factions separate to enforce peace.

GENERAL CAUSES FOR CIVIL UNREST

1-1. In these modern times, demonstrations, civil unrest, public disorder, and riots happen for a number of reasons. Some of these reasons are economic hardships, social injustices, ethnic differences (leading to oppression), objections to world organizations or certain governments, political grievances, and terrorist acts. An event can be triggered by a single cause or a combination of causes. For example, operations in the Balkans involving civil unrest and riots were the result of ethnic hatred, a lack of civil authority, food shortages, a revolution, and religious-based fighting factions.

1-2. Demonstrations may range from simple, nonviolent protests that address specific issues, to events that turn into full-scale riots. Gatherings in protest are recognized rights of any person or group, regardless of where US forces may be operating. This fundamental right is protected under the *Constitution of the US*. During peacekeeping or peace enforcement operations, US forces should never violate basic civil or human rights. Most protesters are law-abiding citizens who intend that their protests be nonviolent, but some protest planners insist that the event involve some kind of violence. Often, in the media, protesters can gain sympathy for their cause by prompting authorities to take physical action against them. Violence is often the result of demonstrators beginning to conduct unlawful or criminal acts and authorities (who are responsible for the safety and welfare of all) enforcing the laws of the municipality, state, or nation. The depth of violence is determined by the willingness of demonstrators to display and voice their opinions in support of their cause.

1-3. Commanders must be aware of the possibility that some individuals or groups within an organized demonstration may have the intent to cause disruption, incite violence, destroy property, and provoke the authorities. The tactical situation and actions of the crowd should dictate control and enforcement options. Agitators and criminal infiltrators within the crowd can lead to the eruption of violence. Inciting a crowd to violence or a greater intensity of violence through the use of brutish enforcement tactics should be avoided. Publicity can be detrimental to authorities and beneficial to crowds because it can further their causes.

1-4. Community unrest results in urban conflicts that arise from highly emotional social and economic issues. Economically deprived inner city residents may feel that they are treated unjustly or ignored by people in power and authority. Tensions can build quickly in a community over a variety of issues, such as hunger, poor employment opportunities, inadequate community services, poor housing, and labor issues. Tension in these areas creates the potential for violence. When tensions are high, it takes a small (seemingly minor) incident, rumor, or act of injustice to ignite groups within a crowd to riot and act violently. This is particularly true if community relations with authorities are part of the problem.

1-5. Significant ethnic differences in a community can create an atmosphere of distrust, even hatred. Unrest among ethnic groups competing for jobs, living areas, and sparse essentials can cause an eruption of civil disorder and/or riots. As emotions run high, violence becomes likely.

1-6. Terrorist organizations may infiltrate groups within a demonstrating crowd. These terrorist groups may intend to embarrass their government or other governments. Terrorist infiltrators can be used to provoke crowds as a diversion, as part of a demonstration, or as cover for terrorist acts.

CROWD DEVELOPMENT

1-7. Crowds are a gathering of a multitude of individuals and small groups that have temporarily assembled in the same place. These small groups are usually comprised of friends, family members, or acquaintances that represent a group belief or cause. Individuals assume a sense of anonymity—they are viewed as just another face in the crowd. People in small groups are known only to companions in their group and to others in the gathering that have come from the same neighborhood or community. Commanders must consider how the individuals assembled and how they are interacting during the gathering process.

GATHERINGS

1-8. The assembly process of a gathering refers to the movement of people from different locations to a common location within a given period. This largely determines who participates. Creating a gathering from a crowd is a process with a beginning, middle, and end. As shown in *Figure 1-1* a crowd has an assembly process that leads to the gathering of the crowd, which is always followed by a dispersal process.

1-9. Gatherings are often assisted by the activities of individuals or groups with a specific agenda, such as yelling catchy slogans and cheers that everyone can easily pick up and join in on. Some groups are so well organized that they can prestage leaders to infiltrate a gathering. This creates unity, even inciting newcomers to join their cause. The first phase of the gathering is the assembly process. It can occur in one of two ways—impromptu or organized.

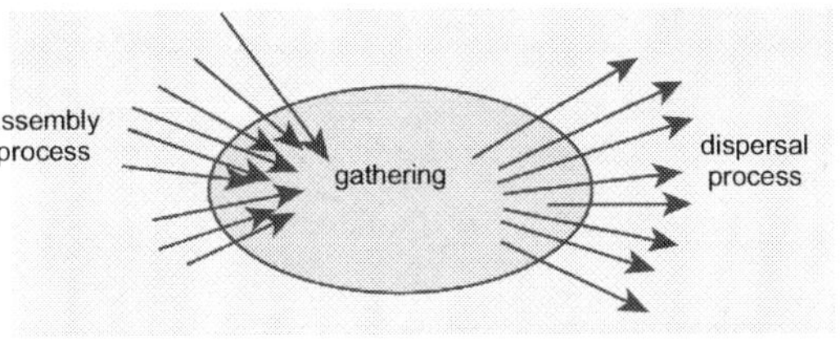

Figure 1-1. Crowd Building

IMPROMPTU GATHERINGS

1-10. Impromptu gatherings usually develop informally and are mostly done by word of mouth (one person to another or one group to another). Participants spread intelligence by telling one another when, where, and what is happening and inviting them to participate. An example of an impromptu gathering would be a gathering at a secured food distribution point after receiving information (by word of mouth) that a large truck carrying much-needed supplies and food is about to arrive. In this case, hunger would be the driving force causing the migration of people to the food distribution facility.

ORGANIZED GATHERINGS

1-11. Like an impromptu gathering, an organized gathering can also involve individuals and groups passing on information to one another. Passing intelligence of police activity and occurrences and on when and where events will take place helps organizers to prestage participants. Rarely is one represented group responsible for pulling together a gathering. Organized gatherings rely heavily on established groups that attract people to gather. Recent examples of these well-organized groups are anarchists, anti-globalization groups, and anti-free-enterprise groups. Groups representing extreme religious faiths and ethnic organizations have been common too. Some gatherings involve groups from rival, or even warring, factions.

1-12. Organized gathering processes rely more on centralized planning and organization. One or more of the groups offer organizers lists of individual names and groups, which they contact as potential participants. Modern technologies are available to these organizers through telephone banks, mailing lists, or e-mail addresses. In preparation for a long-term event, some group organizers have the means to provide logistical support of transportation, food, and water to participants.

> During the preelection period, Bosnian-Serbs (in a bid not to lose any of the territory they had won during the war) began to organize this fiercely contested city. Serbian mobs attacked US forces with bricks, rocks, metal bars, and firebombs. Organization for this altercation started days before, as many witnessed the prestaging of the items to be used in the attack. Wheelbarrows, 55-gallon metal drums, and other large containers were filled with the ammunition that would later be used to attack peacekeepers.
>
> Brcko, Bosnia
> September 1996

CROWD BUILDING

1-13. The second phase of the gathering process is the building of the crowd. Old concepts of crowds based on stereotypes are no longer true. Studies show that—

- Crowds are made of many entities, and all participants are not the same.
- Crowds are not made of isolated individuals but of a minority of individuals and a majority of small groups who may or may not be acquainted with each other.
- Groups and individuals in the crowd are not unanimous in their motivation.
- Groups and individuals in the crowd are usually not anonymous to each other, especially different groups gathered at the same place.
- Individuals within a crowd are not given to unique emotional displays; although known within a group, most individuals would not want to draw attention to themselves.
- Groups within a crowd will often act in unison in an attempt to achieve their agendas.
- Crowds are generally not unique or distinguished by violence or unlawful conduct.

DISPERSAL PROCESS OF A GATHERING

1-14. The last and final phase of any gathering of a crowd is the dispersal process. It involves the movement of people from the common location where they assembled to one or more alternate locations. The dispersal process ends the gathering of a crowd or at least begins its decline. The dispersal can occur on a routine, emergency, or coerced basis.

ROUTINE DISPERSAL

1-15. The routine dispersal may be specified in advance. It can also be included in the assembly instructions given by the organizers of an event.

EMERGENCY DISPERSAL

1-16. An emergency dispersal occurs when people evacuate an area in result of an unexpected crisis, such as a fire, explosion, bomb threat, or terrorist act. Individuals in such an emergency quickly recover from the initial shock. They keep their wits about them and improvise ways of extricating themselves and their companions from the dangerous situation. However, tragedies have occurred in some emergency dispersal situations. Therefore, it is important that forces on the scene are careful not to misinterpret what is occurring.

COERCION DISPERSAL

1-17. Coercion dispersal is caused by the use of force at some level. This is not necessarily the best way to force the dispersal of a crowd. The negotiated management of crowds is the preferred method and has proven to be highly successful in getting crowd organizers to police themselves, especially if the demonstration and/or protest leaders are available and willing to participate.

CROWD DYNAMICS

1-18. Understanding crowds and how individuals and groups form crowds through the gathering process is important because these issues apply to the dynamics of a crowd. Under most circumstances, gathered crowds are orderly and present little or no problems for authorities. Crowds consist of people who, although very motivated and passionate, are also subject to their own need for creature comforts. Rainy, cold, and nasty weather has a way of disheartening all but the few highly motivated and disciplined individuals. If problems exist, they usually fall into the following three categories:

- **Public disorder.** Public disorder is a basic breach of civic order. Individuals or small groups assembling have a tendency to disrupt the normal flow of things around them.
- **Public disturbance.** Public disturbance is designed to cause turmoil on top of the disruption. Individuals and groups assembling into a crowd begin chanting, yelling, singing, and voicing individual or collective opinions.
- **Riot.** A riot is a disturbance that turns violent. Assembled crowds become a mob that violently expresses itself by destroying property, assaulting others, and creating an extremely volatile environment.

1-19. Being part of a crowd of people has certain effects on different people. Each individual in a crowd is susceptible to behaving in a way that is contrary to their normal behavior. Some reasons for these behaviors are as follows:

- Crowds provide individuals with a sense of anonymity. With so many others, an individual realizes that he is just another face in the crowd, giving a sense of invulnerability.
- Crowd and individual behaviors are impersonal by nature. The "them-against-us" attitude affords those within the crowd the ability to freely (without hesitation or reservation) be verbally abusive, throw objects, or attack anyone who gets in the way.

- Crowds provide individuals with the idea that their moral responsibilities have shifted from themselves as a person to the crowd as a whole. Large numbers of people discourage individual behavior, and the urge to imitate others within the crowd is strong. Individuals look to others around them for cues of what to do next, disregarding their own background and training. Often, it is only the strong, well-disciplined person who can resist the prevailing behavior of a crowd.
- Crowd behavior influences the actions of both the disorderly individuals of the crowd and the authorities tasked to control them.

1-20. Individuals within a crowd are at times driven by deep-felt emotions. Emotional contagion is the most dramatic psychological factor of crowd dynamics. It provides the crowd with a temporary bond of psychological unity. Lasting long enough, this unity can push a simple organized crowd into a mob. Normal law and authority are rejected en masse under these conditions, increasing the potential for violence and panic to erupt.

1-21. Panic can erupt quickly, especially when crowds turn into mobs. Individuals within the mob can easily sense that their safety and well-being are at risk, putting them in a "fight-or-flight" mentality. Adding to the panic and confusion is the use of riot control agents (RCAs) by authorities in an attempt to gain control. Individuals in a mob, during the heat of confrontation, may attempt to leave and find that there are no escape routes and that roads are blocked. This can often lead to violent, physical attacks.

CROWD TYPES

1-22. Gaining an understanding of crowds and the dynamics that cause individuals or groups to join together in a crowd should be every leader's responsibility. Active, expressive, acquisitive, and hostile are not acceptable terms used to describe a crowd. They are the motives inferred from the actions the crowd takes. No gathering of a crowd is persistently or exclusively active, expressive, acquisitive, or hostile. For the sake of description, crowds can be identified as casual, sighting, agitated, or mob-like.

CASUAL CROWDS

1-23. Casual crowds are identified as individuals or small groups with nothing in common to bind them together. If they have an agenda, it is their own. They arrive separately and leave separately. Casual crowds are made up of individuals or small groups occupying the same common place, such as a shopping mall where these individuals or small groups meet.

SIGHTING CROWDS

1-24. Sighting crowds are similar to casual crowds with one additional element—an event. People migrate as a crowd to sporting events, are attracted to fires and accidents, and attend music concerts. Individuals or small groups gather at these events for the same purpose. It is the event and/or one's curiosity that compels a crowd to come together.

AGITATED CROWDS

1-25. Agitated crowds add responses that are based on the elements (people, space, and event). Individuals with strong emotional feelings within a crowd can quickly spread and infect the rest of the crowd. As more people within the crowd become emotionally involved, a sense of unity can develop, causing changes in the overall demeanor of the crowd. Yelling, screaming, crying, and profane name-calling are all associated with an agitated crowd.

MOB-LIKE CROWDS

1-26. Mobs have all the elements found in the first three types of crowds, with the addition of aggressive, physical, and sometimes violent actions. Under these conditions, individuals within a crowd will often say and do things they usually would not. Extreme acts of violence and property damage are often a part of mob activities. Mobs consist of (or involve) the elements of people and groups being mixed together and becoming fluid.

CROWD TACTICS

1-27. During a public disorder or disturbance, individuals and small groups within a crowd use any number of tactics to resist authority and disrupt and add turmoil in order to achieve their goals. These tactics can be unplanned or planned and violent or nonviolent. The more organized and purposeful a crowd becomes, the more likely a tactic will be used.

NONVIOLENT BEHAVIOR

1-28. As explained earlier in this chapter, most gatherings of individuals and small groups into a crowd do not involves violent behavior. A public disorder or disturbance usually involves some harmless name-calling, demonstrations to express views, corporate yelling and chanting, and even singing and dancing. In some instances, there may be property damage and the erection of makeshift barricades. All four types of crowds can be nonviolent, but mobs and agitated crowds have the greatest tendency to turn violent. Nonviolent actions of a crowd are disruptive because they are in direct conflict with what authorities want them to do, such as refusing to leave when directed, locking arms, and sitting in front of or around areas and buildings that the authorities are attempting to clear.

VIOLENT BEHAVIOR

1-29. A crowd that becomes a mob can be very violent and destructive. Although some nonviolent activities occur, violent crowds strike out physically at bystanders or others in the crowd, destroying both private and government property, setting fires, and employing bombs (in extreme cases). The only limitations for violent crowds are their own imaginations, the training of their leaders, and the materials readily available.

1-30. Violent crowds may erect barricades and physical barriers to impede the movement of authorities and prevent them from entering certain areas and/or buildings and to hide their activities. Common materials used for barricades include vehicles, felled trees, furniture, felled light poles, fencing materials, or anything else on hand and readily available.

1-31. Setting fires and preventing firefighters access to the structure by blocking it is another violent crowd tactic. Often, fires are set to create confusion in authorities or as a diversion to another activity.

RIOTS

1-32. A riot is one or more groups or individuals who are part of a larger crowd that involves threats of violence against persons or property. In some cases, a crowd will continue to gather until it evolves into a riot.

1-33. Riots vary considerably in both targets and players. A communal riot, for instance, deals with deep-seated ethnic, religious, and language differences. Commodity riots involve an attack on property by acts of vandalism, looting, and arson. Protest riots, such as the riot around the World Trade Organization (WTO) Assembly in Seattle, Washington (30 November 1999), illustrated individuals and groups aggressively and sometimes violently acting out or voicing their opposition to the assembly. The Democratic Convention in Chicago, Illinois, in 1968 reflects a riot that directly targeted police and authority in general. Celebration riots occur across the US as a result of home team victories in sporting events, among other reasons. Celebrating crowds look to make the moment more memorable through raucous acts that demonstrate their joy or happiness, for example, the riots that took place in Chicago in 1992 as a result of the Chicago Bulls winning the National Basketball Association (NBA) Championship.

1-34. Commanders and leaders must be aware that highly organized groups of protesters and/or demonstrators have developed tactics to disrupt the control force. These crowd tactics were published in handbooks for communist organizers during the Cold War, and today the Internet contains sites that are devoted to sharing these tactics with the general public and various protest groups. Although these tactics are somewhat outdated, commanders and leaders must be aware of them and their potential consequences. Many of these groups, such as anarchists, often mock the tactics employed against large crowds; for example, they may identify the police as "Darth Vader Cops" when the control force response is in full riot control gear. Commanders must be aware that well-organized crowds may attempt to engage the control force, surround it, and overpower it by sheer numbers. To avoid this, commanders must do a detailed terrain analysis (intelligence preparation of the battlefield [IPB]) of the area to include all approach and exit routes in the area. The control force flanks have to be protected to avoid being enveloped by the mob. To avoid being enveloped, the on-site commander should keep the crowd at a comfortable distance from the control force formation, which is usually the lethal zone of his NL munitions. Establishing predetermined rally points for the control force is critical in the event that this type of tactic is used by the crowd.

TACTICS USED TO DEFEAT AUTHORITIES

1-35. Organized mobs will try to defeat the control force by employing several different types of tactics. These tactics include the following:

- Constructing barricades.
- Using Molotov cocktails, smoke grenades, rocks, and slingshots.
- Feinting and flanking actions.

BARRICADES

1-36. By constructing barricades, the rioters are trying to protect themselves from assault by the control force. If the control force does advance toward the barricade, rioters hope that the barricade will split the control force and leave it vulnerable to actions by the mob.

MOLOTOV COCKTAILS, SMOKE GRENADES, ROCKS, AND SLINGSHOTS

1-37. Angry mobs will often attempt to disrupt the control force by throwing rocks and other projectiles. As the situation deteriorates, the mob may escalate the violence by using a battery of slingshots that will pellet the control force with a barrage of projectiles ranging from small rocks to marbles. They may also use smoke grenades (homemade or store-bought) to mask their movements. Finally, Molotov cocktails are used against personnel, employed vehicles, and portions of the control force. The mob will attempt to gain a position above an armored vehicle to enable them to drop a Molotov cocktail into an open hatch.

FEINTING AND FLANKING ACTIONS

1-38. Organized mobs may attempt to disrupt the movement of the control force by feinting an assault. When the control force moves to blunt the assault, the mob will assault the exposed flanks in an attempt to split the control force and envelop a portion of it.

WEAPONS

1-39. Mobs will often use various types of weapons against authorities to achieve their agendas and goals. It is extremely important that leaders train soldiers to recognize possible threats of the various weapons used by crowds or mobs, enabling the soldiers to react with the appropriate minimum force.

VERBAL ABUSE

1-40. In almost every instance of a civil disturbance or riot, verbal abuse will be an aggressive tool. Obscene language, racial remarks, taunts, ridicules, and jeers should always be expected. It is apparent that the purpose for using verbal abuse is to anger, demoralize, and provoke a physical response. Undisciplined, untrained soldiers who face such an attack could cause the situation to escalate. Just one provoked action of a soldier could be interpreted as an act of brutality by the media.

PERCEIVED INNOCENT PEOPLE OR THE WEAK

1-41. Women, children, and the elderly have often been used as pawns in civil disturbance operations. On many occasions, the innocent and the weak are moved to the front of the crowd and used as a barrier. With the innocent and the weak directly facing authorities, aggressive and violence-prone individuals or groups behind them attempt to provoke authorities to react. Often, this tactic is to exploit the situation in the eyes of the media for propaganda value. Despite the perception of sometimes being weaker, women, children, and the elderly can be just as committed to the insurgency as those behind them.

BARRICADES

1-42. Barricades are made of anything large or heavy enough to impede or prevent the movement of authorities. Trees, vehicles, fires, and furniture are all examples of items used to erect a barricade.

THROWN OBJECTS

1-43. In many cases, a crowd will throw anything in their reach that can be picked or picked up. Rotten fruit and/or vegetables, rocks, bricks, bottles, sticks, and pieces of lumber are most often used because they are readily available. More sophisticated types of thrown objects are improvised bombs, such as petrol bombs or tennis balls with nails sticking out of them. Never discount that these individuals and groups could acquire hand-thrown explosives, such as sticks of dynamite or grenades.

VEHICLES AND OTHER OBJECTS

1-44. Vehicles and other objects can be placed on higher ground (atop a hill or in a building) to disrupt civil disturbance operations. For example, a vehicle or tire can be set on fire and rolled down a hill, carts or barrels filled with hazardous or flammable materials can be pushed down a hill, or burning furniture and petrol bombs can be thrown from two-story or higher buildings or toward a formation of soldiers.

FIRE

1-45. Rioters have set buildings, vehicles, and other structures on fire to block the advances of authorities and to create confusion and diversions. If property damage is the goal of a crowd, fire is an effective tool. A sniper, for example, may set a building on fire to divert attention or provide maximum target possibilities. Other tactics used, particularly in the Balkans, were to flood the lower floors or basements of structures with flammable liquids or gas and then ignite it in an opportune moment.

FIREARMS

1-46. Sniping or massing fire from within the crowd, in buildings, or other covered positions can (at times) be effective tools for terrorists using a crowd to cover their acts. Firearms are anything that fires a projectile toward a target of opportunity.

EXPLOSIVES

1-47. A bomber's imagination is the only limiting factor in the use of explosives. Explosives can be used as simple diversions to block the advances of formations or provide an escape for rioters. The worst use of explosives is to purposely kill, injure, or demoralize authorities that are protecting and assuring the safety of everyone.

Chapter 2

Control Force Operations

Information is the key to developing civil disturbance plans. One must know who the demonstrators are; when, where, and why they are demonstrating; what their capabilities are; and what their possible course of action (COA) is. A commander's need for current, valid information cannot be overemphasized. The commander must learn as much as he can about the participants (their motivations, strategies, tactics, targets, and dedication). The more knowledge a commander has about the participants, the better equipped he is to counter their actions. He needs sound information to decide how to best use his available resources.

CIVIL DISTURBANCE PLANNING CONSIDERATIONS

2-1. The side that possesses better information and uses that information to gain understanding has a major advantage over its opponent. Forces that have this advantage can use it to bring about changes in attitudes, decisions, and actions in assembled demonstrators. In planning for crowd control or civil disturbance operations, planners must decide what data is needed to develop threat assessments. This decision is based on a multitude of information sources.

STANDARDIZED POSTURES IN COMPANY TACTICAL STANDING OPERATING PROCEDURES

2-2. Within squads, platoons, and companies, equipment may be increased or decreased, as needed. Some examples are as follows:

- Choose the M9 pistol for extraction and apprehension teams. The use of a long weapon (for example, an M16 with an M203, an M4 with an M203, or a 12-gauge shotgun) with NL munitions capability is also recommended, especially for overwatch personnel.
- Add nonstandard weapons such as shotguns for greater NL capabilities.

NOTE: The shotgun is used to protect the M203 gunner as he reloads.

- Add NL munitions to existing organic weapons systems, such as the M203.
- Arm soldiers in the front line of the formation with their standard weapon. If the weapon is a long weapon, it should be carried across the back from left to right or vice versa, with the butt up and muzzle down. Ensure that the weapon is cleared and that the magazine is in the appropriate ammunition pouch.

- Balance the mix of weapons and munitions according to the mission, enemy, terrain, troops, time available, and civilian considerations (METT-TC).
- Establish command relations and determine who has or gives the authority to fire NL munitions.
- Maximize the distance and barriers between the crowd and control formations. Use NL munitions to create a standoff distance.
- Maintain a lethal overwatch of the control force at all times.
- Maintain a reserve force at all times to reinforce the control force, as needed.
- Create NLW range cards for static positions.
- Consider environmental conditions and their effect on the performance of NL munitions. Proper storage, periodic inspections, and the rotation of NL munitions are critical to maintaining the effectiveness and viability of NL munitions.

WEAPON AND AMMUNITION CONFIGURATIONS

2-3. Individuals designated as NL shooters must have the means to transition to lethal rounds, if required. Lethal rounds are carried separately from NL rounds so that the shooter will not confuse them in the heat of confrontation. However, in the heat of confrontation where a lethal option is necessary, NL rounds can be used lethally by adjusting the point of aim to a vulnerable part of the body and closing in on the distance to the target.

2-4. Squad leaders should designate NL shooters within their squad. Ideally, the squad should not change its organization to accommodate NLW.

2-5. Commanders should not dispatch NL patrols. They should plan a combat or security patrol with NL capabilities. Soldiers are never sent in harm's way without lethal protection. NLW are only considered additional tools for the mission and not a mission in itself. There is no such thing as a NL mission.

2-6. Soldiers manning a static position should have NL capabilities depending on METT-TC. Shotguns and the M203 work well at static positions.

2-7. Recovery, apprehension, and/or extraction teams should be established before deployment. Team members should be equipped with personal protective equipment (PPE), an M9, NL munitions and weapons, and Flex-Cufs®. They should have some training in open-hand control, pain compliance, and handcuffs and/or Flex-Cufs.

2-8. Crowd control formations (see *Chapter 6*) should be well trained and well rehearsed. Rapid, coordinated movements of a well-trained and well-rehearsed control force can often be a strong enough deterrent. A lethal overwatch marksman always covers the control force.

2-9. During a NL engagement, the use of designated marksmen (DM) provides confidence and safety to those facing a riot. The DM in an overwatch position scans the crowd to identify threats and designate personnel for recovery and lethal rounds firing (if a lethal threat is presented). They are ideally suited for flank security and countersniper operations.

NONLEHTAL PLANNING

2-10. The fundamental principles of mission planning are not changed by the incorporation of NLW into a plan. NLW are intended to be operational-enhancement tools. The first concern of a commander is the successful completion of the mission. NL munitions provide a commander with a wider range of response options, but these munitions are not a replacement for lethal capabilities. Commanders must always be capable of answering a lethal attack with a lethal response.

2-11. When a commander commits his soldiers and equipment to a crowd control situation, he commits his forces with the additional tools of NL capabilities. This is an addition to the force continuum that the force commander now has available to him. Commanders that properly employ NL munitions and weapons have a tactical advantage over those who rely completely on lethal means.

2-12. When developing a tactical standing operating procedure (TSOP) (see *Chapter 6*), units should consider the following:

- Crowd control formations.
- Extraction teams.
- Apprehension teams.
- Lethal overwatch marksmen and/or observer teams.
- Reserve and/or security forces.

MILITARY WORKING DOG TEAMS

2-13. Military working dog (MWD) teams may be employed with a control force formation as a method of increasing crowd apprehension about approaching or engaging the formation. The teams should be in the rear of the formation in plain sight of the crowd, but in front of the command element and the M33A1 squad RCA disperser. The MWD teams work back and forth behind the formation as an intimidation measure. The presence of the MWD may produce a profound psychological effect on the crowd. These teams may also be used to help control individuals who have been captured by the recovery and apprehension teams.

CAUTION

Do not unleash an MWD on a crowd.

COMBAT CAMERAMEN

2-14. Video and still cameramen should make a photographic record of the individuals in the crowd who are leaders and instigators. Events must be documented to hold personnel, factions, gangs, or groups accountable for acts that violate law, destroy property, or cause physical harm. Electronically recording events aids in the prosecution of such cases and eliminates the sense of anonymity that people in large crowds often feel.

INFORMATION OPERATIONS

2-15. Information superiority helps forces to anticipate problems and requirements. It allows commanders to control situations earlier and with less force, creating the conditions necessary to achieve the optimal end state. Public affairs, psychological operations (PSYOP), and civil military operations are activities that will allow the commander to control situations earlier and with less force.

2-16. Information is available from a multitude of sources. A diversity of sources is the best approach because it prevents biased perspectives. Primary sources are as follows:

- Open sources.
  - Libraries.
  - Newspapers and news periodicals.
  - Radio and television.
  - Internet.
- Law enforcement sources.
  - Local law enforcement agencies.
  - National law enforcement agencies.
- Military sources.
  - Department of Defense (DOD) intelligence community (most restrictive source).
  - Local military intelligence (MI) field offices.

2-17. Police intelligence operations (PIO) is one of the five military police functions. This function is a process of actively and passively collecting information that is of a police, criminal, or combat nature. As military police perform the other four functions of maneuver and mobility support (MS), area security (AS), law and order (L&O), and internment and resettlement (I/R), they are gathering information that supports, enhances, and contributes to the commander's protection program, situational awareness, and battlefield visualization by portraying relevant threat information that may affect operational and tactical environments.

2-18. IPB is a continuous process for analyzing the threat and the environment of a specific geographic area. During the IPB process, the Intelligence Officer (US Army) (S2) or Assistant Chief of Staff, G2 (Intelligence) (G2) uses all available databases, intelligence sources and products, and related MI discipline to analyze the threat and the environment. The PIO function supports this process by providing the S2 with collected police, criminal, and combat information that can directly and significantly contribute to the success of the MI effort. In addition to combat information, the PIO function provides additional information on possible criminal threats and COAs. This is intended to support the S2 IPB process and can be used by the commander to upgrade force protection. (See *FM 19-10*.)

THREAT ANALYSIS

2-19. Crowd control requires its own thought process. Emphasis should be on prevention rather than confrontation. In combat, military forces are taught to fight and eliminate threats. In crowd control, military forces must deal with noncombatants that have internationally recognized rights. These rights must be respected while maintaining public order. This is an issue that law enforcement agencies (LEAs) have been struggling with for years.

2-20. Dealing with crowd control incidents is a matter of using some basic guidelines. It also includes asking the right questions in a logical manner so that key issues are not omitted. The commander must use these guidelines in context with METT-TC and the location of the incident (CONUS or OCONUS).

PREINCIDENT PLANNING

2-21. This phase of planning begins before the incident and is initiated at the operational level with guidance from the strategic level. The planning includes guidance on crowd control and addresses responsibilities, training, organization, operating procedures, use of force (CONUS), and/or rules of engagement (OCONUS). The most difficult and productive decisions are those made in the preincident planning process.

Avoid Confrontation

2-22. Crowd situations are highly unpredictable, but one thing seems certain—confrontation will likely cause crowd resistance. When pushed, people tend to resist opposition to the realization of their purposes.

Focus on Prevention

2-23. Planning should key in on the prevention of unfavorable outcomes. Experience has shown some LEAs attempting to help crowds accomplish their goals within the law that have been beneficial and even led to conceding some violations for the purpose of avoiding confrontation. However, LEAs maintain a law enforcement presence, which signifies social restraint. LEAs also decide when and where they will not compromise and the amount of force to use.

Define Goals

2-24. When defining a goal, deciding what must be accomplished is the first step. Defining the goal is fundamental. However, actually working toward and accomplishing the goal is easier said than done, as the process can drive the situation. Commanders and leaders must be aware of this. The military force must focus on what they are trying to accomplish.

BACKGROUND INFORMATION AND INTELLIGENCE

2-25. Seek to know as much as possible about social protest groups (within the limits of the law, see *Appendix B*) before an incident. This will provide insight into the organization and its functions and provide a warning as to what to expect. When you know as much about these organizations as possible, then it is possible to anticipate their next move. See *Appendix C*.

2-26. Continually assess what is to be accomplished against what others are trying to accomplish. Some groups may not have goals that conflict with those of US forces, but their actions may. Other groups may have goals and activities that do conflict. In either case, the assessment of group goals compared to the goals of US forces helps to understand and avoid potential conflicts.

CIVIL DISTURBANCE OPERATIONS PREPARATION

2-27. Counterdemonstration workgroups are a mixture of traditional and nontraditional staff proponents brought together to coordinate the resources to be employed by task force (TF) units before a potential civil disturbance. Meetings of the counterdemonstration workgroup are held on a regular basis (generally weekly) and are chaired by the Assistant Chief of Staff G3 (Operations and Plans) (G3). The following staff proponents may be involved, and others may be involved as needed:

- Public affairs office (PAO).
- Joint military commission (JMC), if organized.
- Provost marshal office (PMO).
- PSYOP.
- G2, Assistant Chief of Staff G5 (Civil Affairs) (G5), Assistant Chief of Staff G6 (Signal) (G6).
- Civil affairs (CA).
- Army airspace command and control (A2C2).
- Chaplain.
- Engineer.
- Surgeon.
- Fire support element (FSE).

RELATIONSHIPS DEVELOPMENT

2-28. Working relationships between commanders and protest group leaders are increasingly seen as the best means for preventing bad outcomes in crowd situations. This is called the negotiated management model of crowd control. LEAs in large metropolitan cities in the US and Europe practice it.

2-29. Open dialog helps develop working relationships between commanders and protest group leaders, providing an opportunity to communicate clearly. Talking allows group leaders to tell authorities and tactical commanders what they want to accomplish. It also allows authorities and commanders to tell group leaders what they are prepared to do and how they might respond to certain crowd behaviors. Such communications can do much to resolve issues and prevent violence.

2-30. Commanders may not be able to talk to all leaders before a crowd assembles. Some groups do not have recognized leaders and are ad hoc organizations. Other groups may have several leaders, but only some of those leaders will negotiate. In such cases, meetings with some leaders may tell you who the other leaders are and provide critical information. Negotiations may also encourage more moderate leaders to do things that will support the commander.

2-31. Initiate communication with group leaders to work out issues before an event. Commanders should make a concentrated effort to maintain a working relationship with the leaders of protest groups.

2-32. Communication also means persuasion. Commanders should make a concerted effort to win over demonstration leaders. Commanders may tell group leaders that they want to help them complete their mission in a safe manner. They should encourage demonstrators to protest in an acceptable way and, if possible, offer favors to get them to do just that.

NOTE: Commanders must only offer what they are willing and able to deliver.

2-33. Several rules apply in negotiations, and credibility is key. Only communicate necessary information and those actions that authorities and commanders intend to do. Negotiations are made from a position of strength, and by negotiating an agreement may be reached. However, contingencies should be developed in the event that the agreement is violated.

2-34. Working relationships between commanders and protest group leaders often result in protest groups policing themselves. This is one of the basic premises of negotiated management. Protesters are more likely to listen to their own leaders, as opposed to listening to US forces. Protest organizers are likely to buy into a cooperative effort and agree to ground rules set during preprotest negotiations. Many protest leaders seek to stay on the right side of the law.

ACCEPTABILITY

2-35. Deployed US forces will find themselves engaged in crowd control operations under difficult circumstances. The host nation (HN) may have groups that do not accept the presence of US forces, which will make imposing order and the protection of citizens difficult at best. US forces will be under intense media and political scrutiny (an environment much like what LEAs operate in on a day-to-day basis).

2-36. Winning in this environment is not like winning in combat. US forces may appear to be invincible and formidable, but they risk being portrayed as oppressors. Thus, US forces can lose by appearing to win. Groups that perceive themselves as oppressed will readily seek victimhood in an effort to gain the support of public opinion. Winning in this environment is about seizing and holding the moral high ground. US forces must maintain the authority and legitimacy of what they are doing.

2-37. Projecting a favorable image will require outreach to local leaders and citizens. It will also require developing a relationship with the media. For example, the Los Angeles Sheriffs Department often invites the media to accompany their tactical commander during crowd control situations. This shows that they have nothing to hide. It also provides an opportunity for individuals to see the commander's side of an event.

2-38. In this environment, commanders must consider how actions will play among several audiences—local, allied, US, and international. The media effect is inescapable. In some cases, it will be difficult to accommodate all of these audiences. However, the most important audience will be the US public, which is key to the continued support of US forces.

CROWD ASSESSMENT QUESTIONS

2-39. Assessing crowds requires its own set of questions. These questions should be answered before a crowd assembles or as quickly as possible if a gathering occurs without notification. In some cases, all the questions may not be answered. These questions represent a logical way to think about crowds. The questions are as follows:

- **Who are they? What is the overarching identity of the crowd? Are they strikers, ethnic factions, or social protesters? Do they identify themselves as strikers, ethnic groups, religious factions, or protesters against some perceived social injustice?** Understanding who they are will indicate what they may do. It may be possible to determine identities and goals from advance assembling instructions, leaflets distributed to bystanders, placards and banners, and chants and songs.
- **What are their goals?** What the group wants to accomplish by assembling could determine the extent to which they can be accommodated once they have assembled. They may only seek recognition for their cause (being seen and heard). If so, this goal is usually easy to accommodate. However, some groups may have more demanding goals, for example the demonstrators in Seattle who sought to stop the WTO. Goals that cannot be accommodated make confrontation very likely.
- **What is the composition of the crowd and are there any known factions?** Seattle demonstrations against the WTO were comprised of groups that were protesting environmental issues, wages, and child labor laws. Differing goals and the resulting friction were evident between such organizations as the Ruckus Society and the American Federation of Labor-Congress of Industrial Organizations (AFL-CIO). Factions within a crowd represent threats and opportunities.
- **What are they capable of doing?** Protest groups often claim that they will assemble large numbers of people to produce some disruptive action. However, there is often a difference between the claim and reality. An organization may claim that it can mass a 100,000 people, but in reality can only get 40,000. Organizers exaggerate for the following reasons: they want to boost the morale of their own people and they want the media to report that they have strength in numbers. Studying the past activities of a group may provide indications of what they are capable of doing in the future.
- **What are their traditional behaviors or cultural repertoires?** What people do during protests is not universal. It varies with the group and the culture. Social protest organizations and striking unions will carry placards and banners. Other groups will protest in a more quiet way, like the 5,000 women in Sarajevo whose standard Sunday behavior was to sit and block traffic. They were protesting the loss of male relatives in the Bosnian war. Understanding the goals of

the protesters can be helpful in deciding how to respond to their behaviors.

- **When and where will they assemble?** Every protest organizer has a time and place for assembling and perhaps a destination for the crowd to move toward. If the organizer attempts to mobilize large numbers of participants, the time and place for assembling and dispersing must be made known in the instructions. This information may be stated in the mobilization instructions or disseminated by an informal network, such as word of mouth. Such impromptu networks in densely populated areas can enable rapid assembling.
- **Where will they go?** Many crowds have destinations. Organizations may march a specified distance to ensure that their cause gets sufficient attention. Commanders need to know the route to minimize disruptions to the rest of the community. In Los Angeles, LEAs attempt to reroute traffic and prevent congestion caused by protest marches. It is also necessary to provide security along the route to prevent counterdemonstrators from confronting the marchers. This will prevent an even greater problem for the community.
- **What are the possible targets of violence?** Riots, in particular, may focus on target facilities. In the 1992 Los Angeles riots, gun stores were major targets. In developing countries, targets may be more basic. In Somalia, throngs of people stormed food supply facilities. In Haiti, riots broke out over trash dumps where US forces had discarded the remnants of meals, ready to eat (MREs).
- **What is the worst-case scenario?** The worst-case scenario must be recognized for the sake of avoidance. This may be when a peaceful crowd degenerates into a violent riot.
- **When and where will they disperse?** Crowds have a life cycle that includes how they disperse (see *Chapter 1*). Commanders must consider this. It is essential that there be one or more avenues that individuals can use to disperse. These dispersal routes should be clearly marked, visible, and open-ended. There may be a need for US forces to assist the crowd with dispersal. For example, in Los Angeles a 70,000-man march went from east to central Los Angeles. Once at their destination, they engaged in peaceful activities and then dispersed. The LEAs provided transportation back to the assembly area (AA); otherwise, the protesters would have returned to the AA in a disorganized and unsupervised manner.
- **Are there plans for subsequent gatherings?** A crowd may disperse for a short time so participants can take care of personal needs. This could be days or only a few hours.

LEGAL CONSIDERATIONS

2-40. The legal authority in which a peace operation is conducted defines the parameters of the operation. All commanders, leaders, and soldiers must know the legal authority that regulates their operations.

MANDATES

2-41. There are two types of mandates—United Nations (UN) and non-UN. Mandates are the authority under which an operation is conducted. UN mandates are created from resolutions by the UN Security Council (UNSC) or the UN General Assembly (UNGA) that authorize and define an operation. Non-UN mandates are created from treaties, accords, resolutions, or agreements evolving from other international or regional organizations.

STATUS OF FORCES AGREEMENT AND STATUS OF MISSION AGREEMENT

2-42. A status of forces agreement (SOFA) and a status of mission agreement (SOMA) are key documents that define the legal authority and responsibilities of a force and force personnel participating in an operation. A SOFA or SOMA may be a treaty or memorandum of understanding. It is an agreement negotiated between the UN and the host countries, which details the rights, privileges, immunities, and nature of services to be provided to the force and its personnel and their responsibilities and obligations.

INTERNATIONAL LAW

2-43. Legally, international law affects most stability operations and support operations (SOSO). These laws consist mainly of international agreements, treaties, and customary international law, which include the agreements and laws known as the law of war. International agreements prescribe the rights, duties, powers, and privileges of nations relative to particular undertakings. International agreements will affect US participation in SOSO in the—

- Right of US forces to enter a foreign country.
- Status of US forces while in the foreign country.
- Construction and operation of US bases of operations.
- Aircraft overflights and landing rights.
- Processing of claims for any damage to persons and property of the foreign country.

2-44. These agreements or customary international laws govern all aspects of operations carried out. In the absence of a viable HN government or proper international agreement, many aspects of the operations are controlled by domestic (US) or customary international law.

UNITED STATES LAW

2-45. Operations in peace and conflict must comply with US law, whether as a statute, executive order, regulation, or other directive from a branch or agency of the federal government. The *Uniform Code of Military Justice (UCMJ)* applies to questions of military justice. Various statutes, *Executive Order 12333; Section 3, Code of Federal Regulations (CFR);* and DOD and service regulations govern all intelligence-gathering activities. The staff judge advocate (SJA) must actively advise and participate in all activities from initial planning to redeployment.

HOST NATION LAW

2-46. If a viable HN government exists, whether at the national or the local level, all laws of the HN will apply to US forces in that country unless an international agreement provides otherwise. The types of laws that may inhibit US operations are in the fields of immigration, labor, currency exchange, the procurement of goods and services, customs and taxes, and criminal and civil liabilities. A SOFA usually covers these areas before an operation begins. Therefore, planners must understand the law to assess it and determine if it will adversely affect the operation. Assistance may be available from the local US embassy or the command judge advocate; if not, the command may have to rely on other sources for guidance. If local law hinders the operation, the commander must inform the US embassy and request that it negotiate a solution.

CIVIL AUTHORITIES

2-47. There will be times that US forces will deploy to nations where the infrastructure of civil authorities has survived the conflict and is functional. Local authorities and police are responsible for ensuring the safety of their citizens, including the UN High Commissioner for Refugees (UNHCR). US forces assist in this by providing general military security and by facilitating negotiations. In countries where the civil authority infrastructure has collapsed, the UN mandate may require the coalition of forces to help ensure the safety of citizens until their own civil authorities can be rebuilt.

RULES OF ENGAGEMENT

2-48. The rules of engagement (ROE) are the commander's rules for the use of force (RUF). Commanders are responsible for interpreting, drafting, disseminating, and training on the ROE. Ensure that the SJA assists throughout the ROE development process. This will ensure that the ROE will not improperly constrain actions, but are still consistent with domestic and international law and policies and orders of the chain of command.

2-49. Clearly stated ROE are published before Army forces are committed. The highest military authority, with input from subordinate commanders, will continually evaluate the ROE and modify them as appropriate. When the use of NLW has been authorized in the ROE, the decision to use NLW should be delegated to the lowest possible level, preferably the platoon or squad level. This requires that all personnel, not just leaders, have a clear understanding of the ROE and the commander's intent (see *FM 3-22.40*).

2-50. In peace operations, the use of force is restrained while the diplomatic solution is sought through negotiation or mediation. Accordingly, ROE are more restrictive in peace operations, making them more politically sensitive. In today's world, peace operations are usually conducted by a coalition of forces operating under the purview of the UN charter and customary international law. Therefore, the UN may mandate certain restraints. Using overwhelming force could compromise diplomatic efforts to reach a peaceful settlement. Even a single, relatively small engagement could jeopardize the legitimacy of the peacekeeping force and upset negotiations or mediations. Commanders must beware that in any confrontation ROE decisions made by soldiers can have strategic political implications on current and future operations. Therefore, it is vital that leaders and soldiers hold a common understanding of when, how, and to what degree force is to be used. This requires training.

MULTINATIONAL OPERATIONS

2-51. Multinational operations include military forces from different nations. These operations are usually interagency operations. Conducting operations with foreign military partners, like operations with civilian partners, is uncommon to many soldiers, so a clear understanding of this different environment is necessary.

2-52. The majority of US peace operations will be part of a UN peace operation. Their multinational character merits particular attention because national interests and organizational influence may compete with doctrine and efficiency. Commanders can expect contributing nations to adhere to national policies and priorities, which at times complicate the multinational effort.

2-53. When working within a multinational force and with allies, do the following:

- Train coalition partners and allies on tactics and techniques.
- Remember that teamwork and trust are essential.
- Achieve unity of effort.
- Consider equipment and communications capability and interface.
- Strive to eliminate sources of confusion and misunderstanding.
- Remember that military doctrine varies from country to country.
- Remember that national culture influences the way units operate.
- Remember that language barriers represent a significant challenge.

SCALABLE EFFECTS

2-54. Scalable effects concepts are a measured approach in response to a crowd gathering. By recognizing a use-of-force policy, soldiers must be taught and understand that they use the minimum force necessary. Without the appearance of a graduated response, the gathering crowd may consider actions as excessive, causing a possible escalation of hostilities or violence. Do the following to aid in the scalable effects process:

- Try to persuade the crowd to quietly disperse by talking with leaders.
- Use translators as necessary.
- Let the first approach be by local authorities (the mayor or police).
- Pass out handbills requesting that the crowd return home.
- Use video and still cameras to photograph individuals and events for later use at trials.
- Give warnings before moving to the next level of force.

2-55. Ideally, the force should be positioned out of sight of the crowd. Have the troops move into position with few shouted commands. For maximum effect on the crowd, have them form their formation decisively and professionally. Do the following to aid in the formation process:

- Display force in a graduated manner (for example, a UH 60, hardened vehicles, and soldiers with charged weapons).
- Exploit the psychological effects of show of force.
- Demonstrate force (do not use on unarmed civilians).
- Escalate the PSYOP message.
- Highlight the target pointer.
- Demonstrate sniper precision strike capability.
- Employ riot control measures.
- Use NL munitions.
- Use RCAs (if approved).
- Move through the crowd using riot control formations and movement techniques.
- Use the graduated response matrix (GRM)

2-56. With the increased participation of US forces in contingency operations around the world, it accentuates the need to establish procedures for applying graduated military responses to situations that threaten these missions. Numerous GRMs and similar products exist throughout the military. These products graphically portray available responses in a graduated manner. The intent of these products is to give the on-scene commanders a list of options with which to control or defuse a situation before it gets out of hand. Most threats can be eliminated without loss of life or collateral damage by effectively applying the resources available.

PLAN A GRADUATED RESPONSE MATRIX

2-57. The planning for and development of a GRM begins with the mission analysis portion of the military decision-making process (MDMP). Some missions require US forces to enforce treaties or accords to protect the lives of civilians in uncertain or hostile environments (such as noncombatant evacuation operations [NEO]) or to provide large-scale humanitarian assistance. These missions require some sort of graduated response criteria to maintain order and prevent uncertain environments from becoming hostile. The development of a GRM requires the following seven steps:

- Identify the need for a GRM.
- Establish a team to develop the GRM.
- Develop targets.
- Coordinate staff functions.
- War-game.
- Receive command approval.
- Operate deliberately.

Identify the Need for a Graduated Response Matrix

2-58. Not all missions require a GRM. The decision to use a GRM requires careful consideration. Once it is agreed that a GRM is necessary, it requires guidance from the commander regarding the response options available.

Determining the appropriate responses is based on the facts, assumptions, and limitations identified during mission analysis.

2-59. Planners (staff) must agree on the intent of the GRM. The GRM can be used as a training and rehearsal tool. It provides leaders with the most likely vignettes that can be incorporated into a COA analysis, predeployment training, and rehearsals. The GRM can also be used as a handy reference during situations that require graduated responses.

Establish a Team to Develop the Graduated Response Matrix

2-60. Establish a GRM development team with a broad range of skills, knowledge, and professionalism. The following list of individuals makes up a typical GRM team:

- Provost marshal or senior military police advisor.
- FSE.
- Brigade legal officer.
- PSYOP representative.
- Land information warfare or information operations officer.

2-61. Since the GRM is designed to give commanders graduated options for dealing with both hostile and nonhostile threats to the mission, this team composition allows for target selection, application of the ROE, and attack methods using both NL and lethal means.

Develop Targets

2-62. The FSE, in conjunction with the S2 section, develops targets for both lethal and NL attacks. In the case of stability operations, these targets are usually not the conventional specific point or piece of equipment on the ground. They are more situational than specific. The GRM identifies situations or acts that subordinate elements could face during the mission. The sample GRM in *Figure 2-1* shows three possible situations or acts that on-scene commanders could expect to encounter. From the targeting standpoint, these are groups of more specific targets.

2-63. During mission analysis, the fire support officer (FSO) identifies both lethal and NL assets available to the unit. A tactical PSYOP team (TPT) attached to the unit is an example of a NL attack asset that should not be overlooked. The following are examples of what the FSO should look for:

- RCAs.
- TPTs.
- Electronic warfare assets.
- CA teams.
- Information operations (IO) teams.
- Artillery smoke projectiles.
- Aircraft (AH-64s, OH-58Ds, and AC-130s).
- Mortars.

2-64. The lethal assets described could be used in a NL show of force or demonstration to defuse a situation before it requires lethal force. The critical element of this mission analysis is not to focus solely on lethal assets. In

<table>
<tr><th>Crowd Action</th><th colspan="4">Nonlethal Response</th></tr>
<tr><th></th><th>Command Presence/PSYOP</th><th>Show of Force</th><th>Demonstration of Force</th><th>Riot Control Means (if approved)</th></tr>
<tr><td>Unarmed Civilians</td><td rowspan="3">- Ensure that the on-site commander of the TPT directs the broadcast of the dispersal proclamation and/ or passes handbills.
- Escalate the tone of the dispersal proclamation from information to a warning of force.</td><td rowspan="3">- Display force along with escalating the dispersal proclamation.
- Display force in a graduated manner, such as a helicopter hovering over a crowd or soldiers with charged weapons.
- Exploit the psychological effect of a show of force.</td><td>- Do not demonstrate force toward unarmed civilians.</td><td rowspan="2">- Employ RCAs at the point of penetration.
- Use PSYOP to exploit the psychological effect.
- Move through the crowd using riot control formations and movement techniques.</td></tr>
<tr><td>Armed Civilians (knives, clubs)</td><td rowspan="2">- Display force along with escalating the dispersal proclamation.
- Highlight the target pointer.
- Demonstrate sniper precision strike capabilities.</td></tr>
<tr><td>Armed Crowds/ Military (firearms)</td><td>- Do not use RCAs—they may escalate the situation.</td></tr>
<tr><td colspan="5">Hostile intent/hostile act occurs by armed threat.</td></tr>
<tr><th>Crowd Action</th><th colspan="4">Lethal Response</th></tr>
<tr><th></th><th>Sniper Response</th><th>Small Arms Direct Fire</th><th>Aerial Fires</th><th>Indirect Fires</th></tr>
<tr><td>Unarmed Civilians</td><td rowspan="3">- Ensure that target leaders or troublemakers are targeted.
- Use the minimum response necessary.
- Exploit the psychological effect of an attack.</td><td rowspan="3">- Determine that sniper attack is insufficient.
- Consider demonstrating capabilities.
- Exploit the psychological effect of a lethal response.
- Escalate gradually, starting with a small caliber, single round and work up to a large caliber, automatic.</td><td>- Determine that small arms direct fire is ineffective.
- Use the minimum response necessary.
- Use a minimal precision strike initially; use subsequent fires based on the situation.
- Exploit the psychological impact of each strike.</td><td>- Determine that air assets are unavailable or ineffective.
- Use the minimum response necessary.
- Ensure that the response is directed by the on-site commander.</td></tr>
<tr><td>Armed Civilians (knives, clubs)</td><td colspan="2" rowspan="2">- CAS/indirect fires must be authorized by the MACOM commander.
- Consider requesting permission for use when—
• All lesser means have been ineffective.
• There are physical eyes on target.
• Proximity to civilians has been considered.
• Risk to friendly forces/evacuees outweighs the risk of collateral damage.</td></tr>
<tr><td>Armed Civilians/ Military (firearms)</td></tr>
</table>

Figure 2-1. Sample GRM Card

stability operations, the Army wants to prevent acts of hostility first and then be prepared, if necessary, to apply lethal force.

2-65. Graduated responses can range from command presence through the show of force, a demonstration, the use of RCAs, and other techniques (such as the application of lethal force using snipers, small arms, AC-130s, and indirect fires).

Coordinate Staff Functions

2-66. To coordinate staff functions the rest of the GRM team assembles to complete the escalation sequence for each response. PYSOP and legal representatives are critical attendees during the escalation sequencing process. In the area of psychological operations, the TPT must exploit the effects of all responses.

2-67. The legal officer evaluates each escalation of force option and graduated response to ensure that it is consistent with the ROE. The GRM is designed to recommend applications of force consistent with the ROE, yet not limit the leader or individual soldier's right of self-defense. A GRM should clearly show that if a hostile act occurs, lethal options would be first and foremost.

2-68. In the case of lethal responses, the commander's guidance must again be applied. For example, lethal responses are allowed only in self-defense.

War-Game

2-69. Once the types of escalations for each potential graduated response are determined and annotated, the GRM must then be war-gamed. The staff must walk through each act or situation from the on-scene commander's standpoint.

Receive Command Approval

2-70. Once the GRM has been war-gamed, it must be submitted to the commander for approval. This is the final check to ensure that the GRM team has applied the commander's guidance correctly and met his intent.

Operate Deliberately

2-71. Operating deliberately affords commanders ample opportunity to plan and prepare for possible civil disorder situations. Through the effective gathering of information and a working cooperation with local government and police officials, commanders can often be made aware of dates, times, locations, and what groups may assemble before the operation. The purpose is to gather as much information as possible who is involved, where they are assembling, what incident promoted the activity, and what seems to be the prevailing attitude of the assembling crowd.

2-72. Civil disturbance operations are dynamic, ever-changing environments requiring effective communications both up and down the chains of command. They require advance preparation and planning using the established troop-leading procedures (TLP) outlined in *FM 3-19.4*. The eight-step process of TLP aids commanders and their subordinate leaders in planning and preparing for an operation. Most steps can be accomplished concurrently, but rarely is there enough time to go through each step in detail. Despite this, leaders must follow these steps to ensure that nothing is left out of the planning process. As outlined in *FM 3-19.4*, the following are the eight steps of TLP:

- Receive and analyze the mission.
- Issue a warning order.
- Make a tentative plan.
- Initiate movement.
- Conduct a reconnaissance.
- Complete the plan.
- Issue the order.
- Supervise, refine, and rehearse.

CONDUCT HASTY OPERATIONS

2-73. Conducting hasty operations seems to be the norm rather than the exception. Unlike deliberate operations, hasty civil disorder operations are reactionary in nature with little or no time for planning. In these situations, the event of a crowd gathering is already underway. There is very little, if any, advanced warning of the unfolding situation, and commanders are usually put in the position of sending their soldiers into an already volatile, and perhaps hostile, environment.

2-74. Commanders should be in immediate and constant communication with local civil and police authorities. The purpose is to gather as much information as possible about who is involved, where they are assembling, what incident promoted the activity, and what seems to be the prevailing attitude of the assembling crowd.

2-75. It is essential that commanders operating in environments where civil disturbance operations are likely ensure that their soldiers receive training at all levels. Commanders and subordinate leaders need to instinctively rely on the same TLP that they would in a deliberate operation. See *FM 3-19.4*.

CONDUCT TACTICAL OPTIONS

2-76. In the peacekeeping and peace enforcement environment, the use of conventional firearms or the threat of their use may not be the solution to a situation where US forces must separate two belligerent, hostile ethnic groups or prevent a similar group from entering an area that is off limits to them.

2-77. Crowd control options are often combined. Commanders choose their options based on an evaluation of the particular crowd. Commanders select any combination of control techniques and force options they think will influence the particular situation (METT-TC). Commanders must always try to choose the response that can be expected to reduce the intensity of the situation.

Monitoring

2-78. Monitoring an assembled crowd consists of gathering necessary intelligence and watching them to determine progress and development. It is this gathered information that helps commanders select the appropriate actions. Gathered information can often be used to help defuse a situation through persuasion.

2-79. Monitoring is continuous. Without timely information, responses could cause the situation to escalate beyond what it should have or cause a response to be inadequate based on the situation.

2-80. Commanders may task teams with specific missions to monitor crowd activity and note any new developments. Monitoring done by these observation teams gives the commander up-to-the-minute information so he can gauge crowd activity and intent in relation to the overall situation.

2-81. Observation teams monitor crowd activities to gather information. They observe and report on crowd size, location, and mood and on the developing situation. An observation team may consist of a marksman, a radio operator, and an observer equipped with binoculars. They may be posted strategically on rooftops and other high terrain that overlooks the crowd. Sometimes observers use helicopter-mounted observation devices. This also affords security for the operational forces. Commanders must know where their observation teams are located so they are not mistaken for teams of snipers.

2-82. The timely flow of information may allow the commander to influence the outcome of the situation with simple negotiations. Monitoring is appropriate when more decisive action is not feasible because of crowd size or when the intensity of a situation might escalate. It is particularly useful in large, nonviolent demonstrations. Monitoring can serve as an interim measure until more control forces arrive. It includes establishing communication with crowd leaders to convey official interest and intent to the crowd. Monitoring also includes efforts to gain the cooperation of crowd leaders.

2-83. Communication with crowd leaders and participants can help a commander control a situation without more severe measures. If communications exist with crowd leaders, the authorities may be able to divert the leaders or the crowd from their stated or apparent goal. Pressure can be put on leaders to channel the crowd into an area that minimizes disruption to the community and aids crowd control operations. March routes and demonstration areas can be limited to those that will help contain the crowd and reduce their potential for disrupting the community. Pressure can be positive (offering concessions) or negative (with deterrents).

2-84. If the commander can gain the cooperation of crowd leaders, it can decrease the potential for disorder, and if crowd leaders seek cooperation from authorities, officials should try to be accommodating. Crowd leaders can be placed in liaison positions between the crowd and the control force. Crowd leaders can be made responsible for managing the crowd by policing their own activities.

2-85. Taking still pictures or videotaping the faces of individuals within a crowd may prevent or reduce unlawful and violent acts. This could lessen their sense of anonymity. If needed, photographs or videotapes can be used as evidence for prosecution. To be effective, crowd members must see their presence being recorded. The photographer or cameraman should be in uniform to let the crowd know who he or she is. Recorders must be close enough to the crowd to be seen easily, but not close enough to be endangered.

Blocking

2-86. It is not uncommon for protests and demonstrations against the US or its forces to occur outside the main entrances to military installations or US Embassies. Because of the sensitive nature of equipment and information that can be found at these locations, blocking unauthorized entry is critical.

2-87. Blocking is the physical denial of advance upon a facility or an area by a crowd. Commanders may have to task their forces to block a crowd. Blocking physically denies crowd advance. Crowd control formations, particularly line formations, along with barricades can be used to block advancing crowds. Barricades of vehicles, concertina wire, and water-filled barrels are used to block or channel the movement of the crowd.

Dispersing

2-88. Dispersing is taking deliberate actions to fragment an assembled crowd in order to prevent the destruction of property or prevent injury. It is extremely effective against smaller crowds in congested urban environments. However, this may increase and spread lawless activity rather than reduce it. Thus, the commander's forces must control dispersal routes and the areas in which the dispersal will occur. Forces must protect the facilities that could be considered likely targets of opportunity for dispersing individuals or groups. Dispersal of the crowd may require apprehension of small groups still active in the area. Proclamations, shows of force, crowd control formations, and RCAs can disperse crowds.

2-89. Avoid (if possible) the dispersion of crowds into wide-open areas because it gives the crowd the opportunity to grow rapidly in size. The direction of crowd movement is difficult to anticipate and channel.

2-90. Once the crowd has started to disperse, it may occasionally be necessary for the forces to advance, keeping the crowd on the move and in the right direction. A crowd on the move should not be hurried to avoid panic. At no time should the crowd be cornered in a position where there is the perception of no escape. This invokes the "fight-or-flight" syndrome, possibly escalating violent activity.

2-91. Issuing a proclamation (see *Figure 2-2,* page 2-20) can help disperse a crowd. Proclamations officially establish the illegal nature of crowd actions and put the populace on notice that the situation demands special measures. Proclamations prepare the people for the presence of military authority and it tends to inspire respect. Proclamations support law-abiding elements and psychologically bolster military forces trying to restore order. They also demonstrate the gravity of the situation to all concerned and are excellent ways to make a commander's intentions known to a crowd. It is also a good way to reduce crowd size before direct action is taken. Commanders can make a verbal proclamation similar to the following statement: "*In the name of the President of the United States, I command that you disperse and retire peaceably to your homes.*" Such a proclamation may even make direct action unnecessary.

I. DISPERSAL

Unlawful Gathering

"Attention! Attention! This area must be cleared at once! Further unlawful behavior will not be tolerated. Clear this area at once or the necessary force to do so will be used."*

Demonstration

"Disperse and retire peaceably! Disperse and retire peaceably! Attention all demonstrators!"

"The demonstration in which you are participating ends at ______. The permit that was agreed to by the leaders of the demonstration expires at that time. All demonstrators must depart from the ______ by _____. All persons who wish to leave voluntarily may board the buses. These buses will go to the _____. Those who wish to take buses should move to ______. Those demonstrators who do not leave voluntarily by ______ will be arrested and taken to a federal detention center. All demonstrators are urged to abide by the permit."*

Warning of Looting

"Return to your homes! Someone may be looting them at this moment! During a disturbance, criminal activity is at its peak. Your family or your property may be in danger."*

II. EMPLOYMENT OF TROOPS

Employment

"Attention! Attention! Troops are present in this area. They are preparing to advance. Order must and will be maintained. Disperse peaceably and leave this area. To avoid possible injury, leave at once."

"Disperse now and avoid possible injury! Disperse now and avoid possible injury!"*

(Repeat until troops are committed.)

When Troops Are Committed

"Troops are advancing now. They will not stop until this crowd is dispersed and order is restored. To avoid injuries, leave the area at once. Return to your homes as peaceful citizens. Troops have their orders, and they will not stop until the crowd is dispersed. Do not get hurt. Leave this area."*

III. PRESENCE OF CHILDREN

(Used in conjunction with other announcements.)

"Attention! Attention! Do not attempt to cause further disorder. Disperse now in an orderly manner and avoid possible injury to children. Return at once to your homes."*

IV. ADDITIONAL INSTRUCTIONS

*Indicate the method, streets, and direction that the crowd should use when dispersing.

Figure 2-2. Sample Proclamation

2-92. In making a proclamation, a commander may consider imposing a time limit. However, the situation may change, and not imposing a time limit leaves the commander free to choose other actions when he wants. A proclamation must be specific in its instruction.

2-93. If a time limit is stated, it must be a reasonable length of time for the crowd to comply with the instructions. When drafting a proclamation, the commander must consult closely with the SJA. He must use the simplest language possible to maximize the effects of the proclamation. If proclamations must be translated to a local language, the translation must be made with great care.

2-94. The commander at the scene may direct that a proclamation be issued over public address (PA) systems. The force of the words used in the proclamation must be gauged to the composition of the crowd. If the crowd consists of normally law-abiding citizens who are presently assembled to show disagreement with an existing situation, the proclamation requires less force. On the other hand, if the crowd consists of militant rioters, the proclamation requires more force. The text may take a number of forms, depending on the situation.

2-95. A show of force is often a useful measure for dispersing a crowd. When troops arrive, the psychological impact of their arrival can be used. Soldiers can dismount from the helicopters, buses, or trucks in plain sight of the crowd, but they must be far enough away to prevent a provoked attack of thrown objects. The first echelon to dismount from the vehicle secures the surrounding area.

2-96. When small groups are scattered throughout a large disturbance area, a show of force can be made by marching troops conducting motor marches, conducting saturation patrolling, and setting up static posts. Sometimes marching well-equipped, highly-disciplined soldiers in view of a crowd may be all that is needed to convince them to disperse and retire peaceably. On the other hand, a show of force may attract people to an event. It may also provoke a nonviolent crowd into a violent confrontation. Intelligence can help best in preparing a show-of-force response.

Containing

2-97. Containment is the process of limiting a crowd to the area they are presently occupying. It is a suitable option when the crowd must be prevented from spreading to surrounding areas and communities. Additionally, it prevents those outside the already assembled crowd from joining the gathering. Containment is useful when apprehensions become necessary for preventing those in the crowd from escaping. Crowd control formations, perimeter patrols, and barriers are effective methods to accomplish containment. In all instances, caution must be used to avoid the "fight-or-flight" syndrome common to people feeling trapped with no escape.

2-98. Armored vehicles, armored security vehicles (ASVs), and up-armored high-mobility multipurpose wheeled vehicles (HMMWVs) are adaptable to roadblock operations as they can serve as barriers. They also provide added protection for the soldiers inside and those outside manning the roadblocks because they provide an easily accessible barrier for them to crouch behind.

2-99. Military vehicles traveling at close intervals in a column formation next to a crowd are largely a psychological barrier (see *Chapter 6*). They can be used to contain a large, fast-moving crowd. The moving cordon creates a temporary obstacle between the crowd and the line beyond which they will not be allowed to cross. A well-trained driver in a mobile cordon can do a better job than dismounted soldiers. In order to execute safe, quick reversals of direction for mobile cordons on narrow roads, the soldiers executing this formation must be highly-trained vehicle operators.

2-100. By-the-number commands issued over vehicle radios are the most common method for coordinating a cordon movement. Vehicles equipped with PA systems can also prove effective to issue both commands to troops and directives to the crowd. The cordon speed is no more than 5 miles per hour with an interval of at least 20 feet. Blue and red lights, sirens, blinkers, and horns may or may not be used. Armored vehicles can also serve as mobile command posts. When used properly, vehicles provide security, communications, and mobility.

DECIDE TO CONTAIN OR DISPERSE

2-101. Current crowd control doctrine places emphasis on crowd dispersal. Forced dispersal may result in a crowd breaking up into multiple groups that scatter over a large area. This may pose even greater public order problems and may pose a continued threat to control forces. A crowd is often better controlled by means of containment (confining its activities to a given area). A crowd has a limited duration, and its numbers are likely to diminish as individual needs take precedence over those of the crowd.

2-102. During the G8 summit protest in Ottawa, Canada, in June 2002, several thousand people showed up to demonstrate against the evils of world corporate leaders. Canadian Police determined that their best response to this event was a "soft response." Throughout the two-day event, hundreds of officers in standard patrol uniforms operated in four-officer teams, prestaged throughout the city. As the demonstrators began to assemble and eventually took to marching, officers provided escorts both in the front and rear of the enormous crowd along with officers marching to the flanks of the demonstration. This "soft approach" frustrated demonstrators, especially those who had it within their agenda to cause property damage and personal injury. Even more frustrating to the demonstrators was the heavy downpour of rain they endured on both days. Their organizations quickly disbanded out of concern for their own creature comforts. In the end, 10,000 people participated in the protest, with only three arrests made in two days.

USE PATROLS

2-103. Alert, aggressive patrolling of the disturbance area deters the gathering of crowds. The use of saturation patrolling is most effective for this purpose. It allows information to be collected, and it creates the psychological impression of the control force being everywhere at once.

2-104. Standard military transport vehicles provide mobility and communications for area coverage. Soldiers must be deployed with enough vehicles to provide the flexibility to handle all situations in the disturbance area. The vehicles can be modified with sandbags, wire screens, or similar materials to protect against sniper fire and thrown objects. If sniper fire is expected or has been encountered, then up-armored HMMWVs or light armored vehicles should be used for patrolling.

2-105. Armored vehicles patrolling an area of violence provide an added psychological effect and allow troops to maneuver close to snipers to make an apprehension. They may also be used as rescue vehicles to extract hostages or people surrounded by a hostile crowd.

2-106. Varying patrol routes and times for mounted and dismounted patrols keep lawbreakers from being able to pick a safe place or time to act. Patrols are particularly useful in preventing overt arson and violence. Patrol members can also spot and promptly report fires. Whenever possible, military patrols are integrated with civil police patrols. Joint patrols conserve military forces and aid civilian military communications. They also help troops become familiar with an area quickly. In addition, the civil police are available to make apprehensions, if needed. Patrol members must practice proper standards of conduct and fair treatment of civilians at all times. They are performing an important community relation and control function.

2-107. Motorized patrols, because of their speed and mobility, provide timely reconnaissance and broad area coverage. Motorized patrols are in radio contact with TF headquarters (HQ). They can make periodic contact with foot patrols and stationary posts, while patrolling their own areas. Moreover, motorized patrols can respond quickly to calls for help from other patrols and guard posts. Motorized patrols should have at least three vehicles with three men in each. Strength in numbers is necessary for protection. When planning patrol routes, avoid areas where the patrols can be isolated or placed in jeopardy. In addition, motorized patrols are equipped with fire extinguishers to put out small fires, thus reducing the burden on the fire department.

2-108. Besides foot and motorized patrols, air patrols provide a third dimension in directing the overall control effort in the disturbance area. They can perform reconnaissance and surveillance and provide near real time formation over the disturbance area. They are an excellent means of providing timely information to the commander concerning the demonstration area and its surrounding perimeter. They can monitor the actions of rioters, the extent of damage, the status of access routes, locations and conditions of road barriers, and other important conditions.

USE DELAYING TACTICS

2-109. AOs that units are assigned to patrol are often larger than the unit has resources to manage effectively. In addition, units must manage taskings, contingency plans, and other commitments, which will quickly drain available manpower. This problem will manifest itself when factions threaten to riot or gather in large numbers to demonstrate. In order for the commander to stall for time while he shifts unit assets, he may want to delay the arrival of buses and vehicles full of demonstrators. This technique is based on the assumption that demonstrators are moving to the demonstration site by vehicle. Key and secondary routes into the demonstration site must be analyzed by the S2 when he does the IPB.

NOTE: There should be an overwatch when this technique is attempted.

2-110. As demonstrator buses and/or vehicles move toward the demonstration site, soldiers in two HMMWVs block the road just ahead of the approaching vehicles. Once the demonstrators dismount their vehicles and begin to close in on the soldiers, the soldiers leapfrog (see *Figure 2-3*) 1 kilometer farther down the road to repeat the process again. This is repeated as often as necessary. This process, if repeated enough times, should discourage the demonstrators enough that they give up and go home. It should also provide enough time to shift assets to the demonstration site.

2-111. Demonstrators are just as capable of blocking secondary and key routes leading into a demonstration or riot site to prevent relief forces from assisting with riot control as US forces are capable of preventing demonstrators access to the site. One must remember that demonstrators are often well organized and have handheld radios and other items. A great example of this was the 1999 demonstration against the WTO in Seattle, Washington. Police had erected temporary portable fences in the downtown areas to keep demonstrators from disrupting the delegates. Determined to be disruptive, the demonstrators relocated the temporary fences by not allowing police to enter the area. Police were forced to back patrol cars over their own fences to gain access.

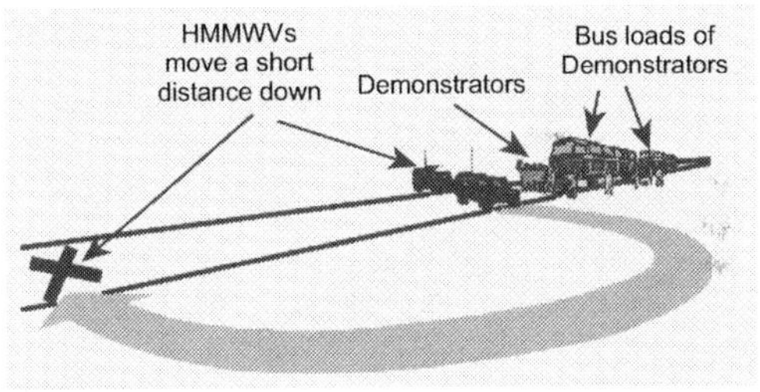

Figure 2-3. Delaying Tactics

Chapter 3

Apprehension, Search, and Detention

During the course of a civil disturbance, some members of the crowd may take part in unlawful activities, such as looting and assault. This will require the civil authorities or attached elements to apprehend, search, and detain people who are participating in the demonstration. Circumstances of the unfolding situation may require US military forces to be called upon to search, take custody of, and detain people who are participating in some form of protest and violating the law. Search, apprehension, and detention operations are conducted to halt these violations and to deter future violations.

LEGAL CONSIDERATIONS

3-1. All apprehensions are made by the civil police force, unless it is not possible for them to do so. Individuals must be known to have committed an illegal act, or there must be probable cause to believe that an individual committed such an act to be apprehended. If it becomes necessary for a military control force to apprehend or temporarily detain such violators, control force members often do so with the approval of the civil authorities. This ensures that all searches, apprehensions, and detentions (if necessary) are conducted within the parameters of the law.

3-2. Commanders operating overseas, similar to Bosnia and Kosovo, may find themselves without an established government or LEAs to work with. Commanders on the ground are then thrust into positions of authority, making them responsible for the actions, conduct, searches, apprehensions, and detentions performed by their troops.

3-3. If US military forces are called upon by the local authorities, certain policies have to be observed during search, apprehension, and detention operations. First, treat all people firmly but with reasonable courtesy and dignity. Remember that participating in a legal demonstration to express views is a right of all people, not just Americans in the US. The action, attitude, and behavior of US soldiers performing these operations are very important.

3-4. Our forces should never be seen as a military policing force on our own soil or as an occupying force in another country. Treating people with contempt, hostility, or excessive force increases the likelihood of resistance and violence. Searching people, placing them under apprehension, and detaining them without probable and just cause or without concern for their constitutional rights create problems and hinder due process. Numerous cases exist where authorities violated an individual's rights, resulting in the prosecution of police authorities or civil suits being waged against the officers and their leaders. This includes the military personnel assisting those officials.

3-5. It is very important that civil law enforcement (if available) be present and supervise all activities. Military personnel conducting a search or making an apprehension must carry out all the procedures carefully within the parameters of their training and the authority given to them. For each search and apprehension performed, the apprehension element must document the specific conduct of the person that violated the law.

3-6. Apprehending officers and supporting US servicemen must provide careful descriptive data for each subject. This description must be sufficient so that, at a later time, the suspected person can be clearly identified as the subject in question. Names and addresses of witnesses to the violation must be obtained. This information can be recorded on *Department of Defense (DD) Form 2708 (Receipt for Inmate or Detained Person)*. Apprehension teams must retain and tag each item of physical evidence, such as weapons or stolen goods, that supports the apprehension on *DA Form 4137 (Evidence/Property Custody Document)*. The team must give the owner a receipt for all property collected as evidence and retained.

TYPES OF SUBJECTS

3-7. There are two types of subjects (cooperative and uncooperative); they are as follows:

- **Cooperative.** Cooperative subjects actively and willingly comply with instructions. Cooperative subjects are usually compliant and offer no resistance to verbal directions or physical actions. However, if they provide any resistance to authority, it is usually passive. Some examples of this resistance are as follows:
  - Yelling.
  - Screaming.
  - Tightening the body.
  - Pulling away.
  - Letting the body go limp.
- **Uncooperative.** Uncooperative subjects will display preattack cues. Uncooperative subjects will actively resist being apprehended. Often, this resistance comes in the form of physical actions, such as—
  - Assaulting verbally.
  - Refusing to be handcuffed.
  - Refusing to be searched.
  - Attempting to push away, forcing a pursuit.
  - Wrestling with the apprehension team.
  - Striking out at the apprehension team with fists, hands and elbows, and feet.
  - Striking out at the apprehension team with weapons, such as knives, clubs, bats, bricks, rocks, or firearms.

APPREHENSION

3-8. US military forces will be in a supporting role to the local civilian authorities or acting as the control force in civil disturbance operations. When making apprehensions, it is best to use apprehension teams. These teams provide quick, organized responses to developing situations. Teams can be organized at the team, squad, or platoon level depending on the number of apprehensions expected. The intelligence information gathered helps to prepare for events where apprehensions may be necessary.

3-9. Each team (if possible) consists of a civilian police or apprehension officer, security element, and recorder. Team apprehension officers make the actual apprehension with the help of the security elements. If there is not a civilian police officer available to act as the apprehension officer, then it may be an officer or noncommissioned officer (NCO).

3-10. Ensuring that the apprehension process is properly conducted and documented, determine who is to be apprehended. Then inform the subject of why he is under apprehension and supervise his handling.

3-11. Security elements help the apprehension officer handle subjects. Security elements move, restrain, and search subjects under the supervision of the apprehension officer. The primary responsibility of the security elements is to watch the crowd and act as a blocking force, keeping the crowd from interfering with the apprehension. Recorders document the event, preferably by video, and complete *DD Form 2708*. If possible, recorders also photograph the apprehension officer with each subject. This aids in the identification process because it links the apprehension officer to the subject and provides the apprehension officer with documentation of the apprehension for use in court.

3-12. Electronically capturing the events as they unfold is essential in all civil disturbance operations. If personnel and equipment are available, videotape the events as they happen or, at a minimum, take still photographs to provide pictorial documentation that may be used later in court. Recording the scene before, during, and after the apprehension provides necessary evidence and can also help eliminate additional hostile and illegal activities. Having control force members take videos or pictures can help control crowd reaction. It reduces the feelings of anonymity that a crowd often enjoys and makes crowd members less prone to unruly or illegal acts.

3-13. Verbal commands given to the subject should be in a voice and manner appropriate for the purpose and the situation. Commands should be simple, concise, and in an understood language. Verbal commands should be convincing and convey an expectation that the instructions will be obeyed. With desperate, frightened people, sympathy combined with an attitude of firm assurance on the soldier's part may avoid additional violent behavior or trouble. An attitude of forceful authority may be needed with an uncooperative person. Commands given by soldiers must be done in a manner and context so that the subject has no confusion as to what the soldier is ordering them to do.

3-14. A subject will need to be transported from the area. In CONUS operations, civil authorities will usually be responsible for providing the transportation of a subject. Exceptions to this may be if the demonstration occurs on an installation. OCONUS operations may require US forces to provide vehicles, such as trucks, buses, or sedans, that will be used to transport a subject. Whenever possible, vehicles used to transport a subject should be modified with barriers to separate them from the drivers. If a barrier cannot be provided, a guard is to be placed in the vehicle. Passenger compartments are checked for items that could be used as weapons. When guards are required, the guard always sits with his weapon away from the prisoner. Prisoners must be placed where they can be best controlled, but they are not fastened to the vehicle with hand irons or Flex-Cufs. When seat belts are available they must be fastened. See *Soldier Training Publications (STPs) 19-95C1-SM and 19-95C24-SM-TG.*

3-15. Female escorts should always be used to transport a female subject. If female escorts are not available, measures must be taken to avoid false charges of molestation or sexual misconduct. Net control stations should be notified of the departure and arrival times and vehicle mileage readings (before and after). At least two male escorts must be present at all times when a female escort is not available. These procedures also apply when the male and female roles are reversed.

FLEX-CUFS

3-16. The two main positions used to place Flex-Cufs on a subject are the standing-supported and the prone.

3-17. When using the standing-supported position, the search man—

- Escorts the subject to a wall or other sturdy vertical surface (see *Figure 3-1*).

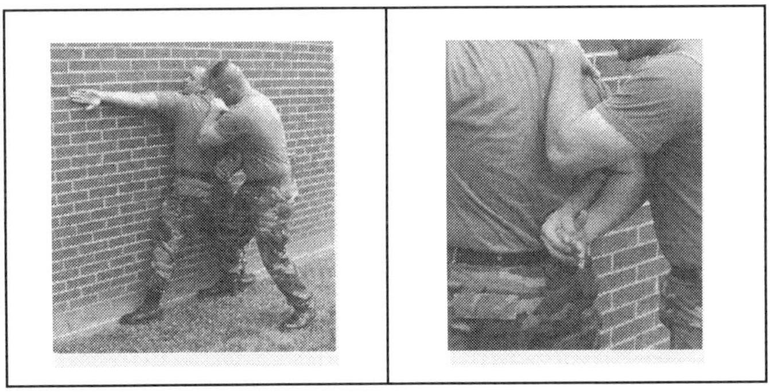

Figure 3-1. Standing-Supported Position

- Instructs the subject (verbally) to do the following:
  - Turn his head away.
  - Place his free arm (palm out, thumb down, and elbow locked) against the surface of the wall.

  - Spread his feet until instructed to stop.
  - Place his heels against the wall.
- Uses the leg closest to the controlled arm to pin the subject's leg (above the knee) to the wall.
- Reinforces verbal commands with pain compliance techniques if the subject fails to follow instructions.
- Maintains a double 90° hold on the controlled arm with his inside hand. With his free hand, the search man grasps the middle portion of the Flex-Cuf, keeping one cuff up and one cuff down.
- Places the first cuff (bottom cuff) on the subject's controlled arm, maintaining wrist manipulation (see *Figure 3-2*).
- Reaches through the free cuff with his outside hand and instructs the subject to place his free hand in the center of his back. With his outside hand, the search man takes the subject's free hand and completes the restraint by securing the cuff. He ensures that there is at least one-finger distance between the cuff and the subject's wrist (see *Figure 3-3,* page 3-6).

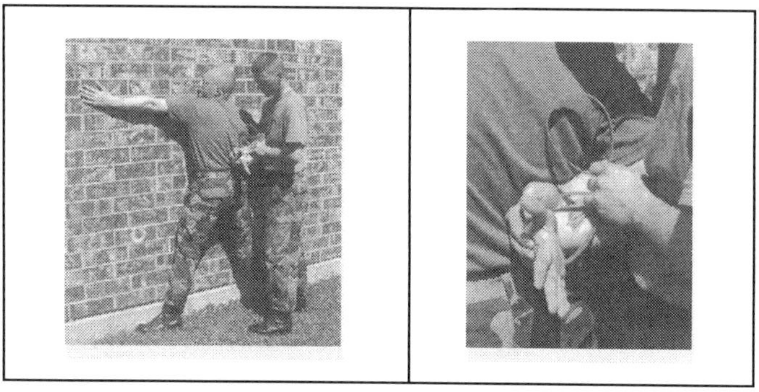

Figure 3-2. Cuffing the First Hand From the Standing Supported Position

3-18. When using the prone position, the search man—

- Grasps the Flex-Cuf with his outside hand, maintains tension on the subject's arm, and places the first cuff on the subject's controlled wrist (see *Figure 3-4,* page 3-6). Next, the search man secures the cuff, keeping at least a one-finger distance between the cuff and the wrist.
- Reaches through the free cuff with his outside hand and instructs the subject to place his free hand in the center of his back (see *Figure 3-5,* page 3-7).
- Reinforces verbal commands with pain compliance techniques if the subject fails to follow instructions. The search man bends the subject's controlled arm (using the natural break of the elbow as a hinge) toward the hand in the middle of the subject's back. With the outside hand, the search man takes the subject's free hand and completes the restraint by securing the cuff.

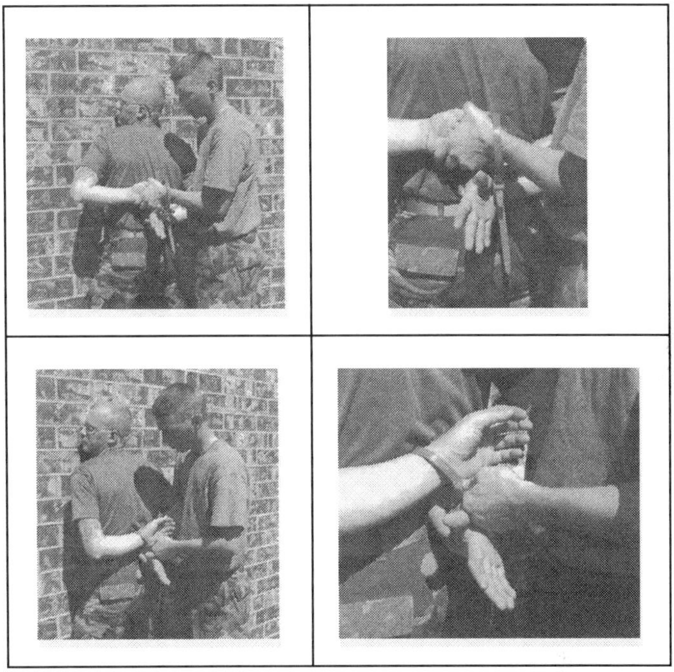

Figure 3-3. Cuffing the Second Hand From the Standing Supported Position

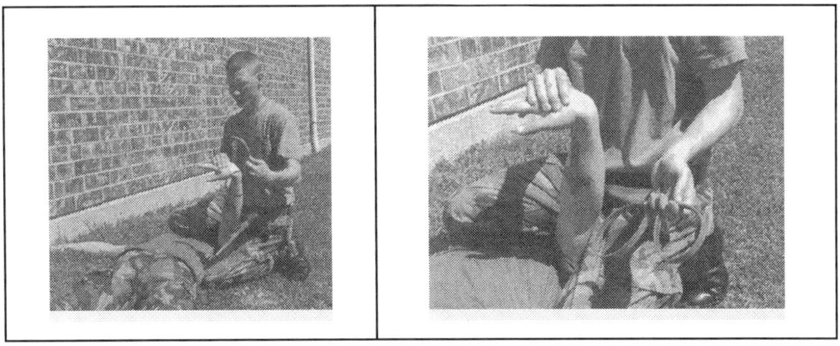

Figure 3-4. Cuffing the First Hand From the Prone Position

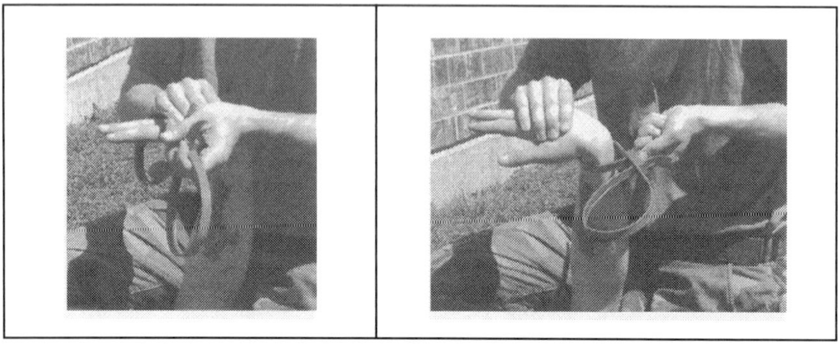

Figure 3-4. Cuffing the First Hand From the Prone Position (continued)

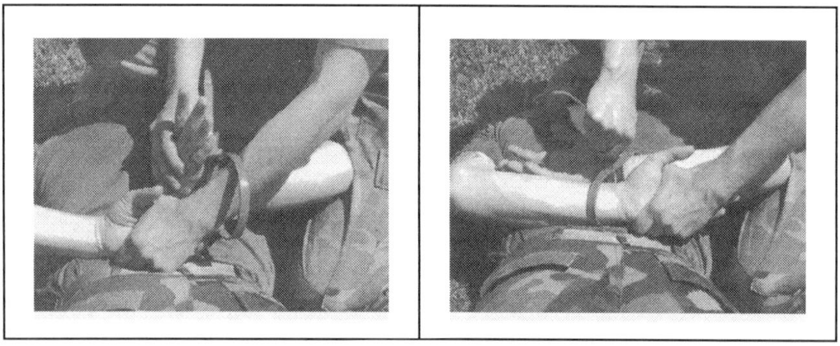

Figure 3-5. Cuffing the Second Hand From the Prone Position

SEARCHES

3-19. Initial contact with the subject is the most dangerous time during any apprehension; therefore, initial contact must be a swift and decisive team effort. Do not give the subject time to think or react. Distract his attention when possible. Apprehension teams are comprised of a search man, threat cover man, and security man (if necessary). Their responsibilities are as follows:

- The search man—
  - Issues verbal commands.
  - Conducts searches.
  - Handles radio communication.
  - Holsters his weapon before approaching a subject and informs the threat cover man that he is holstering his weapon.
  - Performs control procedures and handcuffs the subject.

  - Moves the subject to a safe area.
  - Removes weapons (if any) and sensitive evidence from the subject.
  - Disengages after the subject is restrained.
  - Controls and escorts the subject to a detention area.
- The threat cover man—
  - Provides cover for the search man.
  - Discourages escape and resistance.
  - Discourages hostile interference.
  - Alerts the search man of threats.
  - Resists distractions and maintains target acquisition or scanning.
  - Assesses threats constantly and prepares for the unexpected.
- The security man (if necessary)—
  - Provides cover for the search man.
  - Discourages hostile interference.
  - Alerts the search man of threats.
  - Resists distractions and maintains target acquisition or scanning.
  - Assesses threats constantly and prepares for the unexpected.

Search Principles

3-20. A subject should always be put at a disadvantage and searched immediately, but circumstances may dictate postponing the search until a safer place can be found. At a minimum, a quick pat down can be done en route to a safer location. The longer the apprehension takes, the greater the risk of drawing a hostile crowd and violent acts toward the control force.

3-21. Crowds often sympathize with a subject and try to help him or escalate the activities of the crowd toward violence. A subject's actions may also incite the crowd to violence. The longer a subject is the focal point of an apprehension, the easier it is to stir a crowd to hostile or violent acts. Apprehension teams should choose to move the subject out of the crowd's view. For the safety of the team, always apply Flex-Cufs or handcuffs and escort the subject to an area where a search can be done safely.

3-22. Search team members never operate alone; searches are conducted by a minimum of two team members. Initial steps must be taken to put the subject at a physical disadvantage. Before conducting a search (regardless of rank, sex, position, age, or physical condition), a team member applies the handcuffs or Flex-Cufs. Verbal commands must be short, clear, and distinct.

3-23. A search is conducted on everyone that has been apprehended to ensure that the team and those being apprehended are safe. The search man must abide by the following procedures when conducting a search:

- Search in a clear, secure area (when possible).
- Be ready to counter any resistance or an assault by the subject. Consider the subject armed until proven (by search) otherwise.
- Employ the appropriate level of force immediately, based on the actions of the subject.

- Do not step over the subject when he is at a disadvantage on the ground; go around him and remain out of sight.
- Do a 360° search of the subject's waistline for weapons. Search obvious and inconspicuous areas. Inconspicuous areas such as the groin, armpits, chest, and small of the back must be searched on both men and women; however, do not linger in these areas.

NOTE: Conduct same-gender searches when possible. If mixed gender searches are necessary for speed or security, conduct them in a respectful manner and avoid any action that could be interpreted as sexual in nature. To prevent allegations of sexual misconduct, the apprehension officer shows careful control of the soldiers who perform mixed-gender searches.

- Maintain security.
- Brief subjects quickly on what to do and who to follow.
- Remember that when conducting a search, a subject's body can be divided for a more efficient and thorough search. For example, search from head to toe or from side to side.

NOTE: The search man announces the items found and secures weapons and contraband.

- Military forces (when supporting civil authorities) may perform searches of people and property that are not usually subject to military law during a civil disturbance.
- Remember that it is always better for civilian police to search civilians and civilian property (when possible).
- Remember that if military forces are called upon to support police it is usually a last resort and there are not enough civilian officers available to accomplish the mission.
- Military personnel may search people incident to an apprehension.

3-24. The prescribed methods of searching someone are the frisk, the standing, and the prone. The method of search used largely depends on the situation and the mission. Searches are made not only of the subject and his personal effects such as purses, bags, and wallets, but also of the immediate surrounding area. This prevents the subject being searched from grabbing a weapon or destroying evidence. It may also be necessary for military forces to search private property, including vehicles, if there is—

- Reasonable belief that a person has committed or is committing a violent crime and is hiding in a building or a vehicle.
- Reasonable belief that a vehicle contains weapons or instruments of violence.
- Probable cause for searching a building or a vehicle.
- Probable cause to believe that (unless immediate action is taken) evidence of a crime will be destroyed before a warrant can be obtained.

NOTE: Search teams should always consist of two personnel—one to conduct the search, and the other to provide security and overwatch for the soldier conducting the search. Before being searched, a subject should always be restrained and put at a disadvantage by

securing his hands with handcuffs or Flex-Cufs. Ask the subject if there is anything in his pockets that could hurt the searcher, such as knives or needles.

Frisk Search

3-25. The frisk search (also known as the stop and frisk, pat down, and terry stop) is a search incident to a lawful stop. Subjects are not under apprehension at this point, but if it were an apprehension situation, handcuffs or Flex-Cufs would be applied first. Frisk searches are used when there is reasonable belief that a subject may be armed with anything that could be considered a safety threat.

3-26. Legal and regulatory circumstances for conducting a frisk search are covered in the *Manual for Courts-Martial (MCM)*. When a lawful stop is performed, the person stopped may be frisked for weapons when that person is reasonably believed to be armed and dangerous.

3-27. The primary purpose of a frisk search is safety. Frisk-search techniques are to be used only on a compliant subject. Frisk searches only cover the outer garments. Do not search the pockets of the individual unless a weapon is suspected. A frisk search begins when the search man informs the subject of the intent to search.

- During the preparation phase, the search man—
  - Tells the subject that he must perform a search for weapons for the subject's safety and his own.
  - Tells the subject to turn and face away from him and instructs the subject to keep his hands out to his sides where they are visible.

NOTE: The cover man providing security moves around the subject and positions himself out of reach and in front of the subject.

  - Instructs the subject to spread his feet until instructed to stop (more than shoulder width) and point his toes outward compromising his balance (see *Figure 3-6*).

NOTE: Do not kick at the subject's feet to spread them apart; it is unnecessary and unprofessional.

  - Tells the subject to bring both of his hands back to the small of his back with palms out, thumbs up, and the back of his hands together.
  - Instructs the subject not to move (see *Figure 3-7*).
- During the performance phase the search man—
  - Approaches the subject from behind (if possible) at a 45° angle, off center. The search man approaches with caution to ensure that his feet are to the outside of the subject's stance. See *Figure 3-8*, page 3-12.
  - Stays just inside arms reach and grips the subject's fingers with his support hand. The search man firmly grips at least one finger on each of the subject's hands (both thumbs, both index fingers, or an index and middle finger). An overhand or an underhand grip can be used effectively.

Figure 3-6. Approaching the Subject to Perform a Frisk Search

Figure 3-7. Giving Instructions to Subject Before Frisk Search

NOTE: The primary purpose of gripping the fingers is to control the subject's hands while performing the frisk search (see *Figure 3-9*, page 3-12). If the subject is going to resist or become aggressive, he will attempt to free his hands first. If the subject attempts to pull his fingers from the search man's grip, the search man must respond instantly by increasing his grip. If the grip on the fingers is lost, the search man should immediately push away from the subject and create a safe distance. This will allow the search team to respond with the necessary force to regain control of the situation.

Figure 3-8. Performing a Frisk Search

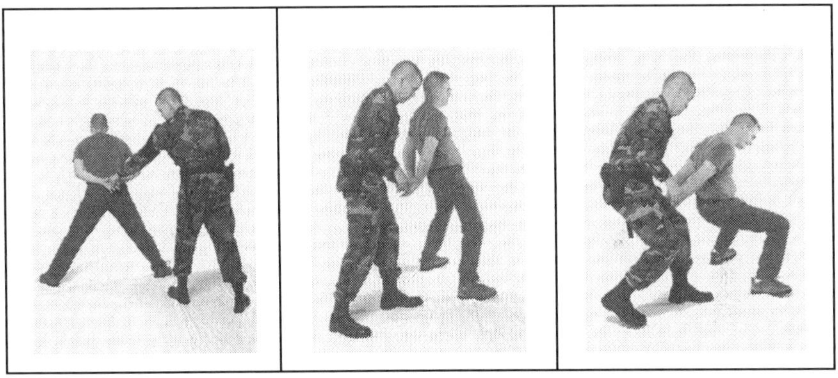

Figure 3-9. Gripping the Subject's Fingers During a Frisk Search

- Ensures that the subject always looks away from him.
- Places his strong hand on the outside of the subject's elbow (similar to the handcuffed escort position).

NOTE: The search man searches one side (the front, back, or side) at a time from head to foot.

- Divides the subject (visually) straight down his middle and does not cross the line while conducting the search.

NOTE: The search man is just feeling or searching for possible weapons, so an open hand should be used.

- Works at arm's reach. It is the sense of touch not the sense of sight that is being used. While performing the search, search men should stay behind and to the side of the subject. This provides the search man with a tactical advantage and prevention from being struck by the subject's elbows or knees if he should begin to resist.

3-28. When searching below the subject's knees, the search man takes a half step back and to the outside. He squats down, but does not drop his knees to the ground.

3-29. If no weapons are found and the subject is to be detained, the hands are secured with Flex-Cufs and the subject is escorted to the holding area. If the subject is to be released, collect information (name and address) and escort him away from the crowd.

Standing Search

3-30. Safety is the primary concern when dealing with a subject under apprehension. Therefore, the search man must protect himself by restraining subjects under apprehension before conducting the standing search. All subjects under apprehension have the propensity to turn violent or hostile. Putting a subject at a physical disadvantage provides safety for the subject and the search man. It limits the subject's ability to hide, destroy, or remove any evidence he may have on him.

3-31. Apply Flex-Cufs to the subject quickly during the apprehension. It is impossible to predict what a subject will do when put under apprehension. A subject may initially be cooperative and compliant, but as the subject becomes aware of his situation he may be overcome with emotion and begin to struggle or pull away.

3-32. There is no legal or regulatory prohibition against opposite-gender searches. Common sense prevails and whenever possible, searches should be done by soldiers of the same gender to avoid obvious questions and complaints. Once a subject is put at a physical disadvantage by cuffing and is more easily controlled, it affords the necessary time to get someone of the same gender there to conduct the search.

- During the preparation phase, the search man—

NOTE: Standing searches always begin with the subject in the handcuffed escort position (see *Figure 3-10,* page 3-14).

  - Takes his outside hand (farthest from subject) and places it on the outside of the subject's elbow while his inside hand (closest to subject) has a reverse-handshake grip of the subject's hand. At this point, both hands are on the same arm (see *Figure 3-10,* page 3-14).
  - Ensures that his position is on the same side of the subject's body being controlled (behind and to the side). When changing sides, change hand position.
  - Instructs the subject to spread his feet until instructed to stop (usually more than shoulder width) and point his toes outward, compromising the subject's balance.

Figure 3-10. Handcuffed Escort Position

NOTE: Do not kick at a subject's feet to spread them; it is unnecessary and unprofessional. If the subject becomes noncompliant or aggressive, pain compliance is usually the best tool to use, but only use enough force to gain compliance.

- Uses his outside hand to conduct the search. His inside hand is always used to maintain control during the search.

- During the performance phase, the search man—

NOTE: Only one side (the front, back, and side) of the subject is searched at a time from head to toe (see *Figure 3-11*).

- Divides the subject (visually) (straight down the middle using the spine as the centerline). He does not cross this line while conducting the search. Searches must be systematic, from head to foot. If there is an interruption, it must be started all over again.
- Searches with an open hand, using a rub down or squeezing method. Do not only pat down like in the frisk search.
- Works at arms reach. It is the sense of touch, not the sense of sight that is being used. To prevent the possibility of being struck by an elbow or knee, the search man stays behind and to the side of the subject.
- Takes a half step back and to the outside when searching below the subject's knees. The search man squats down, but does not drop his knees to the ground. Maintaining positive control is essential throughout the search, so never the reverse handshake is never released.
- Change sides when satisfied that all items have been removed from the side in which the search started. Without losing positive control, the search man conducts the search on the other side of the subject.

Figure 3-11. Performing the Standing Search

NOTE: Changing sides is accomplished by moving the outside hand (the searching hand) to the subject's uncontrolled handcuffed hand located on the outside. The outside hand and forearm will cross over the top of the reverse-handshake grip of the inside hand (see *Figure 3-12,* page 16).

- Crosses over the reverse-handshake grip with the outside hand and grips the subject's uncontrolled handcuffed hand in a reverse-handshake grip.
- Releases (quickly) the first reverse-handshake grip moving that hand to the subject's elbow as he sidesteps to the side yet to be searched. The search man repeats the standing search technique used on the first side of the subject.
- Gives strong verbal commands and uses only the force necessary to control the subject if he resists or becomes aggressive at any time during the search.
- Removes all items found during the search to be evaluated as potential evidence or a weapon. Items not considered a weapon or of evidentiary value will be returned to the subject at a later time.

Prone Search

3-33. Safety is the primary concern when dealing with a person under apprehension. Therefore, the search man must protect himself by restraining subjects under apprehension before conducting this search. Remember that all subjects under apprehension have the propensity to turn violent or hostile. That is the reason a prone search would be necessary. After performing an arm bar takedown on a noncompliant subject, it is especially important for the search man to put him at a physical disadvantage to provide for the subject's safety and his own. By doing so, the subject's ability to hide, destroy, or remove any evidence is limited.

Figure 3-12. Performing the Standing Search (Changing Sides)

3-34. The safest way to handcuff or Flex-Cuf a noncompliant subject is in the prone position, when the search man has greater control of the subject. It is usually the safest position in which to conduct a search

3-35. Handcuff the subject quickly during the apprehension. It is impossible to predict what a subject will do when put under apprehension. Often a subject may initially be cooperative and compliant; however, as the he becomes aware of his situation, he may become overcome with emotion and begin to struggle or pull away.

3-36. There is no legal or regulatory prohibition against opposite-gender searches. Common sense prevails and whenever possible, searches should be done by the same gender to avoid obvious questions and complaints. Once a subject is at a physical disadvantage by cuffing and is more easily controlled, it affords the necessary time to get someone of the same gender there to conduct the search.

NOTE: The subject is put into a physical disadvantage when he is taken facedown to the ground using an arm bar takedown and cuffed.

- During the preparation phase, the search man—
  - Ensures that the Flex-Cufs are not too tight.
  - Rotates his entire body around the subject's shoulder and center at the outside of the subject's hip with his shin over the subject's shoulder and one knee on the ground. At least one knee should remain over the subject's back (near side), with the other on the ground.
  - Pins the subject down by pressing his knee on the side of the subject's back or down on his closest elbow if resistance or aggression is encountered. He simultaneously grips the subject's closest handcuffed hand, grabs it with a reverse handshake, and applies pain for compliance.

- Uses strong, forceful verbal commands, such as "Stop resisting, stop moving around."
- Ensures that the subject's feet and ankles are crossed.

CAUTION

Stay alert. If the subject uncrosses his feet and ankles, it is usually a sign of preattack.

- Removes all items found during the search so they can be evaluated as potential evidence. Items not considered a weapon or of evidentiary value will be returned to the subject at a later time.

• During the performance phase, the search man—

NOTE: The search man must be systematic from head to foot while conducting the prone search. If interrupted during the search, it is important to start from the beginning to ensure that nothing is missed. It is better to go over the same area two or more times than to miss something that can be used as a weapon.

NOTE: The search man searches only one side (the front, back, or side) of the subject at a time until that side is complete.

- Divides the subject (visually) straight down the middle, using the spine as the centerline.

NOTE: He does not cross this line while conducting the search.

- Searches with an open hand, using a rub down or squeezing method for best results. He does not just pat down the subject like in the frisk search.
- Works at an arm's reach, searching only the area of the subject's body that is not in contact with the ground. The sense of touch, not the sense of sight, is being used.
- Stops the search as the subject's knees are reached. He taps one of the subject's legs and firmly commands him to raise his foot up and back, allowing for an inspection of the bottom of his foot. He then searches from the knee to the ankle area. This is performed on both the subject's legs and feet.

NOTE: At this point, the entire back side of has been searched. What remains to be searched is the front side of the subject (the chest and front of the hips, waist, groin, and thighs).

- Pivots back and faces the subject. Using the hand closest to the subject's head, the search man reaches across the subject's back and firmly grips the shoulder, rolling the subject up on the his side toward him to expose the areas not yet searched.
- Rolls the subject back down on his chest once all areas have been searched.

3-37. The search man must be completely satisfied that he has found everything. He must be especially sensitive to areas where weapons are often concealed like the waistband, the small of the back, ankles, pockets, shoes, the tongue of the shoes, and the groin area.

NOTE: Before removing subjects from the area, a brief summary must accompany them. The summary must include the subject's name, what he is suspected of doing, the time and date of the incident, and who saw him do the action.

DETENTION

3-38. Civil authorities must provide adequate detention facilities for all subjects. Authorities must be prepared to detain large numbers of people. They may choose to expand existing detention facilities or set up temporary facilities to accommodate the extra load. If possible, large-scale arrests are delayed until sufficient detention facilities have been set up.

3-39. If US military forces are committed to support local authorities, commanders should coordinate with them to ensure that adequate detention facilities are available and to learn their locations and capacities. If there are more detainees than civil detention facilities can handle, civil authorities may ask the military to provide support by setting up and operating temporary facilities. Army correctional facilities cannot be used to detain civilians. A temporary military detention facility can be set up if—

- Federal troops have been employed according to provisions.
- The TF commander has verified that available civilian detention facilities can no longer accommodate the number of prisoners awaiting arraignment and trial by civil courts.
- The Army Chief of Staff has granted prior approval.

3-40. Use of the temporary facility ends as soon as civil authorities can take custody of the detainees. Military forces are responsible for the custody, health, comfort, and sustenance of all detainees in their facilities until custody is transferred to civil authorities. Temporary facilities cannot be used to confine people arraigned or convicted in civil courts.

3-41. Women are detained in their own facility. Temporary facilities are supervised and controlled by officers and NCOs trained and experienced in military correctional operations. Females guard females. If two female guards are unavailable, use two male guards or one male and one female guard (see *FM 3-19.40*).

3-42. The same operational procedures that apply to the management of an installation confinement facility (see *FM 3-19.40*) apply to the management of temporary detention facilities, except for training, employment, and administrative discipline. Guards and support personnel under direct supervision and control of military officers and NCOs need not be trained or experienced in military correction operations. They must be specifically instructed and closely supervised in the proper use of force, custodial procedures, and the completion of military and civilian forms and reports that may be used. Soldiers that may be tasked to operate a temporary detention facility should be familiar with the forms and reports used for civilian apprehensions too.

3-43. Temporary facilities are set up on the nearest military installation or on suitable property under federal control. Ideally, the facility should be close enough to the disturbance area to minimize transportation and escort needs. However, it should be far enough away not to endanger those being detained. Whenever possible, existing structures are adapted for this use, but construction may be needed to provide segregation for ensuring effective control and administration.

3-44. The basic structure must include the following:

- Search areas.
- Holding areas for incoming men and women.
- A processing area.
- Holding areas for men and women who have committed misdemeanors.
- Holding areas for men and women who have committed felonies or are violent.
- A holding area for detainees' property and evidence obtained during a search.
- A holding area for administrative support and records.
- A medical station.
- Latrines.

For more information on these structures see *FM 3-19.40.*

3-45. Facility personnel must ensure that proper sanitation is maintained. When large numbers of people are detained or processed through a facility, sanitation becomes a problem. Medical personnel must conduct regular health inspections to detect unsanitary practices and conditions.

3-46. Facilities must be organized for a smooth flow of traffic. Processing stations must be set up so there is a linear or circular sequence of movement. These stations are set up out of sight of the holding areas. If possible, they are separated by a door to reduce noise. To reduce distractions, each station may be partitioned. Detainees may be more cooperative if they are out of sight of each other.

3-47. Injured people are given prompt medical treatment and transportation to medical facilities when necessary. A medical aid station for screening detainees and treating minor injuries is set up inside or next to the detention facility. Treatment areas, however, must be out of sight of the processing and holding areas. If possible, access to the medical facility bypasses the holding and processing areas. Facility personnel may also consider setting up a separate holding area for injured detainees.

3-48. Detention facility operation plans must contain emergency procedures. As a minimum, the following areas must be addressed:

- Fire evacuation.
- Disturbance control.
- Facility security.

NOTE: Depending on the situation, other emergency procedures may be needed.

3-49. Upon arrival at the detention facility, the detainee is logged in and searched, regardless of whether a complete search has already been conducted by the apprehending team in the field. Separate search areas are set up for men and women. Weapons, contraband, flame-producing devices, suspected evidence, money, and high-value items are confiscated. Medications are also confiscated. Medical personnel screen detainees that have had medications confiscated. Receipts are provided for any property or evidence that is confiscated. Confiscated items are tagged, and the items are stored in a controlled-property area.

3-50. As a detainee is brought to the facility, a file is initiated. Assigned detainee case numbers are used on all paperwork, such as logs, evidence tags, reports, and visual documentation. All paperwork (including photographs) that began outside the detention facility is marked with the case number. Facility personnel may also use hospital identification tags. Using indelible ink, the case number is written and attached to the tag on the detainee's wrist. Different colors may be used to identify different subject classifications, such as misdemeanors, felonies, or violent offenders. If opposing factions are involved, procedures are established to ensure that members of opposing factions are not processed together or detained in the same holding area.

3-51. After a detainee has been searched and classified, he is taken to a processing station where his paperwork is processed. If subjects passively resist by going limp, they may be moved by wheelchair. This reduces the number of escorts needed and the fatigue to facility personnel. Paperwork is reviewed to ensure that all information is complete, including charges, witnesses, and reasons for the apprehension.

3-52. Equipment such as height charts, scales, fingerprinting equipment, and cameras must be available for completing the police report and identifying the subject. Detainees may use aliases and not carry identification, so pay special attention to obtaining as much information as possible about the detainee's physical characteristics.

3-53. After processing is complete, the detainee is placed in the appropriate holding area. Paperwork for the detainee is forwarded to the administrative section. Files are reviewed for completeness and to determine the disposition of the detainee. Information from the file may be placed in a computer to find repeat subjects. Computers may be used to file criminal information only.

3-54. Custody transfers and release procedures must be coordinated with civil authorities and the appropriate legal counsel. Every effort must be made to arraign subjects quickly. The purpose of a detention facility is not to keep people off the streets, but to aid in processing subjects through the legal system. To speed up the release process, planners should consider issuing citations or subpoenas for minor offenses.

3-55. Military force members may have to respond to writs of habeas corpus. Writs are court orders addressed to a prisoner's custodian. Such a writ directs the custodian to bring the prisoner to court to determine the legality of the prisoner's apprehension and detention. Custodians must ensure that all documentation concerning the case is brought to court with the detainee.

3-56. Military personnel must obey writs issued by federal courts. For writs issued by a state court, the custodian or his legal advisor should respectfully reply that US authority is holding the prisoner. The Office of the SJA can answer any questions and explain the correct procedures.

Chapter 4

Riot Shield and Riot Baton Techniques

A soldier may be required to be part of a riot control formation where his primary weapon is a riot baton and his only protection a riot shield. A riot baton in the hands of a well-trained soldier is an invaluable weapon in a crowd control situation and may also be the most appropriate weapon. A riot baton is a very versatile weapon that can be used as an offensive and/or defensive weapon. The riot shield offers protection from the top of the head to just below the knees.

OVERVIEW

4-1. Use of the riot baton and riot shield is based on METT-TC and the commander's choice-of-force option. Formations committed to a riot with riot batons must have lethal overwatch, lethal weapons, and NLW employed within the formation and reserve forces with additional NL and lethal capabilities. These forces are positioned to rapidly reinforce the control force formation.

4-2. During the termination phase of the disturbance (when the violence has subsided and is expected to remain that way), soldiers are placed in a defensive posture. They may be equipped with a riot shield and a riot baton to perform their tasks. The riot shield is primarily a defensive weapon, which may be used in an offensive mode. It is primarily used as an offensive weapon when the formation is in close contact with rioters. The riot baton can also be used when formations are on the offensive and in contact with the crowd. However, crowd control elements employed with riot shields and riot batons must have lethal weapons and NLW available.

WARNING

Riot control formations are never employed with riot shields and riot batons only. There must always be lethal overwatch elements in position and lethal and NL capabilities within the riot control formation with a ready reserve force armed with lethal and NL capabilities.

RIOT SHIELD

4-3. In the hands of a properly trained soldier, the riot shield has proven to be the first line of defense for a formation. The riot shield may be used as an offensive weapon when in contact with an aggressive crowd.

4-4. The riot shield may be held with one arm. It is secured to the weak side arm with a self-sticking strap system and a handle. This system is designed to make the riot shield an extension of the soldier's arm. The riot shield is held at a slightly inward angle. This allows for debris thrown at the soldier to fall harmlessly to the ground (see *Figure 4-1*).

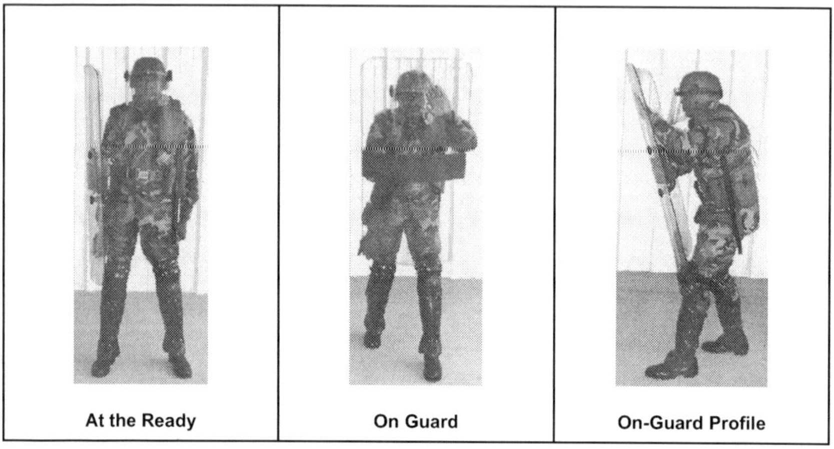

Figure 4-1. Riot Shield Positions

RIOT SHIELD CARRIES

4-5. The two basic riot shield carries are as follows:

- **At the ready.** The at-the-ready position is used while in a column formation or on the move toward hostilities. This position is designed for rapid movement. It should not be used when in contact with a crowd.
- **On guard.** The on-guard position is used when the riot control formation is in contact with the crowd or contact is eminent. It is used primarily in the following formations:
  - Line.
  - Wedge.
  - Echelon.
  - Diamond.
  - Circle.

DEFENSIVE AND RETENTION TECHNIQUES

4-6. The primary purpose of the riot shield is for defense. However, each riot shield holder must be proficient in riot shield retention techniques.

DEFENSE

4-7. The first line of defense for the formation is the riot shield. It is designed to withstand strikes from objects thrown by the crowd. In addition to protecting the riot shield bearer, it also protects other members of the formation from thrown objects.

RETENTION

4-8. When the control force is in contact with the crowd, the crowd will attempt to strip the riot shield from the bearer. When a subject grabs the top of the riot shield (see *Figure 4-2*), the riot shield holder slaps the riot shield with his strong hand near the subject's hands and gives the following commands (while continuing to slap the riot shield until the subject releases it):

- "Get back."
- "Get away."
- "Stop."

Figure 4-2. Riot Shield Retention Technique (Top Attempt)

4-9. If the riot shield is grabbed from the bottom (see *Figure 4-3,* page 4-4), then the riot shield holder forcefully drops to one knee, pinning the subject's fingers and/or hands to the ground. This must be done quickly and with force.

RIOT BATON

4-10. A riot baton in the hands of a properly trained soldier is a formidable weapon. However, it must be used in conjunction with other measures to be most effective. Soldiers must be trained with the riot baton to the point that its various techniques are automatic to them. This training must also include

Figure 4-3. Riot Shield Retention Technique (Bottom Attempt)

learning the vulnerable points of the human body so they can avoid areas that may cause permanent injury or death when struck.

WARNING
The riot baton is never raised above the head to strike a subject in a club fashion. Not only is it likely to cause permanent injury; but it also gives an unfavorable image of the control force. The soldier is also vulnerable to an attack on his rib cage when his arm is raised.

TARGET AREAS OF THE HUMAN BODY

4-11. The target areas of the human body are divided into three areas that represent the primary-, secondary-, and final-target areas. These areas are color coded as green (shown as gray in *Figure 4-4*), yellow (shown as white in *Figure 4-4*), and red (shown as patterned gray in *Figure 4-4*).

Primary-Target Areas

4-12. The primary-target areas are those areas which, when struck or restrained, cause the least amount of trauma to the body. These include the meaty or muscular areas of the body, such as the—

- Foot, shin, and instep.
- Inside and outside of the thigh.
- Lower abdominal region.
- Fore- and upper-arms.

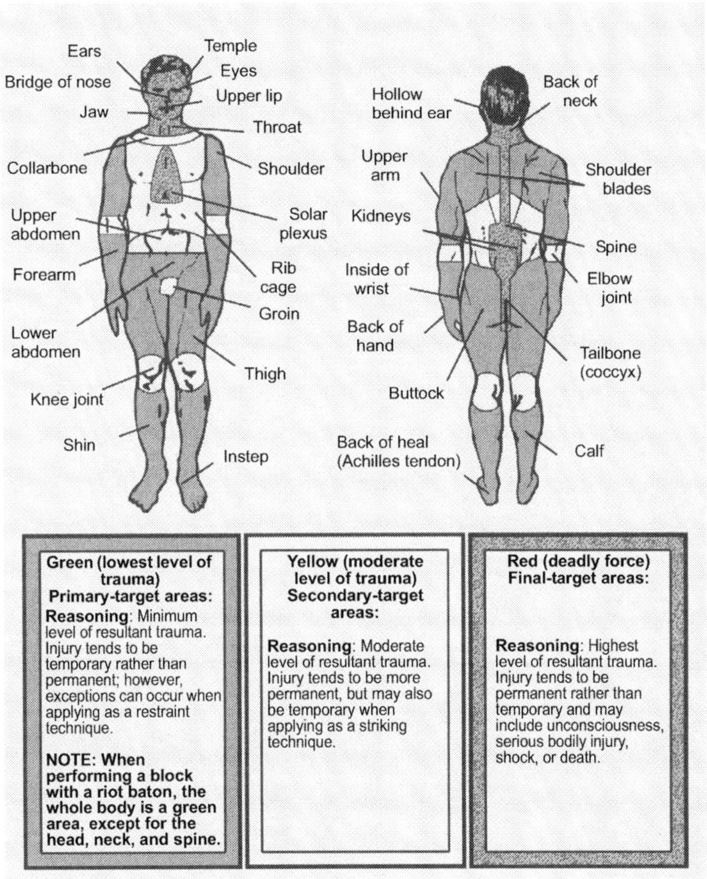

Figure 4-4. Escalation of Trauma Chart

Secondary-Target Areas

4-13. The secondary-target areas are those areas of the body which, when struck or restrained, cause a moderate level of trauma to the body. This type of trauma tends to be longer lasting because the time to heal is longer than with other parts of the body. If these areas are struck with a riot baton, serious injury may occur. However, some of these areas are used for control or restraining holds that tend to lessen the threat of injury. Some secondary targets are the—

- Collar bone and shoulder blades.
- Elbows and knees.
- Upper abdominal region.

Final-Target Areas

4-14. The final-target areas are those areas which, when struck with a riot baton, cause serious trauma that tends to be fatal or critical. However, there are some areas that (when used for restraining only), will not cause death or serious injury. Some final-target areas are the—

- Front and back of the head (eyes, ears, nose, upper lip, and hollow behind the ear).
- Neck and throat.
- Upper chest.
- Xiphoid process.
- Spinal column.
- Groin area.

JOINTS OF THE HUMAN BODY

4-15. The joints of the human body are often described as hinges and, when manipulated in a certain fashion, allow for better control of the subject being restrained, especially if he is aggressive. When applied correctly, minimal pressure is needed to control the subject. When a subject attempts to resist, he creates additional pain for himself. By anatomically locking out these hinges, the subject will be forced to comply. When restraining a subject, emphasis must be placed on locking out two of the three joints of the upper body or two of the three joints of the lower body. The joints are as follows:

- **Shoulder.** Apply pressure to the shoulder to prevent the subject from freely moving the upper portion of his body.
- **Elbow.** Apply pressure on the elbow or just above the elbow on the ulner nerve.
- **Wrist.** Apply pressure by pancaking the hand to twist or lock out the wrist. Apply pressure at the break of the wrist and push the subject's thumb toward a stationary object (the wall or the ground) to lock the hand out.
- **Hip.** Apply pressure to the hip to knock the subject off balance or alter his center of gravity.

- **Knee.** Apply pressure to the knee by pushing outward on it to cause the subject to be placed off balance, or use the knee as a hinge for pain compliance.
- **Ankle.** Apply pressure to the ankle by clasping and turning it in a counterclockwise manner to lock the hinge out.

TYPES OF RIOT BATONS

4-16. The Army is currently fielding two types of riot batons, wooden and expandable. The most common riot baton is the 36-inch hickory riot baton with thong. There is also the 24- to 36-inch expandable riot baton, which has been added to the nonlethal capabilities set (NLCS). Regardless of which riot baton the soldiers are equipped with, all techniques for blocking and striking are identical. Similar to the riot shield, the riot baton offers soldiers both offensive and defensive capabilities.

4-17. Soldiers must be properly trained in all blocking and striking techniques. Improper use of the riot baton by an untrained soldier has the potential for creating a greater problem than what already exists.

NOMENCLATURE OF THE RIOT BATON

4-18. The riot baton is divided into four basic parts: the long end, long portion, grip portion, and grip end (see *Figure 4-5*). Although the expandable and wooden riot batons differ, their nomenclature is quite similar.

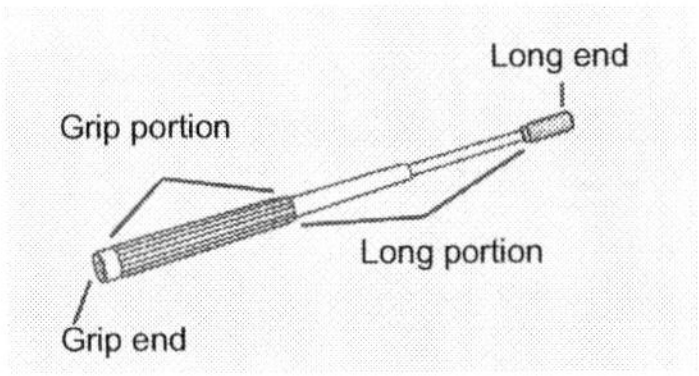

Figure 4-5. The Riot Baton

TYPES OF RIOT BATON CARRIES AND POSITIONS

4-19. As with any other weapon that the soldier will use, there are certain carries and positions from which the riot baton may be used offensively or defensively.

Two-Hand Carry

4-20. The two-hand carry is the most common carry for the riot baton. It is intended for executing all blocks and returning with strikes. When the expandable riot baton is in the collapsed position (at 24 inches), the long end of the riot baton is angled downward approximately 45°. When it is extended to 36 inches the long end is angled upward approximately 45°.

4-21. The wooden riot baton is secured with the soldier's hand by means of a leather thong. He hooks the thumb of his strong hand (palm down) in the thong, pulls the thong across the back of his hand, and grips the riot baton. This provides him with a very secure grip (see *Figure 4-6*).

4-22. The two-hand carry is intended for executing all blocks and returning with strikes. The soldier grips the riot baton with both hands. His strong hand grips the short end of the riot baton, and his weak hand grips the long portion of the riot baton, 2 to 4 inches from the end. His palm is down with his strong hand pulled in close against his strong side hip.

4-23. Wherever the long portion of the riot baton is pointed is where the soldier will strike or jab. The riot baton should be kept at a 45° angle to the ground. All blocks and two-hand jabs and strikes are executed from the two-hand carry.

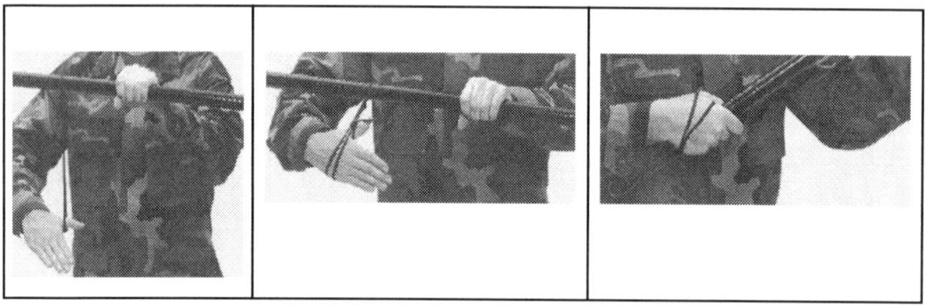

Figure 4-6. Two-Hand Carry

Outside-Arm Carry (Expandable Riot Baton Only)

4-24. The outside-arm carry (see *Figure 4-7*) is used with the expandable riot baton in the collapsed (24-inch) position only. Assuming a wide-based defensive stance (knees slightly bent and feet shoulder width apart), a soldier draws the grip portion of the riot baton with his strong hand. Without striking out or widely swinging the riot baton, he brings the long portion up against the outer part of his strong arm (the tricep area). The soldier's weak hand should be raised in a position to protect and block strikes from an aggressor. It is from this carrying position that a soldier can effectively use both the front and rear strikes. Front and rear strikes should never be performed with an expandable riot baton in the expanded position or with a 36-inch wooden riot baton.

> **CAUTION**
> The expandable and 36-inch wooden riot batons provide a greater risk of injury to the soldier's wrist.

Figure 4-7. Outside-Arm Carry

Vertical High-Profile Carry

4-25. Assuming a wide-based defensive stance, the soldier draws the riot baton with his strong hand by the grip portion. Once the riot baton has been drawn from the carrier, the soldier swings the riot baton in a downward motion, fully extending the arm and holding the riot baton down and parallel to the strong side leg. The soldier's weak hand is raised in a position to protect and block strikes from an aggressor (see *Figure 4-8*, page 4-10). Although this carry can be used with both riot baton systems, it is best suited for the expandable riot baton collapsed to 24 inches. The vertical high-profile carry clearly shows the soldier with a drawn riot baton in an effective defensive posture and ready for attack. A drawn riot baton in the hands of a trained soldier may offer just the kind of deterrence needed to stop the activity of an aggressor.

Vertical Low-Profile Carry

4-26. Assuming a wide-based defensive stance, the soldier draws the riot baton with his strong hand by the grip portion. Once the riot baton has been drawn from the carrier, the soldier swings the riot baton in a downward motion, fully extending his arm and holding the riot baton down and parallel to his strong side leg. His weak side hand is raised in a position to protect and block strikes from an aggressor (see *Figure 4-9*, page 4-10). Although this carry can be used with both types of riot batons, it is best suited for the expandable riot baton collapsed to 24 inches. This position portrays the soldier in an effective defensive position ready for an attack, but with a possible hidden weapon. The major difference between the vertical low-profile and high-profile carry is that in the low-profile carry the soldier hides the riot baton behind his strong side leg, allowing for some element of surprise to a potential aggressor. A drawn riot baton in the hands of a trained soldier may offer the deterrence needed to stop the activity of an aggressor.

Figure 4-8. Vertical High-Profile Carry

Figure 4-9. Vertical Low-Profile Carry

RIOT BATON BLOCKING TECHNIQUES

4-27. There are five basic types of riot baton blocking techniques that are executed from the two-hand carry position.

High Block

4-28. The high block is very effective in blocking a downward, vertical strike directed at the top of the head and shoulders. This type of downward, vertical strike may be from a club, pipe, or similar object. The high block is a two-step movement. To perform a high block the soldier performs the following steps:

Step 1. Raises the riot baton straight across his body until it is parallel to the ground (see *Figure 4-10*).

Step 2. Brings the riot baton straight-up in front of his body until the long portion of the riot baton is in front of his forehead (approximately 2 inches from his head). The riot baton should be in a horizontal position above his head and slightly angled down toward his weak hand, as it absorbs the shock from the strike (see *Figure 4-10*). The fingers of his weak hand should be open and behind the long portion of the riot baton when blocking the object to protect his fingers from being smashed by the strike.

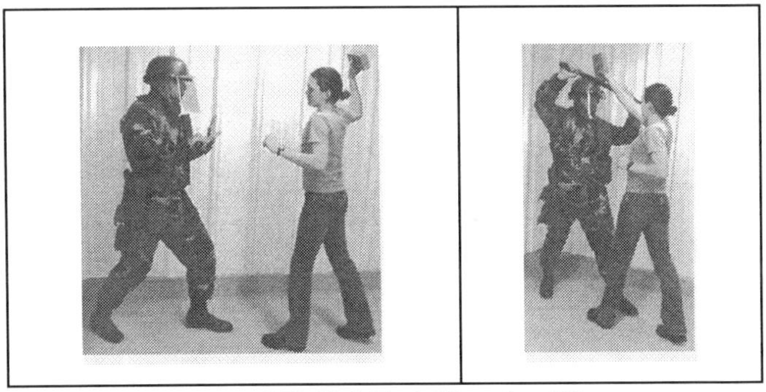

Figure 4-10. High Block (Steps 1 and 2)

Low Block

4-29. A low block is very effective in blocking an upward vertical strike directed at the groin, lower abdomen, chest, or chin. This upward vertical strike may be from an individual's foot, knee, or fist (for example, an upper cut to the chin). To perform a low block, the soldier performs the following steps:

Step 1. Raises the riot baton straight across his body until it is parallel to the ground (see *Figure 4-11*, page 4-12).

Step 2. Brings the riot baton straight down in front of his body, bending at his knees and keeping his body upright until the long portion of the riot baton is just below his knees (see *Figure 4-11*). The riot baton should be horizontal and parallel to the ground as it absorbs the shock from an upward strike. His weak hand fingers should be open and behind the long portion of the riot baton when blocking the object to protect his fingers from being smashed by the strike.

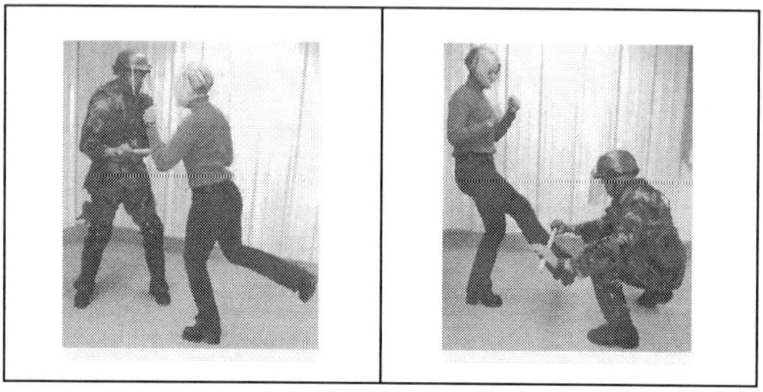

Figure 4-11. Low Block (Steps 1 and 2)

Strong Side Block

4-30. A strong side block is very effective in blocking a horizontal strike directed at the strong side of the head, neck, chest (or flank), or hip area. The horizontal strike may be from an individual's foot, fist, elbow, knee, or weapon. To perform a strong side block, the soldier performs the following steps:

Step 1. Raises the riot baton straight across his body in a vertical position where the grip portion is in his strong hand and the long portion is in his weak hand with the fingers straight up (see *Figure 4-12*).

Step 2. Turns his body laterally by rotating his hips toward his strong side and moving the riot baton across his body, keeping the riot baton vertical (see *Figure 4-12*).

4-31. The strong side block provides protection for a soldier from the front to the rear of his body. A riot baton held in the vertical position absorbs the shock of a strike coming from the strong side. The fingers of his weak hand should be open and behind the long portion of the riot baton, when blocking the object to protect his fingers from being smashed by the strike.

Weak Side Block

4-32. A weak side block is very effective in blocking a horizontal strike directed at the weak side of the head, neck, chest, or hip area. The horizontal strike may be from an individual's foot, fist, elbow, or knee. To perform a weak side block, the soldier performs the following steps:

Step 1. Moves the riot baton in a vertical position toward his weak side, ensuring that the grip portion is pointing straight down and the long portion is pointing straight up.

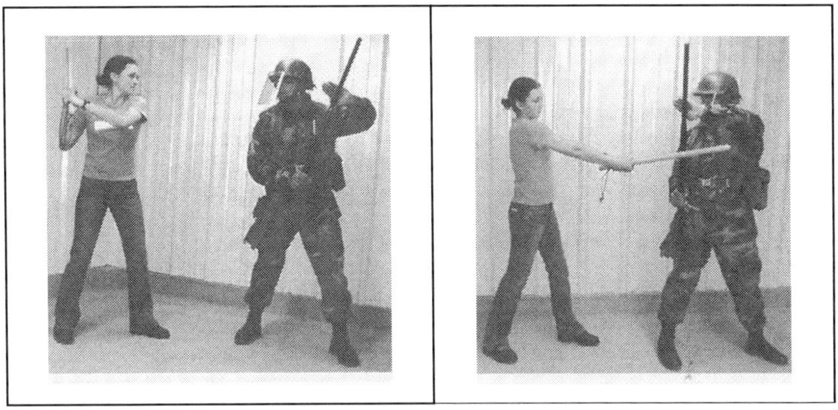

Figure 4-12. Strong Side Block (Steps 1 and 2)

Step 2. Keeps the fingers of his weak hand open and behind the long portion of the riot baton when blocking the object to protect his fingers from being smashed by the strike. At the moment of contact with the opposing strike, the blocking surface of the long portion should be at a right angle to the opposing object.

Step 3. Assess the situation following a weak side block, and take appropriate follow-up action, as needed.

Middle Block

4-33. A middle block is very effective in blocking a front horizontal strike directed at the face, throat, chest, abdomen, or hip. The horizontal strike may be from an individual trying to tackle near the waist or shove or punch at the face. To perform a middle block, the soldier performs the following steps:

Step 1. Moves the riot baton toward the middle of his body at a 45° angle, keeping the long portion pointed up and slightly forward of the grip portion (see *Figure 4-13*, page 4-14).

Step 2. Keeps the fingers of his weak hand open and behind the long portion of the riot baton when blocking the object to protect his fingers from being smashed by the strike. However, his weak hand may have to grip the long portion when pushing an individual away or forcing an individual down on the ground if the individual is trying to tackle.

RIOT BATON STRIKING TECHNIQUES

4-34. Seven strikes can be executed using the riot baton. Two of these strikes come from the outside-arm or vertical carry. The other five strikes come from the two-hand carry.

Figure 4-13. Middle Block

One-Hand Forward Strike

4-35. The one-hand forward strike is a very quick and effective offensive strike. It is usually employed as a countermeasure to an attack from the front that has been blocked effectively.

4-36. Target areas can be as high as the outside upper arm, down to the outer thigh region. An effective one-hand forward strike must be aggressively quick, with the purpose of distracting, disabling, or altering behavior. Therefore, those employing this technique must be sensitive to what part of the riot baton they use to strike the individual. Strike the target with the last 2 to 4 inches of the long end for optimal results.

4-37. A one-hand forward strike may be employed in one of the following three ways. When selecting available targets, use the principles in *Figure 4-4*, page 4-5.

- A horizontal manner.
- A downward diagonal manner to destabilize an individual.
- A vertical manner in front of the body to clear an aggressor's hands.

NOTE: When the one-hand forward strike is used in a vertical manner, keep the long portion straight up, not angled forward. By keeping the long portion up rather than angled forward, the chance of striking the individual in the face or head is significantly reduced. For the counterstrike to be effective, follow through with the technique when the baton makes contact with the aggressor and/or his object to destabilize the him. A pattern of movement such as a

forward shuffle, forward pivot, strong side step, or rear pivot may enhance this technique.

> **WARNING**
>
> **Do not execute the one-hand forward strike with the wooden or expandable riot baton in the extended position. It has been shown to cause injury to the wrist of the user.**

4-38. When using the vertical or outside-arm carry position (see *Figures 4-7,* page 4-9 and *4-8,* page 4-10), the soldier quickly moves the riot baton across his body using the strength and power of his hips, moving from his strong side to his weak side (see *Figure 4-14,* page 4-16). Ending this strike leaves the riot baton positioned under the weak side armpit in preparation for the one-hand reverse strike (see *Figure 4-14*). When delivering this strong side strike technique (in a horizontal or downward diagonal manner), ensure that the palm is facing up. In a vertical delivery, the palm will be toward the chest rather than up. Pause following the one-hand forward strike and assess the situation. Take appropriate follow-up action, as needed. If the situation does not call for additional strikes, avoid striking again.

One-Hand Reverse Strike

4-39. The starting position for the one-hand reverse strike is under the weak side armpit (see *Figure 4-14*). The one-hand reverse strike is used as a follow-up strike and is quick and effective.

4-40. Target areas can be as high as the outside upper arm, down to the outer thigh region. An effective one-hand reverse strike must be aggressively quick, with the purpose of distracting, disabling, or altering behavior. Therefore, those employing this technique must be sensitive to what part of the riot baton they use to strike the individual. Strike the target with the last 2 to 4 inches of the riot baton for optimal results.

4-41. A soldier may employee a reverse strike in one of the following three ways:

- A horizontal manner.
- A downward diagonal manner to destabilize an individual.
- A vertical manner.

4-42. In front of the body to clear an individual's hands. When the one-hand reverse strike is used in a vertical manner, keep the long portion straight up, not angled forward. By keeping the long portion up rather than angled forward, the chance of striking the individual in the face or head is significantly reduced. When selecting available targets, use the principles in *Figure 4-4,* page 4-5.

4-43. The one-hand reverse strike is considered a follow-up strike to the one-hand forward strike. As a follow-up strike, it is important that soldiers are taught the discipline of assessing the target. An aggressor may not require another strike, based on his condition and actions after enduring the one-hand forward strike.

Forward Strike | Target Assessment | Reverse Strike

Figure 4-14. One-Hand Forward and Reverse Strikes

4-44. When using the one-hand reverse strike, the solder quickly moves the straight baton across his body, using the strength and power of his hips (from weak side to strong side). When the strike is completed, the soldier's arm will be back in the outside-arm carry position (see *Figure 4-7,* page 4-9). The soldier ensures that his palm is facing down when delivering the counterstrike in a horizontal or diagonal manner. This will prevent injury to the user's wrist. In a vertical delivery, the palm will be toward the chest rather than up. Following the one-hand reverse strike, assess the situation and take appropriate follow-up action, as needed. If the situation does not call for additional strikes, the soldier should avoid striking again.

4-45. When the one-hand reverse strike is used in a vertical manner, keep the long portion straight up, not angled forward. By keeping the long portion up rather than angled forward, the chance of striking an individual in the face or head is significantly reduced. For the counterstrike to be effective, follow through with the technique when the baton makes contact with the aggressor and/or his object to destabilize him. A pattern of movement such as the forward shuffle, forward pivot, strong side step, or rear pivot may enhance this technique.

WARNING

This strike will not be executed with a 36-inch wooden or expandable riot baton in the extended position. It has been shown to cause injury to the wrist of the user.

Two-Hand Strong Side Horizontal Strike

4-46. Starting from the two-hand carry position, the two-hand strong side horizontal strike can be aggressively quick. The purpose of this strike is to create distance, distract, disable, or alter behavior. Strong side horizontal strikes are quick and effective offensive strikes, usually employed as a countermeasure to close frontal attacks that have been effectively blocked. When selecting available targets, use the principles in *Figure 4-4,* page 4-5.

4-47. The two-hand carry position enables soldiers to use riot batons at the 24- or 36-inch length. To effectively employ this strike, use the strength and power of the hips to thrust the riot baton in a horizontal manner. Simultaneously, pull back with the weak hand as the strong hand drives the grip end toward the target, striking a rib or the abdominal region of the aggressor (see *Figure 4-15*, page 4-18). Following the two-hand strong side horizontal strike, assess the target before following up with another strike. An aggressor may not require another strike, based on his condition and actions after enduring the one-hand forward strike.

NOTE: A pattern of movement such as a forward shuffle or forward pivot may enhance this technique.

Two-Hand Weak Side Horizontal Strike

4-48. An effective two-hand weak side horizontal strike must be aggressively quick to create distance, distract, disable, or alter behavior. The weak side horizontal strike is a quick and effective offensive strike usually employed as a follow-up strike to the strong side horizontal strike. It is a countermeasure designed to close frontal attacks that have been effectively blocked. When selecting available targets, use the principles in *Figure 4-4*. Maintaining a strong defensive stance after employing a strong side horizontal strike puts the soldier in position to execute a two-hand weak side horizontal strike.

4-49. Use of the two-hand carry position enables soldiers to use riot batons at 24 or 36 inches. To effectively employ this strike, use the strength and power of the hips to thrust the riot baton in a horizontal manner. Simultaneously, pull back with the strong hand as the weak hand drives the grip forward toward the aggressor, striking a rib or the abdominal region of the target (see *Figure 4-15*). Following the two-hand weak side horizontal strike, it is important to assess the aggressor. An aggressor may not require another strike, based on his condition and actions after enduring the one-hand forward strike.

NOTE: A pattern of movement such as a forward shuffle or forward pivot may enhance this technique.

Two-Hand Front Jab

4-50. The two-hand front jab starts from the two-hand carry position (see *Figure 4-16,* page 4-18). It is aggressively quick and creates distance, distracts, disables, and/or alters behavior. Front jabs are effective offensive strikes, usually employed as a countermeasure to charging and overpowering frontal attacks. When selecting available targets, use the principles in *Figure 4-4*.

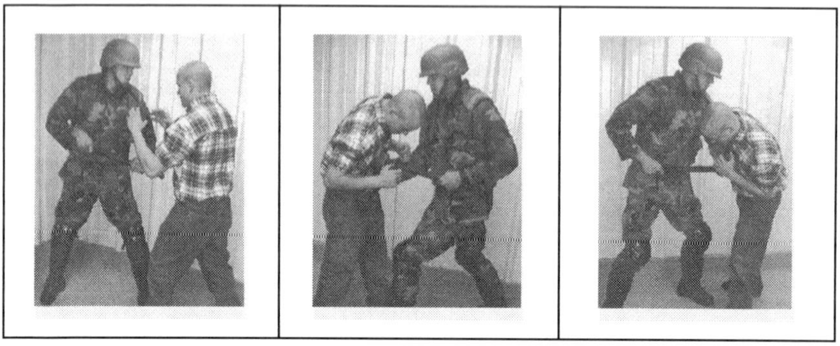

Figure 4-15. Two-Hand, Strong Side, and Weak Side Horizontal Strikes

4-51. Use of the two-hand carry position enables soldiers to use riot batons at 24 or 36 inches. Use the strength and power of the hips to effectively employ this strike in a slightly downward or horizontal manner. Simultaneously, use both arms to shoot the long end of the riot baton straight out from the body to the target area. Immediately, pull back the riot baton to the two-hand carry position (see *Figure 4-16*).

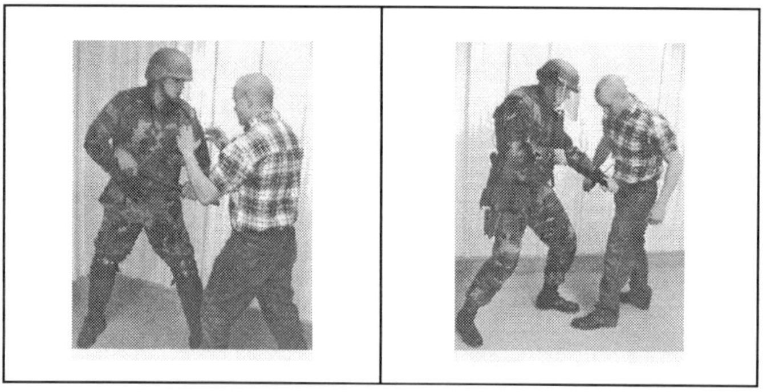

Figure 4-16. Front Jab

4-52. Forward jabs drive the long end of the riot baton toward the aggressor, striking a rib, a hip flexor, or the abdominal region (see *Figure 4-4*, page 4-5). Following the two-hand front jab, it is important that soldiers assess the aggressor before following through with another strike, as the aggressor may not require another strike.

NOTE: A pattern of movement such as a forward shuffle will enhance the power of this technique.

Two-Hand Rear Jab

4-53. Starting from the two-hand carry position, the two-hand rear jab is aggressively quick and creates distance, distracts, disables, or alters behavior. Rear jabs are quick and effective offensive strikes usually employed as a countermeasure to charging and overpowering attacks from the rear. When selecting available targets, use the principles in *Figure 4-4,* page 4-5.

4-54. Use of the two-hand carry position enables soldiers to employ riot batons at 24 or 36 inches. To effectively employ this strike, use the strength and power of the hips to thrust the riot baton in a slightly downward or horizontal manner. Simultaneously, turn the head to the rear to see and assess the target. Use both arms to shoot the riot baton straight back from the body to the targeted area and immediately pull back the riot baton to the two-hand carry position (see *Figure 4-17*).

NOTE: A pattern of movement such as a rear shuffle will enhance the power of this technique.

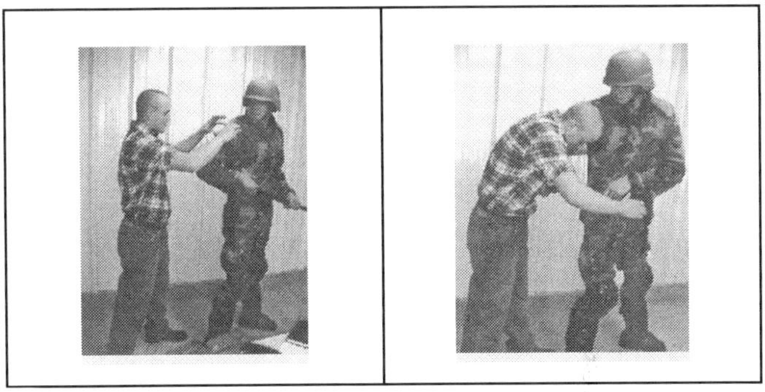

Figure 4-17. Rear Jab

Two-Hand Middle Strike

4-55. The two-hand middle strike starts from the two-hand carry position. It is aggressively quick and creates distance, distracts, disables, or alters behavior. It is proven to be an effective follow-up technique to a two-hand weak or strong side block or to destabilize an individual by pushing him back and away. The two-hand middle strike is used as a two-count movement. When selecting available targets, use the principles in *Figure 4-4*.

4-56. To perform a two-hand middle strike (see *Figure 4-18*, page 4-20) the soldier does the following:

Step 1. Brings the riot baton up to chest level and slightly off parallel to the ground.

Step 2. Steps forward aggressively, simultaneously thrusting the long portion of the riot baton forward in a horizontal manner. Both arms are fully extended at the end of the movement with a quick snappy return to the step 1 position. Use the entire body to generate the power.

NOTE: It is important that soldiers assess the aggressor before applying a follow-up strike. If the soldier steps forward as he employs the middle strike, it will afford greater impact and power.

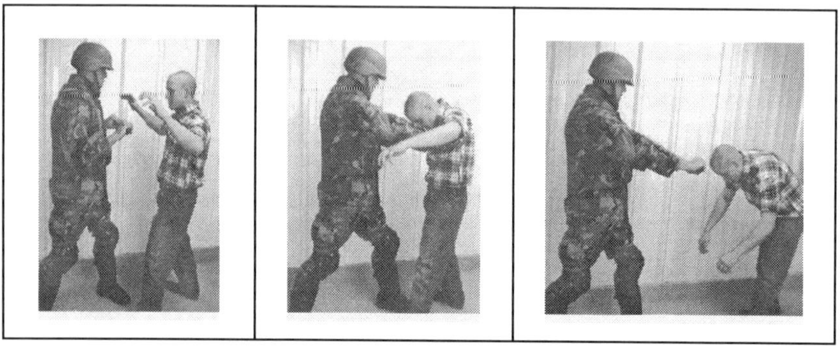

Figure 4-18. Middle Strike

Riot Baton Retention

4-57. A common technique an aggressor may use is grabbing for a soldier's riot baton or, in some cases, even taking the riot baton and using it against the soldier. This causes a disruption or distraction.

4-58. Maintaining physical control of the riot baton is essential for the safety of the soldier and the control force as a whole. Aggressors will generally attempt to grab the riot baton where they can get the strongest hold. Soldiers must resist the instinct of getting into a tug-of-war match over the riot baton, which often proves to be ineffective.

4-59. Using the momentum of the aggressor to pull the riot baton away, the soldier simply steps into or closer to the aggressor with the weak side foot. With the momentum created, the aggressor's hold on the riot baton is lessened. The soldier then rapidly uses the tracing-C technique to recover the riot baton.

4-60. A soldier using the tracing-C technique does the following:

Step 1. Pushes up on the riot baton.

Step 2. Pulls the riot baton around in an upward swing as if tracing a C in the air from bottom to top with the long end.

NOTE: This movement seems to wrap the aggressor's arms around themselves (see *Figure 4-19*).

Step 3. Drives (immediately with a quick and forceful movement) the long end of the riot baton straight down, as if slicing the C in half (see *Figure 4-19*).

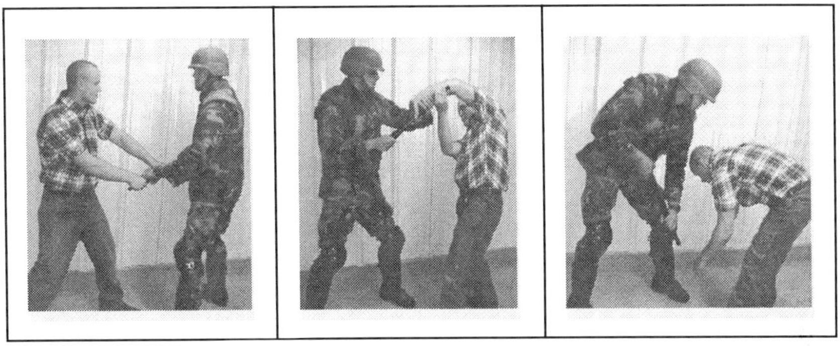

Figure 4-19. Riot Baton Retention

4-61. Once the soldier reaches the top of the C and the aggressor is tied up, it is impossible for the aggressor to maintain hold of the riot baton. Further actions such as a strike may or may not be necessary. Soldiers must be able to assess the situation and react accordingly.

WORKING AS A TEAM

4-62. Soldiers never have to rely solely on their own skills or abilities in riot control operations. Although individual skills were discussed in this chapter, it is important to remember that well-trained soldiers employing offensive, defensive, riot shield, and riot baton techniques will perform the collective task of crowd control well.

4-63. Riot control operations are dynamic and often chaotic. Soldiers armed with riot shields provide a first line of defense while those armed with riot batons and firearms provide overwatch as the formation moves toward engaging a hostile crowd. To work effectively and in concert, soldiers must be properly trained, drilled, and exercised. If an aggressor approaching a line of skirmishers violently grabs a riot shield, help is required to maintain control (see *Figure 4-20*, page 4-22). A soldier armed with a 36-inch wooden riot baton or an expandable riot baton (extended to 36 inches) can be of assistance. The soldier simply moves forward into a gap between the riot shields where he can quickly strike the top or side of the riot shield, slicing the long end across or straight down onto the hands and fingers of the aggressor (see *Figure 4-18*).

4-64. Riot batons are not the only way to effectively knock an aggressor's hands from a riot shield. Soldiers armed with 12-gauge shotguns or M16/M203s can just as effectively smash the hands or fingers of a persistent aggressor with the barrel of their weapons using the technique described above.

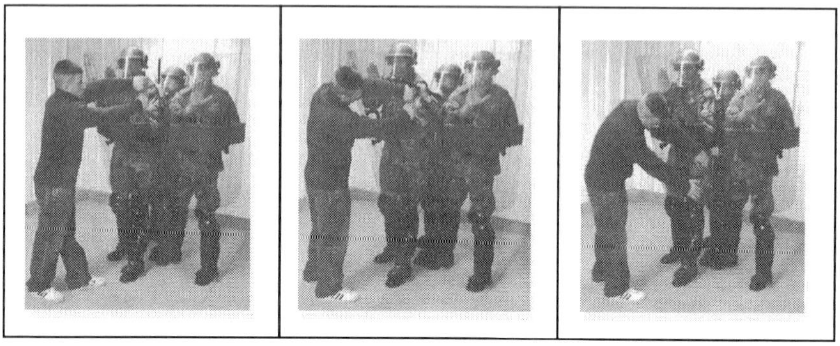

Figure 4-20. Clearing a Riot Shield

NONLETHAL MUNITIONS

4-65. With the inclusion of NL munitions, commanders now have additional tools that afford excellent standoff capabilities. This is critical because commanders now have intermediate options for dealing with a hostile crowd. No longer will a commander issue a proclamation, have it ignored by the crowd, and then be forced to use lethal munitions to backup the proclamation. Now he can use NL munitions to gain the same standoff distance but without the catastrophic results of lethal munitions.

4-66. NL munitions include (but are not limited to) a 12-gauge shotgun and 40-millimeter rounds. These blunt munitions have the capability of effectively engaging targets as close as 5 meters and as far as 50 meters. Range is dependent upon each round. These munitions are used from a distance to dissuade, discourage, and disperse the crowd. If the crowd disperses from 50 meters and soldiers are in close contact with the crowd, then the mission has been successful.

TAP-DOWN TECHNIQUE

4-67. Commanders must be familiar with the characteristics and capabilities of all rounds to use them effectively. When 12-gauge and M203 weapons are used within a line of skirmishers (usually located behind the riot shields), it is imperative that a method is developed, trained, and practiced to effectively operate as a team. Operators of these weapons may encounter problems in riot control situations when trying to identify and effectively engage legitimate targets. The tap-down technique is used to overcome target identification and engagement with NL munitions. If ordered to engage targets or the situation authorizes an engagement, soldiers armed with NL munitions may step forward, level their weapons, and fire upon legitimate targets (see *Figure 4-21*).

4-68. A soldier that is armed with a NLW and is using the tap-down technique must first tap on the shield holder's shoulder. Tapping on the shield holder's shoulder alerts him that the gunner is preparing to fire. In response to the tap, the shield holder drops to one knee while keeping the riot shield firmly affixed to their front for protection. The gunner takes a step forward and fills the gap. He then leans into the riot shield bearer's back with his knee and fires. This technique ensures that the long end of the weapon is extended beyond the riot shield.

WARNING

The long end of the weapon must be projected beyond the riot shield before firing. This is a critical safety step, which will prevent NL projectiles from hitting the riot shield and injuring soldiers.

4-69. Once soldiers firing munitions have fired, they raise their weapon back to a high port arms. The soldiers clear their weapons and tap the riot shield bearer a second time, signaling them to rise to a standing position.

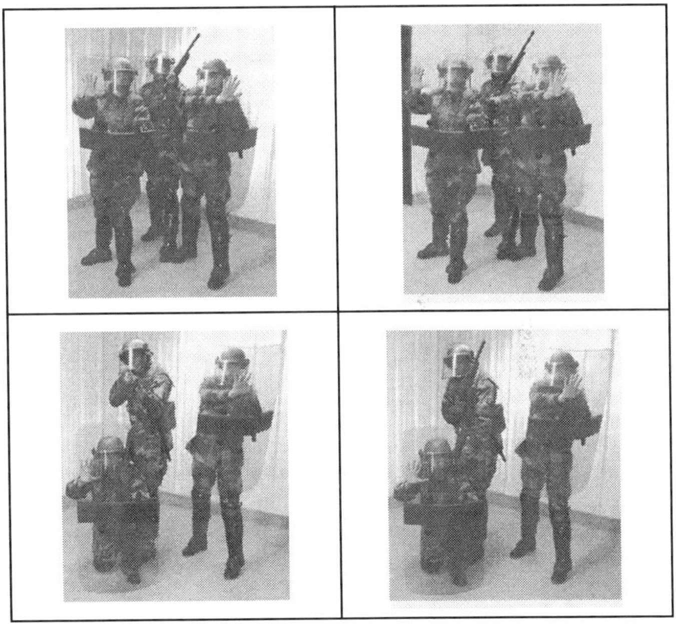

Figure 4-21. Tap-Down Technique

Chapter 5

Nonlethal Capabilities Set and Employment Considerations

During the past several decades, US forces have regularly performed peacekeeping and peace enforcement missions at a greater tempo than in the past. In the peacekeeping and peace enforcement environment, the use of conventional firearms or the threat of their use may not be the solution to a situation where US forces must separate two belligerent, hostile ethnic groups or prevent a similar group from entering an area that is off-limits to them. However, a show of force using NLW and NL munitions will cause crowds to disperse, separate, or leave the area with minimal casualities. The NLCS is a well-rounded, versatile package of both commercial off-the-shelf (COTS) and government off-the-shelf (GOTS) NL equipment and munitions. NLCS contents are divided into four distinct categories: personnel protectors, personnel effectors, mission enhancers, and training devices.

PERSONNEL PROTECTORS

5-1. Personnel protectors provide the individual soldier with added protection to the sensitive and vital areas of his body. They provide excellent protection for the individual soldier from trauma often inflicted by thrown objects (such as rocks, bricks, sticks, and bottles).

NONBALLISTIC RIOT FACE SHIELD

5-2. A nonballistic riot face shield (see *Figure 5-1,* page 5-2) has two adjustable positions and is constructed of hard durable plastic that securely attaches to the Kevlar® helmet. It may be worn with an M17 or M40 protective mask and is designed to protect the soldier's head, face, and neck from thrown objects. Nonballistic riot face shields will not protect from attacks with ballistic weapons, such as firearms. All soldiers operating in a civil disturbance environment should be outfitted with riot face shields.

NONBALLISTIC FULL-LENGTH SHIELD

5-3. A nonballistic full-length riot shield (see *Figure 5-2,* page 5-2) is a body shield constructed of hard durable plastic. Left- and right-handed personnel can easily use the nonballistic full-length riot shield. It is designed to protect the soldier's face, torso, and upper legs from thrown objects. Nonballistic full-length riot shields will not provide protection from ballistic weapons. Only the front line of skirmishers and those providing lateral protection to both flanks of the formation will be outfitted with nonballistic full-length riot shields.

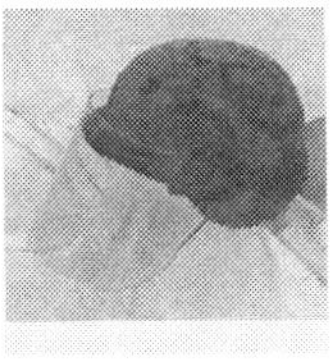

Figure 5-1. Nonballistic Riot Face Shield

Figure 5-2. Nonballistic Full-Length Riot Shield

NONBALLISTIC SHIN GUARD

5-4. The nonballistic shin guards (see *Figure 5-3*) are constructed of hard durable plastic and are easily fastened to a soldier's legs. They are designed to protect the soldiers, feet, shins, and knees from thrown objects. They will not provide protection from ballistic weapons, such as firearms. All soldiers operating in a civil disturbance environment should be outfitted with shin guards.

BALLISTIC RIOT FACE SHIELD

5-5. The ballistic riot face shield (see *Figure 5-4*) has two adjustable positions constructed of acrylic and bullet-resistant materials that provide the wearer with Level IIIA protection. The ballistic riot face shield easily attaches to the Kevlar helmet and is capable of being worn with either an M17 or M40 protective mask. It will protect the soldier from hand-thrown objects. Because of its excessive weight (3.4 pounds), the ballistic riot face shield should not be worn for standard riot control operations. Ballistic riot face shields are best suited for use within specialized teams associated with riot control operations; for example, teams responsible for the search and apprehension of identified criminals or building- and alley-clearing operations.

Figure 5-3. Nonballistic Shin Guards

Figure 5-4. Ballistic Riot Face Shield

BALLISTIC FULL-LENGTH RIOT BODY SHIELD WITH LIGHT KIT

5-6. The ballistic full-length riot body shield (see *Figure 5-5,* page 5-4) is constructed from Spectra Shield® bullet-resistant material that provides a soldier with Level IIIA protection. It will protect the soldier from thrown objects. Because of its excessive size (18 pounds and 20 by 36 inches), the ballistic full-length riot body shield should not be used as part of a standard riot control operation. Ballistic full length riot body shields are best suited for use within specialized teams associated with riot control operations; for example, teams responsible for the search and apprehension of identified criminals or building- and alley-clearing operations.

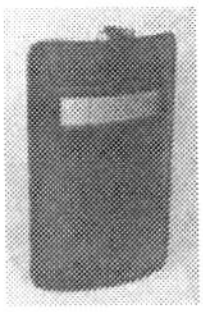

Figure 5-5. Ballistic Full-Length Riot Body Shield

BALLISTIC SHIN GUARDS

5-7. Ballistic shin guards (see *Figure 5-6*) are constructed of Kevlar KM2 materials and provide a soldier with Level IIIA protection to the feet, shins, and upper legs. They protect the soldier from hand-thrown objects. There are three different sizes, and because of their excessive weight (7 to 10 pounds), ballistic shin guards should not be deployed as part of a standard riot control operation. Ballistic shin guards are best suited for use within specialized teams associated with riot control operations; for example, teams responsible for the search and apprehension of identified criminals or building- and alley-clearing operations.

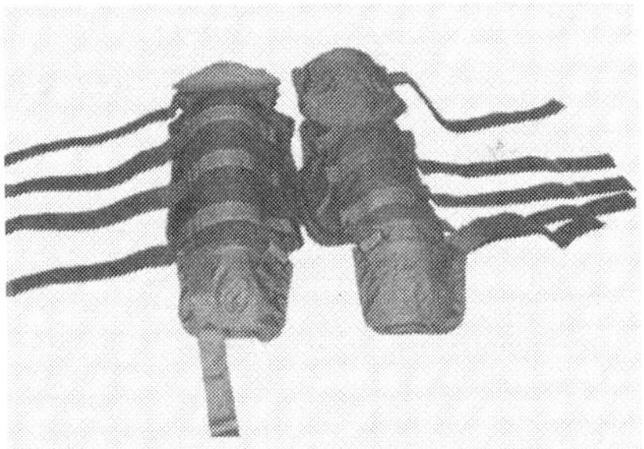

Figure 5-6. Ballistic Shin Guards

PERSONNEL EFFECTORS

5-8. Personnel effectors are those items that provide soldiers with an NL response without the necessity of always coming into direct physical contact with aggressors. Personnel effectors are 12-gauge and 40-millimeter ammunition that inflict blunt trauma. These munitions are compatible with the shotgun and the M203 grenade launcher already in the Army inventory. There are also diversionary and rubber ball grenades, which distract and inflict blunt trauma.

WOOD RIOT BATON WITH BELT RING (36 INCHES)

5-9. The 36-inch wood riot baton (see *Figure 5-7*) is made from solid oak. When held properly, the riot baton thong provides greater retention capabilities. Intended primarily for self-defense, the 36-inch wood riot baton can be used effectively as an offensive tool if it becomes necessary to keep rioters out of arm's reach of soldiers conducting crowd control operations.

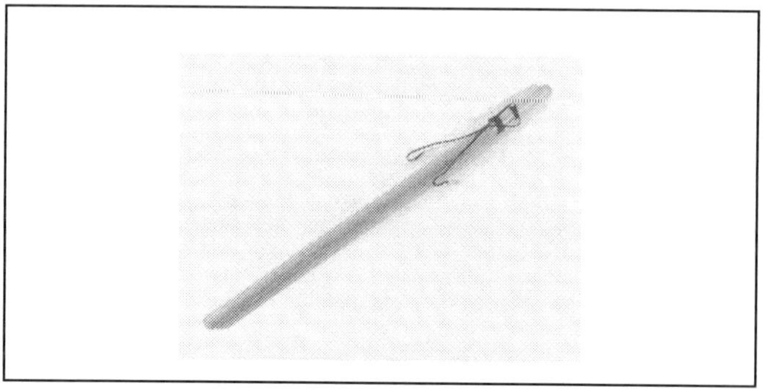

Figure 5-7. Wood Riot Baton

EXPANDABLE RIOT BATON WITH CARRIER (24 TO 36 INCHES)

5-10. The expandable riot baton (see *Figure 5-8*) provides the soldier with an NL means of crowd control and self-defense. The riot baton comes with a mounting device that attaches the riot baton to a belt. The riot baton is 24 to 36 inches in length and is primarily intended for self-defense. The expandable riot baton can be used effectively as an offensive tool if it becomes necessary to keep rioters out of arm's reach of soldiers conducting crowd control operations.

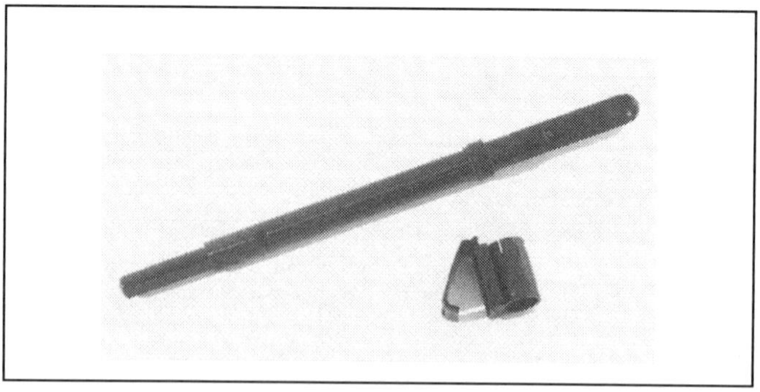

Figure 5-8. Expandable Riot Baton and Carrier

500 (3 INCH) 12-GAUGE SHOTGUN

5-11. The 12-gauge shotgun (see *Figure 5-9*) is a pump action shotgun currently in NLCS inventory. The pump action shotgun is chambered to take up to 3-inch shells. The 3-inch chamber allows for the use of M1012 and M1013 NL munitions. This shotgun also provides a visually distinct alternative to standard military weapons that may be desired based on mission considerations (force signature and acceptability).

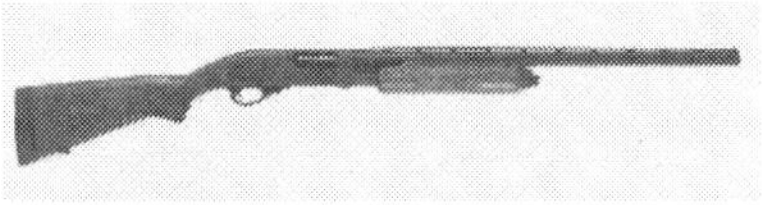

Figure 5-9. 12-Gauge Shotgun

M7 66-MILLIMETER LAUNCHER WITH M315 INSTALLATION KIT

5-12. The M7 (see *Figure 5-10*) is a 66-millimeter vehicle-mounted, NL, grenade-launching device that is mounted on a HMMWV. It is an indirect fire support system that can deliver the M98 distraction grenade that creates a flash-bang effect, L96A1 antiriot grenade, or M99 blunt trauma grenade that creates a sting-ball effect. The M315 installation kit is used to install an M7 discharger on the turret ring of appropriate HMMWV variants. An adjustable bracket allows the launch angle to be depressed for engaging targets at ranges of 50, 75, and 100 meters. The system enforces standoff distances and deters potential threats.

Figure 5-10. M7 Launcher

MODIFIED CROWD CONTROL MUNITION-GROUND EMPLACEMENT

5-13. Similar in nature to the claymore mine, the modified crowd control munition-ground emplacement (MCCM-GE) (see *Figure 5-11*) is an NL munition that can deliver six hundred .32-caliber rubber pellets and a flash-bang effect. The MCCM-GE will stop, confuse, disorient, and/or temporarily incapacitate area targets and/or personnel at ranges out to 30 meters.

WARNING
Avoid employing the MCCM-GE within 10 meters of personnel as it may cause serious bodily harm or death.

Figure 5-11. MCCM-GE

12-GAUGE, FIN-STABILIZED POINT, NONLETHAL CARTRIDGE (M1012)

5-14. The M1012 (see *Figure 5-12*) is a single projectile round made of hard rubber that is shaped like a bomblet and designed to be fired a at a single target. With a muzzle velocity of 500 feet per second, the M1012 has the effective range of no closer than 5 meters and no further than 30 meters. Engagement inside 5 meters could cause serious injury or death. Beyond 30 meters the kinetic energy dissipates to the point where the round becomes ineffective.

Figure 5-12. M1012

12-GAUGE, CROWD DISPERSAL, NONLETHAL CARTRIDGE (M1013)

5-15. The M1013 (see *Figure 5-13*) is a multiple projectile round with .23 caliber hard rubber pellets that is designed to be fired at and employed with the purpose of affecting multiple targets. With a muzzle velocity of 900 feet per second, the M1013 has an effective range of no closer than 5 meters and no further than 30 meters. Engagement inside 5 meters could cause serious injury or death. Beyond 30 meters, the kinetic energy dissipates to the point where the rubber pellets become ineffective.

Figure 5-13. M1013

40-MILLIMETER SPONGE GRENADE (M1006)

5-16. The M1006 (see *Figure 5-14*) is a single projectile round made of pliable foam rubber with a hard plastic back. The M1006 is designed to be aimed and fired at a single target. With a muzzle velocity of 265 feet per second, the M1006 has an effective range of no closer than 10 meters and no further than 50 meters. Engagements inside 10 meters could cause serious injury or death. Beyond 50 meters, the kinetic energy dissipates to the point where the round becomes ineffective.

Figure 5-14. M1006

40-MILLIMETER CROWD DISPERSAL CARTRIDGE

5-17. The crowd dispersal cartridge (CDC) (see *Figure 5-15*) is a multiple projectile round with .48 caliber hard rubber balls that is designed to be fired and employed with the purpose of affecting multiple targets. With a muzzle velocity of 450 feet per second, the CDC has an effective range of no closer than 10 meters and no further than 30 meters. Engagement inside 10 meters could cause serious injury or death and beyond 30 meters the kinetic energy dissipates until the rubber balls become ineffective.

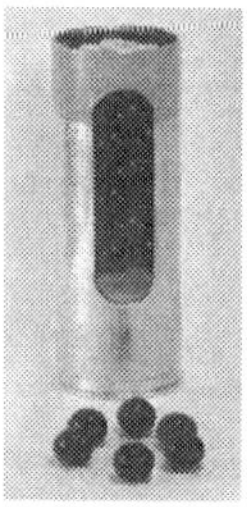

Figure 5-15. CDC

RUBBER BALL GRENADE (GG04)

5-18. The GG04 (see *Figure 5-16,* page 5-10) is a multiple projectile, flash-bang grenade with 100 .25-caliber hard rubber pellets. Each grenade has a fuse delay of 2.8 to 3 seconds with a flash measuring approximately 1 million CP and 180 decibels at 3.5 feet. At detonation, rubber pellets are dispatched at 360° with an effective range of 2 to 3 meters and a maximum engagement range of 15 to 20 meters. The rubber ball grenade is designed to be hand-thrown or muzzle-launched from a 12-gauge shotgun.

RUBBER BALL GRENADE LAUNCHING CUP AND AA30 12-GAUGE LAUNCH ROUND

5-19. The launch cup (see *Figure 5-17,* page 5-10) firmly attaches to the 12-gauge shotgun without any additional tools. It works in concert with the AA30 12-gauge launch round to propel the rubber ball grenade approximately 100 meters. The shooter simply inserts the full body and the safety lever of the grenade into the launching cup, pulls the pin, loads a 12-gauge round, and fires (approximately) at a 30° angle. The desired effect is to have the grenade explode approximately 7 to 10 feet above the target.

RELOADABLE FLASH-BANG GRENADE (M84)

5-20. The M84 (see *Figure 5-18,* page 5-10) is a reloadable flash-bang thrown grenade. Each grenade has a 1- to 2-second fuse delay, with a flash capability of 1.5 to 2.5 million CP and a bang of 168 to 175 decibels. Although it is part of the NLCS, use of the M84 is not recommended for use in crowd control situations. Specialty teams, such as security response teams, and search teams may use this device effectively.

Figure 5-16. GG04

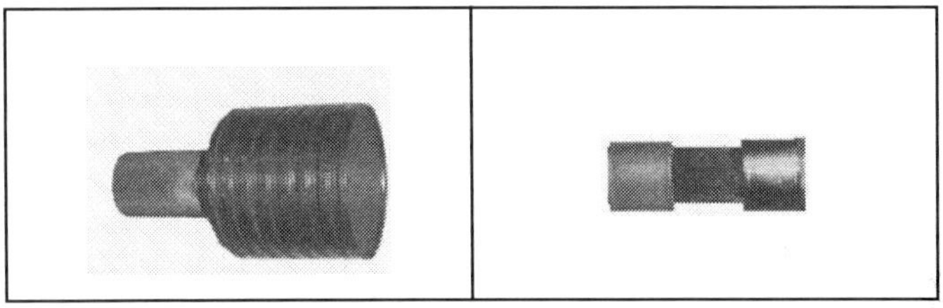

Figure 5-17. Launching Cup and AA30

Figure 5-18. M-84

FLEX-CUFS

5-21. Flex-Cufs consist of a tough pliable plastic band with a self-locking mechanism in the center for each end. When the Flex-Cuf (see *Figure 5-19*) is threaded the restraint bands circle around the wrists or ankles, impeding movement and securing the individual. They are lightweight and disposable and require a cutting tool or other instrument to remove.

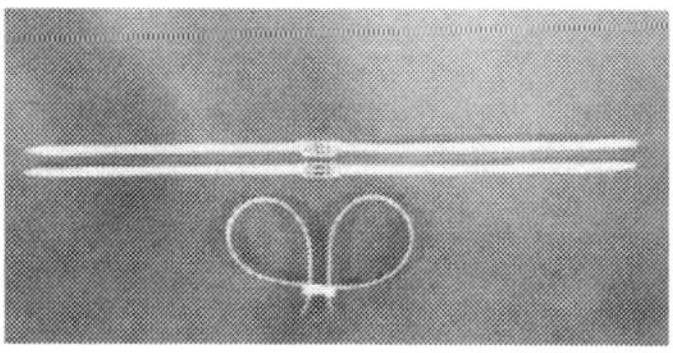

Figure 5-19. Flex-Cufs

INDIVIDUAL RIOT CONTROL AGENT DISPERSER (M36)

5-22. The M36 (see *Figure 5-20*) contains a Dibenz (F, -1 4-oxazepine [CR] solution. It can deliver 25 one-second bursts out to 12 feet. Individual disposable RCA dispersers are intended primarily for self-defense or to keep rioters out of arm's reach of soldiers conducting crowd control formations or soldiers engaged in missions where a noncombatant exists.

Figure 5-20. M36

MIDSIZE RIOT CONTROL AGENT DISPERSER (M37)

5-23. The M37 (see *Figure 5-21*, page 5-12) is the size of a standard fire extinguisher that uses compressed air to force the RCA out to a range of 30 feet. It has the capacity to employ 18 bursts for 3 seconds per charge. It is excellent for providing a wide coverage of RCAs onto a hostile crowd while maintaining excellent standoff capability. The M37 can be refilled and is rechargeable. It can be filled with CR solution (liquid agent) or CS1 (dry agent). For training purposes, CR can be substituted with water and CS1 can be substituted with talcum powder.

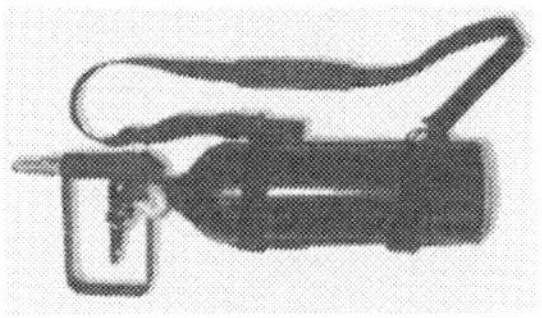

Figure 5-21. M37

SQUAD RIOT CONTROL AGENT DISPERSER (M33A1)

5-24. The M33A1 (see *Figure 5-22*) is designed to provide crowd control and protection at the squad level. It is capable of projecting a ballistic stream of RCAs beyond 25 feet in up to 25 half-second bursts. It consists of a frame and harness assembly, compressed-gas cylinder (agent container assembly), air pressure assembly, gun and hose assembly, multijet spray unit, and check valve assembly. The M33A1 can be refilled and is rechargeable. For training purposes, CR can be substituted with water and CS1 can be substituted with talcum powder.

Figure 5-22. M33A1

MISSION ENHANCERS

5-25. Mission enhancers aid the commander in various tactical situations. Mission enhancers include various types of devices used for illumination, vehicle denial, traffic control missions, and force protection enhancement.

PORTABLE BULLHORN

5-26. The portable bullhorn (see *Figure 5-23*) provides squad leaders with the capability to project their voices above the noise and commotion created by crowds and mobs. The megaphone has a built-in siren for alarm purposes that allows the operator to use the device as an ordinary megaphone and siren alarm during emergencies. The bullhorn is portable and easy to carry. The acoustic range of the megaphone for a clear voice output is ¾ to 1 mile. The power output of the megaphone is rated at 15 watts with a maximum rating of 20 watts. The dimensions of the megaphone are 14.5 inches with a 9-inch diameter bell. It weighs 3.5 pounds without batteries. Battery requirements for the megaphone are eight dry-cell size C batteries.

Figure 5-23. Portable Bullhorn

GROUND-MOUNTED BULLHORNS

5-27. Ground-mounted bullhorns (see *Figure 5-24,* page 5-14) are a critical communication enhancement device for conducting crowd control tactics. The bullhorn can facilitate communication with the crowd in conjunction with PSYOP support and assistance with the communication of commands to troops engaged in the crowd control process. The power output of the ground-mounted bullhorn is rated at 22 watts. The dimensions of the bullhorn are 13.75 by 9 by 14.75 inches, and it weighs 15 pounds.

INDIVIDUAL VOICE AMPLIFICATION SYSTEM (M7)

5-28. The M7 (see *Figure 5-25,* page 5-14) is a critical communication enhancement device for conducting crowd control tactics using RCAs while wearing the M40 protective mask. The M7 is fitted to the M40 protective mask to facilitate oral communication and increase the user's ability to communicate using radios and other devices.

Figure 5-24. Ground-Mounted Bullhorns

Figure 5-25. Individual Voice Amplification System

INDIVIDUAL HIGH-INTENSITY SEARCHLIGHT WITH HOLSTER

5-29. The individual high-intensity searchlight (see *Figure 5-26*) is used primarily for illumination in crowd control operations at night. An individual can use it for general illumination of the operational area to pinpoint agitators and threats, reduce the ability of the rioters to see troop formations and actions, enhance tactical deception and actions, and enhance tactical deception techniques for units conducting crowd control operations.

HIGH-INTENSITY LIGHT

5-30. A high-intensity light (see *Figure 5-27*) is intended for use in low light or night conditions. A high-intensity light can project a beam that will enable the user to identify an individual person up to 1,900 yards away. It has an adjustable beam spread of 1° to 15° and will run continuously at maximum power for up to 45 minutes. The internal power supply requires 4 hours for a full charge.

Figure 5-26. Individual High-Intensity Searchlight with Holster

Figure 5-27. High-Intensity Light

HIGH-INTENSITY LIGHT SYSTEM WITH 12-GAUGE SHOTGUN ATTACHMENT KIT

5-31. This 12-guage high-intensity light system (see *Figure 5-28,* page 5-16) is used to illuminate possible targets in crowd control situations where there is little or no light. It serves to blur or blind the aggressor's ability to see. It mounts easily to the end of the 12-gauge shotgun barrel.

40-MILLIMETER CARRYING POUCH

5-32. The 40-millimeter carrying pouch is made from durable canvas material, (see *Figure 5-29,* page 5-16) and can be easily affixed to load-bearing equipment (LBE) and/or a load-bearing vest (LBV) using Velcro straps. It is capable of carrying up to six 40-millimeter rounds.

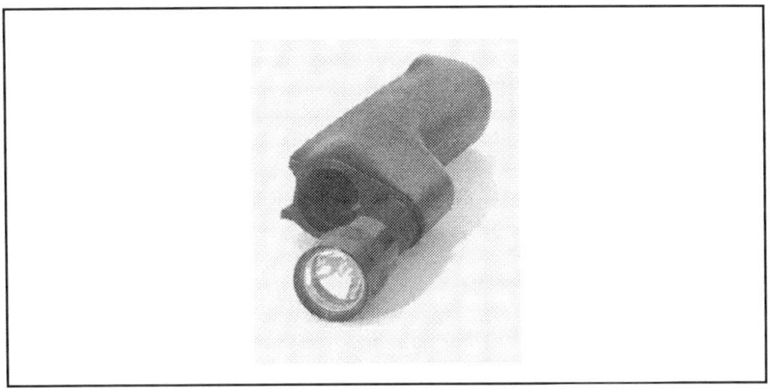

Figure 5-28. 12-Gauge High-Intensity Light System With 12-Gauge Shotgun Attachment Kit

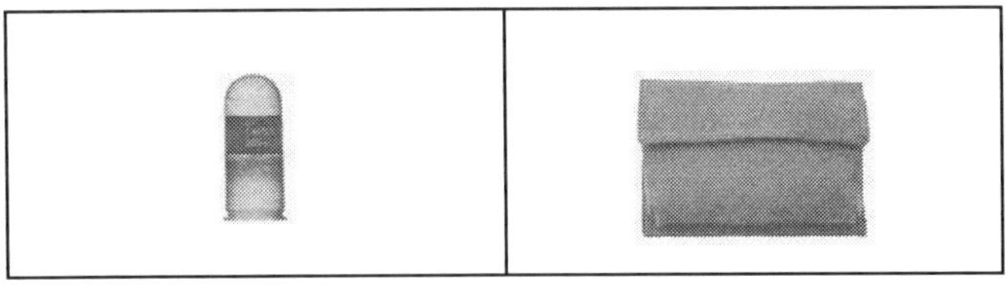

Figure 5-29. 40-Millimeter Carrying Pouch

12-GAUGE UTILITY POUCH

5-33. The 12-gauge utility pouch is made from durable canvas material (see *Figure 5-30*). It can be easily affixed to an LBE and/or LBV using metal clasps. The 12-gauge utility pouch is capable of carrying up to twenty-five 12-gauge shotgun rounds.

Figure 5-30. 12-Gauge Utility Pouch

12-GAUGE BUTTSTOCK CUFF

5-34. The 12-gauge buttstock cuff (see *Figure 5-31)* firmly attaches to the shotgun by sliding over the stock. Elastic straps aid in preventing the buttstock cuff from sliding or rolling. It holds up to seven 12-gauge cartridges, which give the soldier quick access to additional rounds.

Figure 5-31. 12-Gauge Buttstock Cuff

DIVERSIONARY AND/OR RUBBER BALL GRENADE POUCHES

5-35. The diversionary and/or rubber ball grenade pouch (see *Figure 5-32*) is an ammunition-carrying pouch made of durable canvas material. It is partitioned in a manner to provide the soldier with a means of holding up to six rubber ball grenades or six flash-bang grenades in separate compartments.

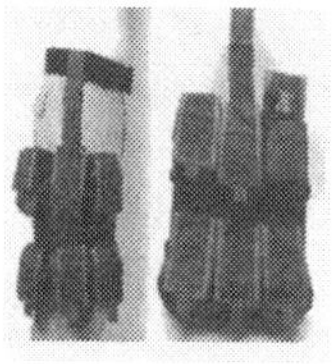

Figure 5-32. Diversionary and/or Rubber Ball Grenade Pouch

CALTROPS

5-36. The caltrop (see *Figure 5-33,* page 5-18) is a field-expedient area denial system. Securing approximately 50 caltrops to 550-type military cord allows a soldier to quickly toss and recover the system. Caltrops are used to deny vehicle and pedestrian movement in designated areas. The four prongs are approximately 2 inches tall and will puncture tires or boots. A soldier can easily employ the caltrop by scattering it in a designated area hidden by camouflage.

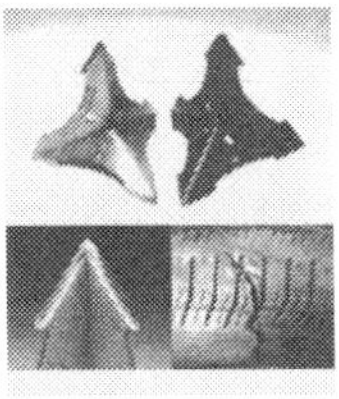

Figure 5-33. Caltrop

PORTABLE VEHICLE ARREST BARRIER

5-37. A portable vehicle arrest barrier (PVAB) (see *Figure 5-34*) is a lightweight, portable, manually-emplaced and recoverable-barrier entrapment device that is used (on command) to create a barrier that will stop vehicles being driven by known or suspected hostile forces with minimal damage to the vehicle and its occupants. Two individuals can easily emplace it in approximately 4 to 8 hours. It has the capability of stopping a light truck (up to 7,500 pounds) at speeds of up to 45 miles per hour with a stopping distance of less than 200 feet. It is not to be used as a stand-alone system. The PVAB may be used at dismount points, checkpoints, and roadblocks to prevent unauthorized access by wheeled vehicles into or out of areas.

Figure 5-34. PVAB

TRAINING DEVICES

5-38. Training devices that commanders can use to effectively prepare their soldiers are included in the NLCS. Training devices allow soldiers to learn and build confidence in the development of personal defensive skills. These are critical pieces of equipment because they allow soldiers to train at almost full speed, which provides reinforcement of realistic motor skills. Training in this manner has proven to be more effective than just striking out at the air.

IMPACT TRAINING SUIT

5-39. Impact training suits (see *Figure 5-35*) are made of closed-cell shock-absorbent foam that allows soldiers to hone their riot baton skills through realistic hand-to-hand training. Impact training suits absorb strikes inflicted by the current expandable and wooden riot batons. They provide protection for the head, face, hands, and legs without significant degradation of the wearer's mobility.

Figure 5-35. Impact Training Suits

TRAINING STRIKE BAGS

5-40. Training strike bags (see *Figure 5-36*) give the soldier a durable, yet absorbing surface to strike or kick for training. It also helps develops proficiency in open-hand control and riot baton techniques.

Figure 5-36. Training Strike Bag

PRACTICE GRENADE BODY (GG05)

5-41. The GG05 is blue in color and is easily identified as a practice grenade (see *Figure 5-37*). It is designed to replicate the live GG04 rubber ball grenade. It is made of durable rubber and can be hand-thrown or muzzle-launched from the 12-gauge shotgun. It can be reloaded.

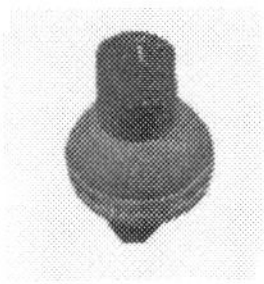

Figure 5-37. GG05

PRACTICE GRENADE FUSE

5-42. The practice grenade fuse (see *Figure 5-38*) is identified by its blue spoon and brown stripe. It easily screws into the body of the GG05. Once used, unscrew the practice grenade fuse from the body and reload it with a new fuse.

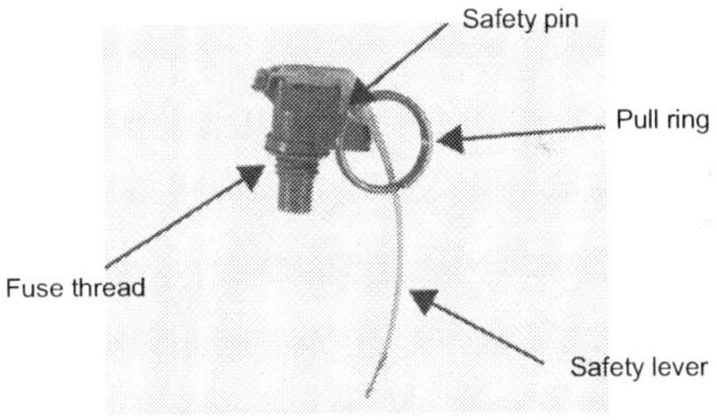

Figure 5-38. Practice Grenade Fuse

CORE CAPABILITIES

5-43. Core capabilities are those fundamental competencies that enable the commander to achieve the desired operational outcome. In the case of NLW, this includes providing a flexible means of response in order to protect friendly forces and/or influence the actions of potential adversaries and noncombatants. With the availability of the NLCS (see *Figure 5-39*) these goals can be achieved without resorting to lethal force and in a manner that will minimize collateral damage. The core capabilities associated with NL effects fall into the following,two major categories:

NOTE: There are nine platoon equivalents within a light infantry battalion.

PLT Organization	Personnel and Weapons	Nonlethal Weapons & Equipment
PLT	7 personnel (5 M16/M4 and 2 M60/ M240)	3 riot face shields 3 sets of shin guards 3 sets of full-body restraints 3 individual voice amplification systems 3 individual RCA dispensers with carriers 1 portable bullhorn 1 ground-mounted bullhorn 1 high-intensity searchlight 1 set of caltrops 3 ballistic riot face shields 3 sets of ballistic shin guards 3 ballistic riot body shields 1 riot baton training suit 1 riot baton strike pad
SQD SQD SQD	3 personnel (3 M16/M4)	3 riot face shields 3 sets of shin guards 3 riot body shields 3 expandable riot batons with carriers 9 sets of cinch-strap restraints 3 individual voice amplification systems 3 individual RCA dispensers with carriers 3 individual high-intensity search lights with carriers (see *Figure 5-26*)
TM TM TM TM TM TM	24 personnel (12 M16/M4 and 6 M16/ M4 with M203 6 SAW)	24 riot face shields 24 sets of shin guards 100 sets of disposable flex-cuffs 24 individual voice amplification systems 24 individual RCA dispensers with carriers 6 expandable riot batons with carriers 11 grenade pouches 12 forty-millimeter grenade pouches 6 twelve-gauge ammunition pouches 6 twelve-gauge bitstock ammunition carriers

Figure 5-39. NLCS Distribution

- **Counterpersonnel.** NL counterpersonnel capabilities enable the application of military force with a reduced risk of fatalities or serious casualties among noncombatants or, in some instances, among enemy forces. There are several specific NL counterpersonnel capabilities to be explored. These include the means to influence the behavior and activities of a potentially hostile crowd, and the capability to bring a mob engaged in a riot under control. While there are many similarities in these two scenarios, each involves unique challenges, which may require radically different solutions. For more information, see *FM 3-22.40*.
- **Countermaterial.** NL countermateriel capabilities enhance operations by rendering equipment and facilities unusable without complete destruction. An NL countermateriel capabilities enable the

application of military force to defuse potentially volatile situations under circumstances in which more destructive conventional military means might prove counterproductive. For example, pre-emptive strikes against troublesome and aggressive nations may be politically unacceptable when only conventional weapons with the high risk of casualties are involved. With NL countermateriel capabilities, the enemy's ability to threaten its neighbors could be curtailed with less political risk by attacking only their weapons of war and their supporting infrastructure. For more information, see *FM 3-22.40*.

ADDITIONAL CAPABILITIES

5-44. NLW are only NL when fired within their parameters.

DANGER

If NLW are fired outside their parameters, they may cause serious injury or death.

5-45. If a situation turns from NL to lethal, an NL gunner only needs to change his point of aim and distance to inflict serious bodily harm and/or death to the threat. The closer the soldier is to the target, the greater the kinetic energy and the more severe the effects are to the threat.

EMPLOYMENT CONSIDERATIONS

5-46. NL munitions and equipment are not completely NL; they are NL by intent. The term NL does not guarantee zero mortality or nonpermanent damage. NLW can add flexibility to combat operations and enhance force protection by providing an environment in which friendly troops can engage threatening targets with a reduced risk of noncombatant casualties and collateral damage.

5-47. If the tactical situation dictates a NL response, a soldier who levels a weapon loaded with NL munitions must be trained in distance to the target and appropriate points of aim. Based on the soldier's understanding and training on the characteristics and capabilities of the round, they can easily discern and engage targets to the desired effect. If the tactical situation dictates a lethal response, soldiers can easily adjust the point of aim to a more vulnerable target location on the body and move in closer to the target. NL munitions can be used as a lethal response if the situation dictates and the intent of the soldier is to inflict death.

NONLETHAL WEAPONS CAPABILITIES IN FORMATIONS

5-48. The mix of NLW within the formation is based on METT-TC. For example, the first rank of the formation will have riot shields, riot batons, and their personal weapon (slung across their back [left to right] with the butt up and muzzle down or holstered). The second row will have a mix of shotguns and M203s. Shotguns and M203s may be used as area denial or point target weapons depending on the specific munitions used. The shotgun provides cover for the slower loading M203. The commander may move these weapons within the formation to meet the needs of the mission and to create distance between the formation and the crowd (see *Appendix D*).

Chapter 6

Civil Disturbance Formations

Control force formations, when properly employed and executed against a crowd, are some of the most practical methods of crowd control. The riot baton has been the weapon of choice for the control force. With the development of NLW, the commander now has more options at his disposal. Before NLW, the only other choices were chemical irritants, MWDs, and lethal force. Now commanders can create a standoff distance with NLW that reduces the risk of close fighting with the mob. During any crowd control operation, the commander must protect the control force with lethal overwatch. This overwatch may be in the form of marksmen and observer teams (sniper with observer) and a response force with lethal capabilities.

OVERVIEW

6-1. With advances in NLW, it is possible to create an effective standoff distance of 15 to 100 meters between the crowd and the control force formation. It is now possible to advance and disperse the crowd without coming into direct contact with it. Some NLW may also be used to assist in the identification and apprehension of instigators.

6-2. Control force formations are used to disperse, contain, and/or block a crowd. These formations are more effective in urban areas than in open areas, such as parks or fields. When control force operations are employed in urban areas, it is easier to split a crowd into smaller segments, isolate instigators, or funnel the crowd into the desired location by using buildings and other man-made structures. Commanders must be mindful of the fact that well-organized crowds may attempt to move to the flanks of a formation and/or get behind the formation to gain a tactical advantage.

6-3. Commanders must be aware of the limitations of control force formations. They are not the answer to all civil disturbance situations. Do not expose the formation to sniper fire and/or unnecessary violence simply for a show of force. Control force formations should be used when it is decided that a crowd poses a threat and must be dispersed or moved to a specified area. Before dividing a large crowd, the commander must consider that this might not solve the problem. It may worsen the problem by creating smaller elements that may engage the control force in small-mob tactics, such as sniping, looting, burning, and attempting to envelop the control force. Commanders must anticipate that breakaway crowds may engage in small-mob tactics. These tactics can be defeated by area control measures, such as building searches, saturation patrolling, and other tactics. After dispersing the crowd, the control force must not assume that there will be an immediate return to peaceful activities. The use of formations is only part of the total dispersal effort.

6-4. Commanders must assess and secure the areas of operations (AOs). This may be accomplished with the use of helicopter overflights and by visually securing rooftops, high elevations, and advantage points. Designated marksmen teams may be used to secure the rooftops and provide lethal overwatch for the control formation. The control force commander must be aware that designated marksmen teams are positioned on rooftops.

6-5. Crowd control formations and their support teams have more capabilities than just crowd dispersal. They also have the capability to apprehend and detain certain members of the crowd that the commander feels may instigate the crowd to further violence. Commanders must analyze the threat, determine the mission of the control force, and decide which control force formations will accomplish the mission according to METT-TC.

6-6. During CONUS operations, there are usually no enemy forces. However, commanders must analyze and prepare for any potential opposition. The formations discussed in this chapter are guidelines and may be adapted to fit any mission or situation. Whatever the modification, soldiers must stay in position and on line. Only through training and rehearsals will soldiers become proficient in basic formation movements and be able to adjust to changing situations. Soldiers must also be prepared to employ NLW. *Figure 6-1* is used as a key to identify soldiers within crowd control formations.

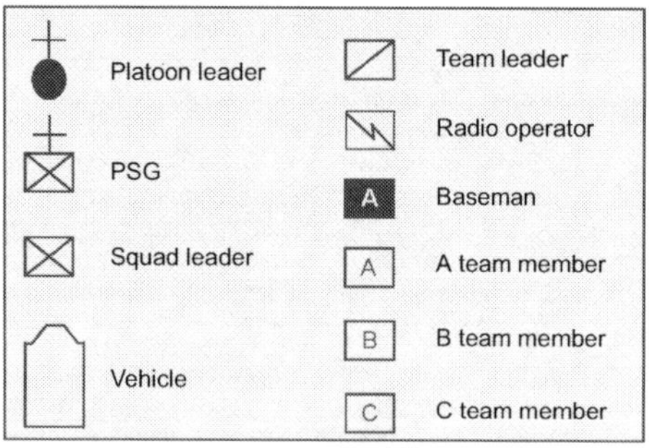

Figure 6-1. Symbols for a Formation Element

ACTIONS BEFORE MOVEMENT

6-7. Before moving the control force to the civil disturbance area and into a potentially hostile urban environment, the commander and his staff must review the preincident plans and expand them (as needed) based on current information. Because of the employment of forces against ambiguous threats,

IPB becomes critical. Information is the key to developing an appropriate response to civil disturbances.

6-8. Civil disturbance training must be part of the annual unit training plan. If a soldier is deployed to an area where civil disturbances are a recurring event, sustainment training must occur more frequently. Before moving the control force to the rally point, an intense rehearsal must be conducted by the leadership and the control force. A top-down review of the rules of engagement (ROE/rules for the use of force (RUF) must be accomplished. The ROE/RUF must be part of the unit training plan and each soldier must be trained to standard (see *Chapter 8*). The use of NLW (type specific) must be granted to the on-site commander from the approval authority.

6-9. Information is the key to developing plans for appropriate responses to a civil disturbance, and IPB is crucial to the success of the mission. During the planning phase (see *Chapter 2*), the information provided by the IPB will aid the staff in selecting various routes and rally points. The selection of ingress and egress routes must include a variety of ways in and out of the area. Direct and indirect routes and those with cover and concealment are necessary. The rally point must be cleared and secured by an advanced party capable of controlling the area.

RALLY POINT ACTIONS

6-10. Once at the rally point, the commander makes contact with the local police, civil authorities, or military officer. He then assesses the situation, decides the next COA, and issues orders, as appropriate. The platoon sergeant (PSG) forms the platoon into a column formation and places team leaders directly in front of their teams (see *Figure 6-2,* page 6-4). Team and squad leaders make last minute checks of their soldiers and await orders.

6-11. The commander places his lethal overwatch (designated marksmen [DM] teams) in areas that overlook the control force and the mob. They may be to the flanks and slightly to the front of the formation. It is their responsibility to protect the formation from lethal fire by constantly scanning the crowd.

6-12. When the commander receives word that DM teams are in place, he forms the control force into a crowd control formation and moves them quickly from the rally point to where the crowd is assembled. Depending on the situation, movement should be in a column formation or one of the three basic control force formations. Reserve forces are left at the rally point until ordered forward as additional lethal overwatch or to reinforce the formation.

CONTROL FORCE FORMATIONS

6-13. Control force formations, when properly employed and executed against a crowd of limited size, are one of the most practical methods of crowd control. Experience has indicated that the most frequently used control force

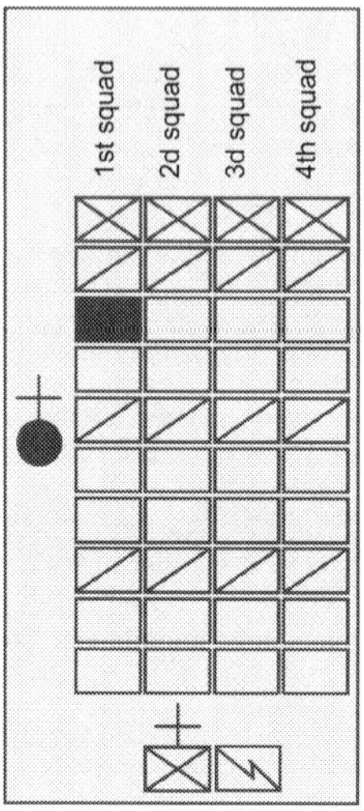

Figure 6-2. Column Formation

formations are the line, echelon, and wedge. The two lesser-used formations are the diamond and circle. Descriptions of all the formations are as follows:

- **Line formation.** The line is the basic formation and it is used more often because of its offensive and defensive applications. As an offensive formation, the line is used to push or drive crowds straight back, across an open area, or up a city street. As a defensive formation, the line is used to hold the crowd or to deny access to areas.
- **Echelon formation.** The echelon is an offensive formation used to turn or divert groups in open or built-up areas and to move crowds away from buildings, fences, and walls.
- **Wedge formation.** The wedge is an offensive formation that is used to penetrate and split crowds into smaller groups.
- **Diamond formation.** The diamond, when used as an offensive formation, is used to enter a crowd and is the formation of choice for extraction teams. As a defensive formation, the diamond is used when all-around security is required in open areas. The decision to use this formation is based on the conformation of the crowd.
- **Circular formation.** The circular formation is used for the same purpose as the diamond formation. The decision to use this formation is based on the conformation of the crowd.

6-14. There are many suitable variations of control force formations that may be employed, but appropriate commands and signals must be created to execute the formations. Because of their somewhat complicated nature and the coordination required for these formations, new variations must be trained, practiced, and rehearsed before they are used in a civil disturbance.

VEHICLES AND FORMATIONS

6-15. Vehicles may be employed with troops in control force formations, especially when a large riotous crowd is on hand. Although vehicles add strength to formations, certain precautions must be taken. Covering the windshield with sturdy, close mesh fencing and the standard safety glass will offer some level of protection to the occupants. Shields and/or mobile barriers may be built by mounting a wooden or metal frame strung with barbed wire across the front of a vehicle. Members of the formation should walk as near to the front corners of each vehicle as possible to keep rioters from attacking the sides and rear of the vehicles.

6-16. Up-armored HMMWVs, ASVs, or armored vehicles may be employed with the control force formation when determined by METT-TC. They present a strong psychological effect and offer protection for the occupants. Wheeled armored vehicles are best because they do not deface the pavement. Whenever vehicles are used in a crowd control formation, the commander must ensure that he is still able to see and control the formation. This may require him to occupy a position in a similar vehicle behind the formation or command subordinate leaders to move away from the front line and into line-of-sight (LOS) control. This will require additional hand-and-arm signals to allow the commander and subordinate leaders to communicate. Commanders can effectively use vehicles as primary communication tools by using horns and lights as signals to organize movement or actions. However vehicles are used, it takes a good measure of training and practice to be effective.

ELEMENTS OF A CONTROL FORCE FORMATION

6-17. Four elements make up the basic crowd control formation. They are as follows:

- **Base element.** The base element is made up of two ranks. The first rank is shield holders, while the second rank contains the NLW. This is the front line of the formation.
- **Support element.** The support element forms up in a column formation behind the base element and has a variety of uses. It may be used to replace base element members as needed, provide lateral or direct support, or perform extraction team operations. The support element helps the base element by performing the following three essential tasks:
  - **General support.** The general-support element is formed from an uncommitted squad in the platoon (generally the 2d squad). When a company is tasked as the control force element, one of the platoons becomes the general-support element. The element is in a column formation centered on and behind the main formation. From this formation, the general-support element can move to lateral or direct support, as needed.
  - **Lateral support.** The lateral-support element is used to protect the flanks of the formation. This is done by moving a set number of teams forward from the general-support element or by using the

end teams in the formation; however, this will make your formation smaller. Once in position, these teams become part of the formation, with their riot shields facing the flanks of the formation.

- **Direct support.** When direct support is ordered, the general-support element moves forward as they form their own formation. While moving forward, the squad and team leaders from the control force formation step back and allow the riot shield holders from the support element to step behind riot shield holders of the original formation. This allows the support riot shield holders to form up between and behind the riot shield holders of the original formation. Now there are two lines of riot shield holders between the crowd and NLW firers. This formation is the strongest of the formations and requires more planning and practice to master.

- **Command element.** The command element contains several different members. A general configuration for the command element is the platoon leader, the PSG, a radio operator, a video recorder operator (if required), and an interpreter (if required). This element does not have a fixed location within the formation and moves about as needed.
- **Lethal overwatch element.** The lethal overwatch element is a team formed from reserve security forces.

6-18. The reserve support element is not part of the control force formation until it is brought forward from the rally point to join the formation. It remains until released by the formation commander. Lethal overwatch teams are formed from the reserve force. Once lethal overwatch teams are deployed and in position, they are under the control of the formation commander. Communication between the lethal overwatch team and the formation commander is a priority because he approves target selection and engagement.

WEAPON POSITIONS

6-19. At the commander's discretion, members of the control force formation may carry their weapons in any of the positions prescribed in *FM 3-21.5*. However, the primary method of carrying weapons in the control force formation is at the safe-port position (see *Figure 6-3*). It allows the soldier to control both ends of the weapon while moving in and out of the formation and advancing on the crowd.

6-20. High-port position (see *Figure 6-3*) is a position of complete readiness. It is used whenever troops are in contact with a crowd that is showing resistance or not withdrawing. This is a tiring position and is hard to maintain for extended periods. Commanders must rest their troops at every opportunity by using less tiring positions.

Safe-port arms | High-port arms

Figure 6-3. Weapons Positions

COMMANDS

6-21. Commands are given to the control force formation orally or with hand-and-arm signals. Oral commands for the on-guard position are given in one count. All other commands are given in two counts, a preparatory command followed by a command of execution. However, oral commands cannot be relied on completely, so commanders must plan to use hand-and-arm signals too. When vehicles are part of the formation, radios become an alternate method for relaying commands. To improve communication, commanders may use hand-and-arm signals (see *Figure 6-4,* page 6-8) with oral commands. The basic commands used in control force formations are outlined in *Table 6-1,* page 6-9. These commands can be added to or combined to form more complex commands, therefore forming more complex formations.

MOVEMENTS

6-22. Control force formations move in the same manner as regular formations. This allows the commander to more precisely maneuver the formation by commanding the number of steps it should move and in what direction, such as "Five Steps, Forward March." The normal rate of march for entering and leaving a control force formation is double-timing. The half-step march is used when the formation is in direct contact with the crowd. This slows the formation down and allows for better command and control. The squad and team leaders echo preparatory commands and provide cadence while in control force formations.

INTERVAL AND DISTANCE

6-23. Interval is the lateral space between elements. Distance is the space between elements in a column. The usual interval and distance between soldiers in control force formations is 30 inches. The interval and distance can be adjusted based on METT-TC. If the crowd has to be physically pushed back, a close interval is preferred. This allows the riot shields to be overlapped, creating a stronger wall. During peaceful demonstrations, the interval could increase to double arm width, allowing the formation to cover more area and still control the crowd. The interval and distance could change several times during the course of operations before the crowd is dispersed.

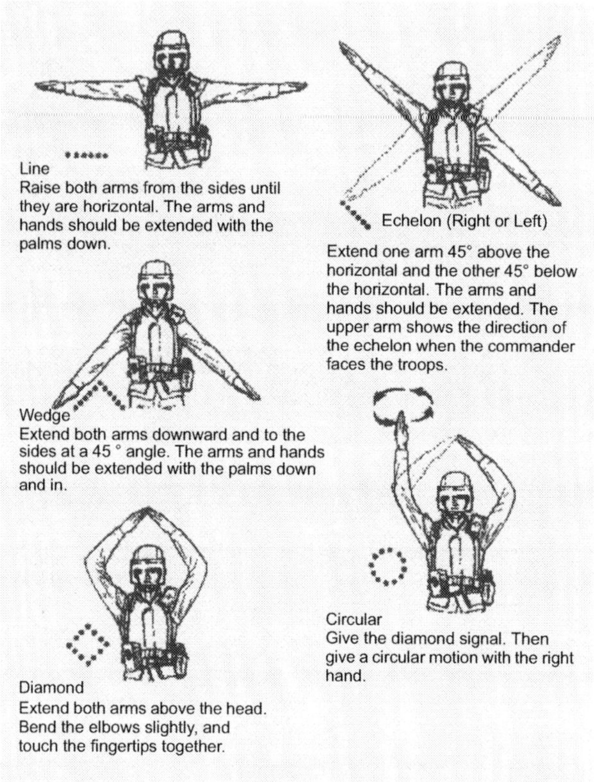

Figure 6-4. Hand-and-Arm Signals

Table 6-1. Basic Commands

Commands for	Command	Given by	Purpose
Formations	Platoon On Line	Platoon leader and/or commander	To form a line formation.
	Platoon Wedge	Platoon leader and/or commander	To form a wedge formation.
	Platoon Echelon	Platoon leader and/or commander	To form an echelon formation.
	Platoon Diamond	Platoon leader and/or commander	To form a diamond formation.
	Platoon Circle	Platoon leader and/or commander	To form a circular formation.
	Move	Platoon leader and/or commander	As a command of execution. **NOTE: The platoon leader and/ or commander also identifies the location for the formation by pointing his arm to the desired location for the formation.**
Support	General Support	Platoon leader and/or commander	To place a specified unit in the rear of the base element.
	Lateral Support	Platoon leader and/or commander	To place support elements on the left or right flank of the base element.
	Direct Support	Platoon leader and/or commander	To move support elements forward to strengthen the base element.
	Open	Extraction team leader	To open a space in the formation.
	Up	Extraction team leader	To inform the extraction team to start moving back to the main formation.
Fire	Weapon System Gunners, Number of Rounds, Type of Round, and Prepare to Fire	Platoon leader and/or commander	To ready NL gunners to fire weapons; for example, M203 gunners, three rounds, area target, prepare to fire.
	Shield Down	NL gunners	To tell riot shield holders to get down on one knee and lower the riot shield to the ground.
	Shield Up	NL gunners	To tell riot shield holders to stand and raise the riot shield.
Weapons	Port Arms	Platoon leader and/or commander	To use when not in contact with the crowd.
	High Port	Platoon leader and/or commander	To tell NL gunners what position to hold their weapons in while the formation is in physical contact with the crowd. This allows for rapid targeting and firing of NL munitions.

EXTRACTION TEAM OPERATIONS

6-24. The extraction team is a squad (see *Figure 6-5*) that is generally formed from the support element, but could also be formed from reserve security forces. Once the team is formed, it is OPCON to the formation commander until the team's mission is complete and it returns to the rear of the formation. This team provides the riot control formation with the means to employ NL and lethal cover forward of the formation. It also provides the formation with the ability to remove and detain rioters downed by munitions.

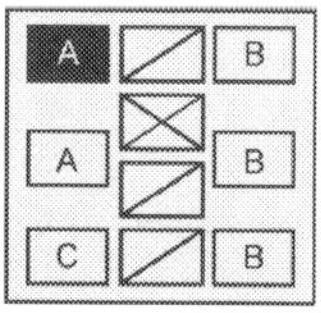

Figure 6-5. Extraction Team Formation

The extraction team—

- Extracts vehicles or personnel that are in immediate danger from the crowd.
- Detains and escorts downed rioters to the rear of the formation.
- Restrains and searches subjects.
- Covers confined or congested areas where a full riot control formation cannot be inserted.

6-25. The extraction team is usually a squad. While the recommended distance for deploying the extraction team from the control force formation is no farther than 10 meters, this may change based on METT-TC. The control formation protects the extraction team and provides immediate support, if necessary.

6-26. The extraction team may deploy from anywhere within the formation. When given orders to deploy, the squad leader of the extraction team gives the formation commands, identifies at least two personnel in the squad to conduct the search and apprehension of the subject, and leads the squad to the base line of the formation. He then extends his arm between the two riot shield holders and commands, "Open," while tapping the riot shield holders on the side of the shoulder. The two riot shield holders take one step backward and one step to the right or left. This clears a path for the extraction team. The extraction team leader then states the number of personnel leaving the formation, such as "Ten security personnel leaving." As the last extraction team member exits the formation he states, "Last Man." The two riot shield holders then return to their original position in the formation. The adjacent base element squad and/or team leaders pass the number of personnel in the team through the formation.

6-27. As the extraction team approaches an instigator or incapacitated demonstrator, the riot shield holders envelop the target and face in the direction of the nearest threat. NL gunners on the left and right cover their respective areas. The search and apprehension team immediately controls and restrains the target. To avoid confusion, one team member concentrates on controlling the subject while another team member applies a restraining system. When the search and apprehension team has the subject under control, the extraction squad leader sounds off with, "Up." Upon hearing the "Up" command, it is repeated by all extraction team members. Team members then grab a shoulder of the team member in front of them to backtrack into the formation. As the extraction team approaches the formation, the nearest base element leader will extend his arm and command, "Open," while tapping the riot shield holders on the side of the shoulder. The riot shield holders move one step backward and to the right or left to open the formation. As he enters the formation, the extraction team leader sounds off with the number of personnel entering the formation, such as "Ten security personnel and one demonstrator." The adjacent base element leader counts the number of personnel as they enter the formation. When the last man enters the formation, he states, "Last Man." The riot shield holders then return to their original places in the formation. The adjacent squad and team leaders in the base element pass information through formation that the extraction team has returned.

LETHAL OVERWATCH TEAMS

6-28. During a NL engagement, the use of a DM team provides confidence and safety to those facing the mob. If a lethal threat is presented, the DM team in an overwatch position (armed with a standard infantry rifle that is mounted with a high-powered scope) can scan the crowd, identify agitators and riot leaders for apprehension, or fire lethal fire if so ordered or warranted. It is also ideally suited for flank and countersniper operations.

6-29. The DM team sets the security overwatch and provides real-time intelligence and reporting that is vital to mission success. It provides coverage during the entire approach to the crowd, and its ability to select positions and provide cover fire (if and when warranted) is critical to the safety of the control force. Successful execution requires training, practice, and rehearsal with the control force formation.

6-30. The DM team is organized from a standard three- or four-man team. Each individual on the team is equipped with a standard-issue weapon. One individual has an infantry rifle with a scope, another has a set of binoculars, and two other individuals provide security for the team. Each team is equipped with a radio for communication with the HQ element of the control force formation.

6-31. The teams must have a visual advantage over the crowd to provide lethal protection to the formation. This is best accomplished by placing them on nearby rooftops, in the upper floors of buildings, and on hilltops. They must have the optical equipment to identify mob leaders, instigators, and/or individuals with weapons. Team integrity must be maintained. In organizations with three-man teams, one individual becomes the marksman, another the observer, and another the security. When teams have a 4th member, that individual becomes the video camera operator. This will help identify individuals for later apprehension and provides a record of events.

6-32. Two teams should be deployed to cover the flanks of the formation. Each team should be deployed to a position that is in front of the formation and in an overwatch position. As the formation moves forward, the teams leapfrog forward from the formation.

6-33. The reserve forces, located at the rally point, can provide another lethal force of some size that can be brought forward, as needed. Each soldier in the control force formation should have a sidearm (9-millimeter pistol) or an M16 or M4 rifle. The rifle is slung diagonally across his back with the rifle butt over his left shoulder and the muzzle below his right hip. Lethal munitions are in a separate ammunition pouch that is isolated from NL munitions. This is to prevent the mixing of NL and lethal ammunition.

6-34. Once the lethal overwatch teams are deployed, they are under the OPCON of the formation commander. The lethal overwatch element must have direct and secure communications with the formation commander. The number of teams deployed is based on METT-TC. Teams will leapfrog to keep ahead of the advancing control force formation with one team in the overwatch position at all times.

RESERVE FORCES

6-35. A large reserve of soldiers should be maintained during civil disturbance operations. Knowing that a large reserve force is available provides confidence and safety within the control force and helps prevent them from overreacting to provocative acts with disorderly and criminal elements in the crowd. When determining the number of reserve forces required, it should be resolved in favor of a large number. Lethal overwatch teams, forces to augment the control force, and apprehension teams come from the reserve forces.

6-36. Apprehension teams are a must when it is likely that a large number of people will be apprehended. These teams provide an organized response to what could be a chaotic situation. The teams can be organized at the team or squad level, depending on the number of apprehensions expected. These teams usually operate behind the base element. During peaceful demonstrations, these teams can be sent forward to remove demonstrators who refuse to disperse.

6-37. Each team consists of an apprehension, movement, and security element (added if the control force cannot provide security); a recorder; and a video or still photographer. The apprehension element is an officer or NCO and a civilian police officer (CONUS or OCONUS) or US marshal. It decides who will be apprehended, informs the subject of the reasons, supervises the handling of the subject, and ensures that the apprehension is properly conducted and documented. The apprehension team is responsible for the overall apprehension process, limiting the number of individuals appearing in court to document the apprehension. The movement element helps the apprehension element handle the subject. They move, restrain, and search the individuals being apprehended under the supervision of the apprehension element. The security element prevents the crowd from interfering with the mission. The recorder documents the apprehension on *DD Form 2708*. When possible, the video and/or still photographer recorder photographs each subject with the apprehension element. This provides the apprehension element with the necessary documentation for use in court. One individual may need to conduct many of the duties of an apprehension team based on METT-TC.

6-38. The video and/or still photographer record the situation before, during, and after the apprehensions. These images may also be used as pictorial documentation for use in court. It is possible that the presence of video and still cameras actively recording the situation can cause crowd members to be less prone to unruly or criminal acts.

6-39. The behavior of the selected individual to be apprehended is observed to decide if the subject is violent, cooperative, or passive. Unless there is good reason to believe otherwise, the subject must be considered dangerous. The mood of the crowd is observed to estimate what it might do. Is it only curious onlookers or does it support the individual targeted for apprehension? Will it interfere?

6-40. When apprehending an individual, the actions must be kept as simple as the situation allows. The element of surprise can give the apprehension team an advantage over the apprehended individual. It is best not to use surprise if the individual might panic and injure someone. Select an area for the apprehension that is to the advantage of the team and not the individual. Avoid apprehensions in crowded areas as these areas provide many avenues of escape and contain supporters of the apprehended individual.

6-41. All verbal commands given to the apprehended individual must convincing authority with short, clear, and distinct commands. Voice, attitude, and deportment must convey an expectation that the apprehended individual will comply.

6-42. The individual should be searched immediately; however, circumstances (violence or sympathizers for the individual being apprehended) may dictate temporarily postponing the search until the individual can be moved to a safer area. Flex-Cufs or hand irons may be applied.

SQUAD FORMATIONS

6-43. Civil disturbance training is a collective task. Using team and squad concepts, team leaders instruct their teams and squad leaders instruct their squads. Safety is key when using force on force training for the control force formation. Soldiers must understand that serious injuries may occur if they do not follow guidelines set by the commander.

6-44. The smallest formation is a squad formation, which is used in back of the main formation to cover side streets. Squad members must know their positions regardless of which control force formation they are in, squad to company.

6-45. Training must enforce soldiers' understanding of the need to stay informed for individual protection. Individual soldiers are much more vulnerable to attack when they break ranks and chase after crowd members. When an individual soldier breaks ranks he not only puts himself in danger, but the entire formation.

6-46. When moving a squad into a crowd control formation from a column formation, the squad leader takes one or more steps to the right of the squad and faces it. He gives the command and hand-and-arm signal for the formation he wants. Then he indicates where he wants the formation located by pointing in that general direction. The A team leader commands, "Follow Me." He places the baseman at the position indicated by the squad leader. The squad forms on the baseman position.

LINE FORMATION

6-47. The command for forming a squad in a line formation is "Squad Line Formation, Move." On the command of execution, the A team leader leads the baseman to the place indicated by the squad leader. The squad members align themselves in sequence with the baseman at normal intervals (see *Figure 6-6*). Based on the situation, the squad leader may designate a specific interval. He does this when he gives the preparatory command. If no interval is specified, the squad automatically forms using the normal interval.

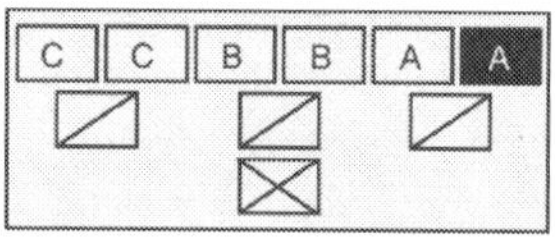

Figure 6-6. Squad Line Formation

ECHELON FORMATION

6-48. The command for having a squad form in an echelon formation is "Squad Echelon Right (or Left), Move" (see *Figure 6-7*). On the command of execution, the A team leader places the baseman at the location indicated by the squad leader. The squad members align themselves in sequence with the baseman, one pace to the side and one pace to the rear.

WEDGE FORMATION

6-49. The command for moving a squad formation into a wedge formation is "Squad Wedge, Move." On the command of execution, the A team leader moves the baseman to the place indicated by the squad leader. The A team lines up to the left of the baseman, one pace to the left and one pace to the rear of each preceding man. B team members align themselves with the baseman, one pace to the right and one pace to the rear of each preceding man. C team members align themselves in the same way to the left of the baseman (see *Figure 6-8*).

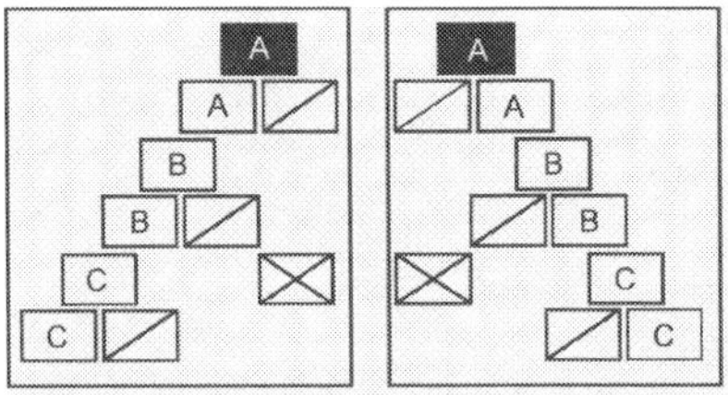

Figure 6-7. Squad Echelons Left and Right Formations

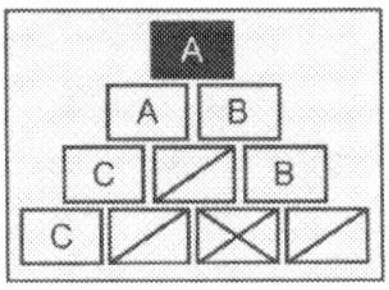

Figure 6-8. Squad Wedge Formation

ASSEMBLY

6-50. When assembling a squad from the line formation, the squad leader takes a position a sufficient distance to the rear of the squad and commands, "Squad Assemble." At the same time, he raises his right hand in the air and makes a circular motion. Squad members, who hear the command, automatically come to the port arms position. The A team leader and the baseman do an about-face movement.

6-51. The other squad members face toward the baseman. On the command of execution "Move," the squad leader points to the place where he wants the squad to assemble. The A team leader double-times to the designated spot, and the other members of the squad follow him. Team leaders fall into the file in their designated positions as they move to the AA. To assemble from a squad echelon formation, the steps are the same as a line formation.

6-52. To assemble from a squad wedge formation, the squad leader takes the same steps as for the line and echelon formations. B team members do a half left face, and the A team and C team members do a half right face. The C team pauses at the baseman position and allows the B team to clear the formation before moving to the AA.

DIAMOND AND CIRCLE FORMATIONS

6-53. The diamond and circle formations are used during extraction team operations. They are small formations used to penetrate the crowd or cover small areas. The decision on which formation to use is based on METT-TC and the ability of the squad to perform.

6-54. Commands for forming a squad in a diamond or circle formation is "Squad Diamond (or Circle), Move." On the command of execution, the A team leader moves the baseman to the place indicated by the squad leader. The A team lines up to the left of the baseman, covering the 8 to 12 o'clock positions. B team members align themselves to the right of the baseman, covering the 12 to 4 o'clock positions. C team members align themselves covering the 4 to 8 o'clock positions (see *Figure 6-9*).

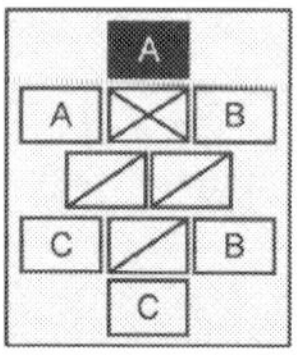

Figure 6-9. Squad Diamond or Circle Formation

6-55. To assemble from a squad diamond or circle formation, the squad leader takes the same steps as he would for a line formation. The A team leads the squad to the AA followed by the B team and then the C team. Teams fall into a column formation as they file back to the AA.

THREE-SQUAD PLATOON FORMATIONS

6-56. The platoon HQ consists of the following individuals: the platoon leader, the PSG, and a messenger or radio operator. When directed by the platoon leader, other individuals such as the selected marksman, interpreter, RCA disperser operators, and firefighters may augment the platoon HQ. In forming all control force formations from the column, the platoon leader moves to the right of the platoon and faces them when giving his commands. He gives the hand-and-arm signal and the preparatory command. On the command of execution "Move," he points to the approximate location where he wants the platoon to form. The platoon leader pauses between the preparatory command and the command of execution to allow each squad leader to issue a preparatory command to his squad.

Platoon Line Formation

6-57. The command for forming a platoon line is "Platoon on Line, Move." Immediately following the platoon leader's preparatory command, the squad leaders of the 1st and 3d squads give the command, "Follow Me." The leader of the 2d squad gives the command, "Stand Fast." On the platoon leader's command of execution, the 1st and 3d squads move forward to the point designated by the platoon leader. The 1st team member of the 1st squad is the baseman for the platoon formation. The 1st squad forms a squad line to the left of the baseman while the 3d squad forms a squad line to the right of the baseman and dresses with the 1st squad. After the 1st and 3d squads have cleared the platoon, the 2d squad leader gives the command, "Follow Me." The 2d squad splits and the A and B teams form a line to the right of the 3d squad.

The C team (with the squad leader) forms a line to the left of the 1st squad (see *Figure 6-10*).

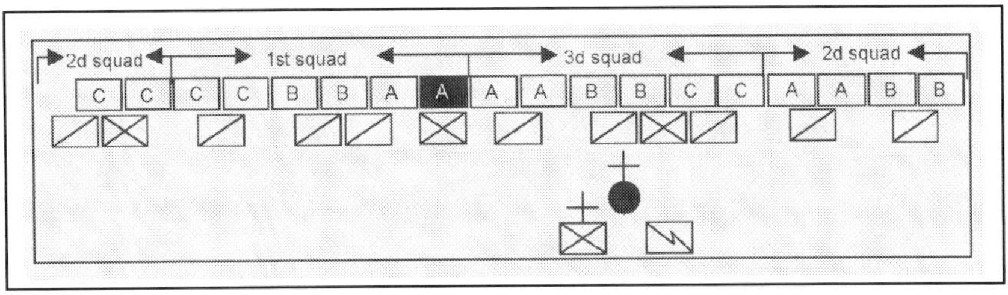

Figure 6-10. Platoon Line Formation

Assembling From a Platoon Line Formation

6-58. When assembling the platoon from the platoon line formation, the platoon leader and his HQ personnel take a position to the rear of the platoon. He then gives the command, "Platoon Assemble," while raising his right arm and making a circular motion above his head. Platoon members automatically assume the port arms position. The 1st squad leader and the baseman do an about face. The 3d squad leader moves to the head of his squad. All other members do an about-face movement toward the baseman. On the platoon leader's command of execution, the platoon leader points to where he wants the platoon to assemble. The 1st squad leader gives the command, "Follow Me," and moves his squad toward the AA. The squad moves to make a single file line to the baseman position followed by the C team of the 2d squad. As the 1st squad clears the baseman position, the 2d squad leader comes to a halt at the baseman position. The 3d squad leader gives the command, "Follow Me," and leads his squad to the AA followed by the A and B teams of the 2d squad. The A team leader of the 2d squad comes to a halt at the baseman position. Once the 3d squad has cleared the baseman position, the 2d squad leader gives the command, "Follow Me," and leads his squad to the AA. The 2d squad reassembles into their original column formation as they move to the AA. As the 1st squad comes to a halt (facing the crowd) in the AA, the 3d squad dresses with the 1st squad and leaves space for the 2d squad to return to its place in the formation. Once the platoon is formed into a column formation, HQ personnel take their position in the formation. The PSG then gives the command, "Countercolumn March," and moves the platoon to the rally point.

Platoon Line Formation With General Support

6-59. Whenever the command for establishing a crowd control formation contains the phrase "In Support" (without modification), it means that the support element is to remain in general support.

6-60. The command for forming a platoon line with the 2d squad in general support is "Platoon Line, 2d Squads in Support, Move." The 1st and 3d squads of the platoon line with general support execute the line, while the 2d squad remains in the column (see *Figure 6-11*). To assemble a platoon with a squad in general support, the procedure is the same as the platoon line, except that the 1st and 3d squads dress with the 2d squad.

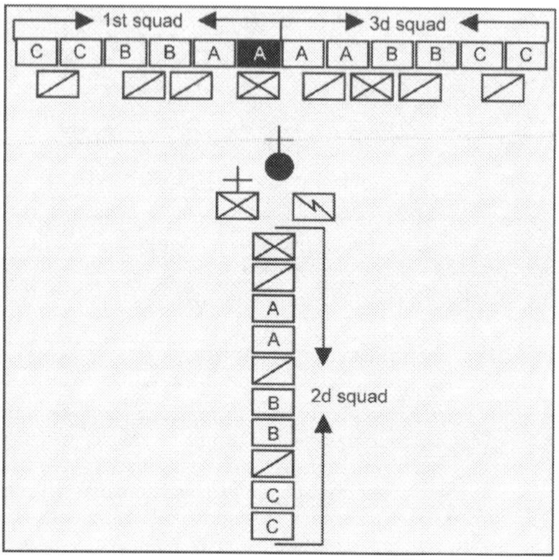

Figure 6-11. Platoon Line Formation With General Support

Platoon Line Formation With Lateral Support

6-61. The command for forming a platoon line with one squad in lateral support is "Platoon Line, 2d Squad in Lateral Support, Move." The 1st and 3d squads execute the line as before, and the 2d squad stands fast. After the line has been formed, the 2d squad leader gives the command, "Follow Me." The A and B teams move to the right flank while the squad leader and the C team move to the left flank of the formation. The 2d squad forms columns behind the last men on the line formation. If no direction is given, the 2d squad splits to cover both flanks. A squad may be used to cover one flank by giving the command, "2d Squad Right (or Left) Lateral Support" (see *Figure 6-12*).

6-62. The 2d squad may be moved from general to lateral support at any time by the platoon leader. The platoon leader commands, "2d Squad, Lateral Support, Move." To have the 2d squad join the line from either support position, the platoon leader commands, "2d Squad, Extend the Line, Move." The 2d squad leader commands, "Follow Me." On the command of execution, the squads set up individual lines, closing and dressing on the existing line. To assemble the platoon from a platoon line with lateral support, follow the same steps as for the platoon line.

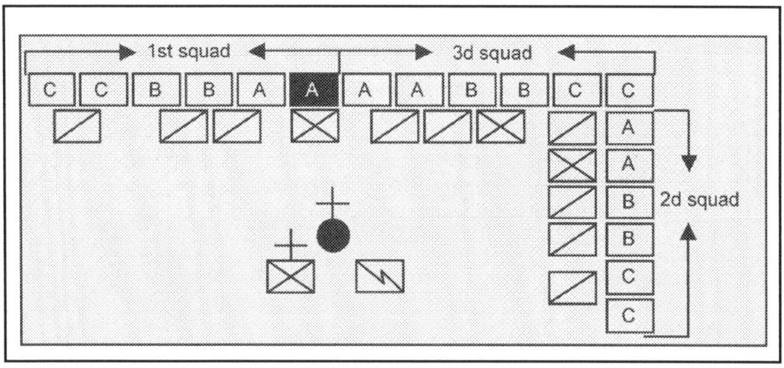

Figure 6-12. Platoon Line Formation With Lateral Support (Right)

Platoon Line Formation With Direct Support

6-63. The command for forming a platoon line with one squad in direct support is "Platoon Line, 2d Squad in Direct Support, Move." The 1st and 3d squads execute a line as before. The 2d squad executes a squad line directly to the rear and centered on the 1st line. The men in the supporting line cover the intervals between the men in the 1st line. To do this, the squad and team leaders of the 1st and 3d squads must take two steps to the rear to allow space for the 2d squad to move into place. After seeing the squad and team leaders move back from the line, the 2d squad leader commands, "Follow Me," and leads his squad into position (see *Figure 6-13*).

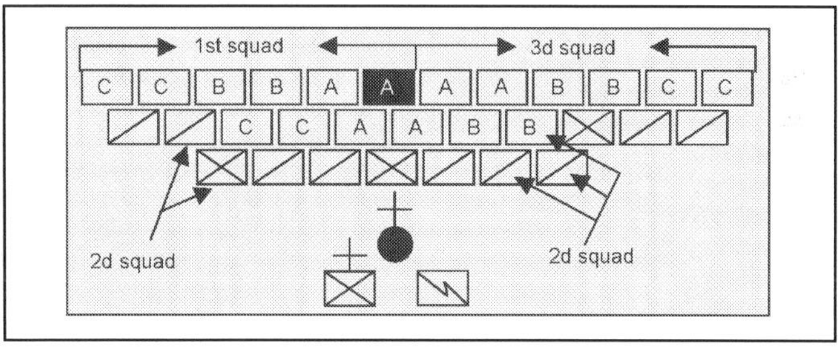

Figure 6-13. Platoon Line Formation With Direct Support

6-64. To assemble a platoon with direct support, the procedure is the same as the platoon line, except that the 2d squad moves to the AA first. To assemble the support squads from any position to general support, the platoon leader commands, "2d Squad in Support, Move." The 2d squad then returns to a column behind the line formed by the other two squads.

Platoon Echelon Right and Left Formations

6-65. The command for forming a platoon echelon right is "Platoon Echelon Right, Move." After the platoon leader's preparatory command, the squad leader of the 1st squad commands, "Follow Me." The squad leaders of the 2d and 3d squads command, "Stand Fast." On the command of execution, the 1st squad moves out and executes an echelon right at the location designated by the platoon leader. As the 1st squad clears the column, the 3d and 2d squads extend the echelon. The 2d squad performs support to this formation in the same ways it supports a line formation.

6-66. The command for forming a platoon echelon left is "Platoon Echelon Left, Move." The platoon echelon right is formed in the same manner as the echelon left (see *Figure 6-14*). *Figures 6-15; 6-16,* page 6-22; and *6-17,* page 6-23 show support formations in a left echelon.

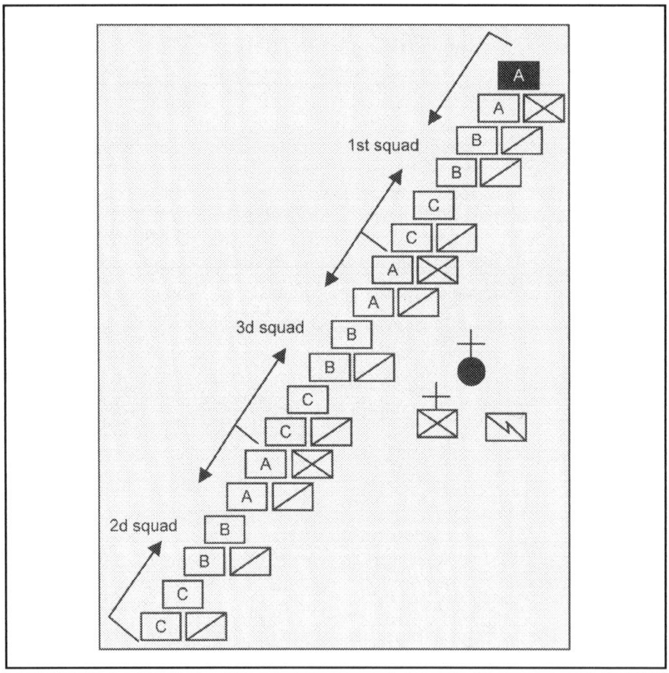

Figure 6-14. Platoon Echelon Left Formation

6-67. To assemble the platoon, the commander takes his position to the rear of the formation with the platoon HQ. On the preparatory command, "Platoon Assemble," the 1st squad leader does an about-face movement while all other

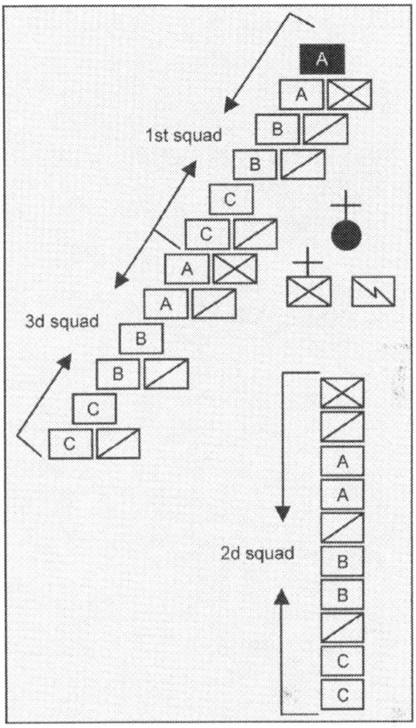

Figure 6-15. Platoon Echelon Left Formation With 2d Squad in General Support

members assume the port arms position and face the baseman. On the command, "Move," the 1st squad moves first, followed by the 3d and 2d squads. Each squad dresses with the 1st squad as it reaches the AA. The assembly of the echelon right formation is the same as the echelon left.

Platoon Wedge Formation

6-68. The command for forming the platoon wedge formation is "Platoon Wedge, Move." After the platoon leader's preparatory command, the squad leaders of the 1st and 3d squads command, "Follow Me." At the same time, the 2d squad leader commands, "Stand Fast." On the platoon leader's command of execution, the 1st and 4th squads move directly to the front. The 1st squad executes an echelon left on the baseman while the 3d squad executes an echelon right. When the last man of the 1st and 4th squads clear the 2d squad, the 2d squad leader commands, "Follow Me," and moves out to the left and right. The 2d squad splits and the A and B teams go to the right and the C

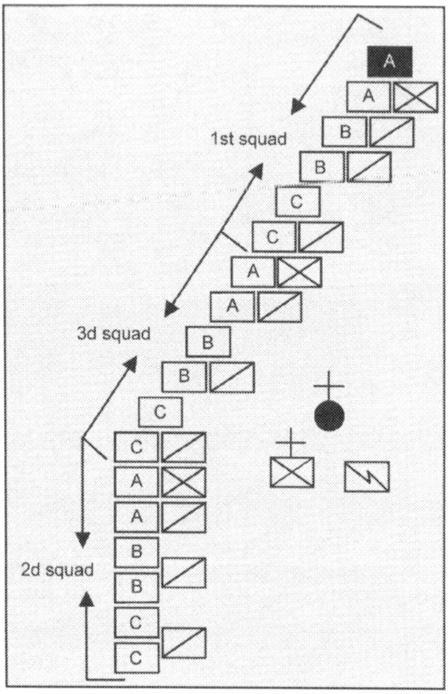

Figure 6-16. Platoon Echelon Left Formation With 2d Squad in Lateral Support

team goes to the left to extend the formation (see *Figure 6-18*). Assembly of the wedge formation is the same as the platoon line.

Platoon Wedge Formation With General Support

6-69. The command for forming a platoon wedge formation with support is "Platoon Wedge, 2d Squad in Support, Move." The 1st and 3d squads execute the wedge while the 2d squad remains in a column formation (see *Figure 6-19,* page 6-24).

Platoon Wedge Formation With Lateral Support

6-70. The command for forming a platoon wedge formation with lateral support is "Platoon Wedge, 2d Squad in Lateral Support, Move." The 1st and 3d squads execute the wedge while the 2d squad stands fast. After the 1st and 3d squads have cleared the column, the 2d squad leader commands, "Follow Me." If no direction is given, the 2d squad splits to cover both flanks. A squad

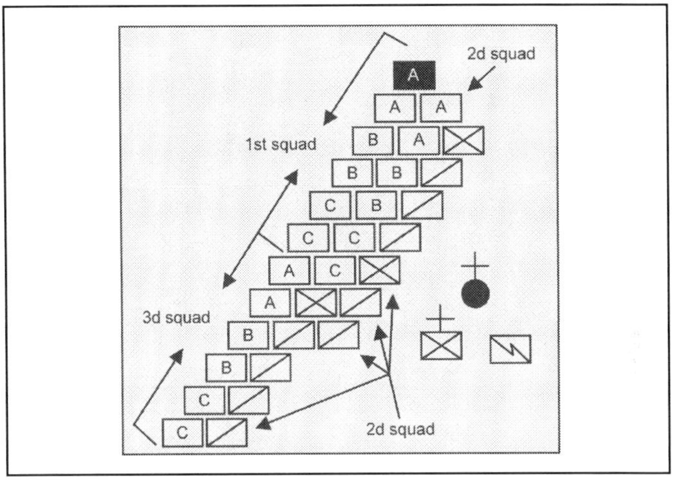

Figure 6-17. Platoon Echelon Left With 2d Squad in Direct Support

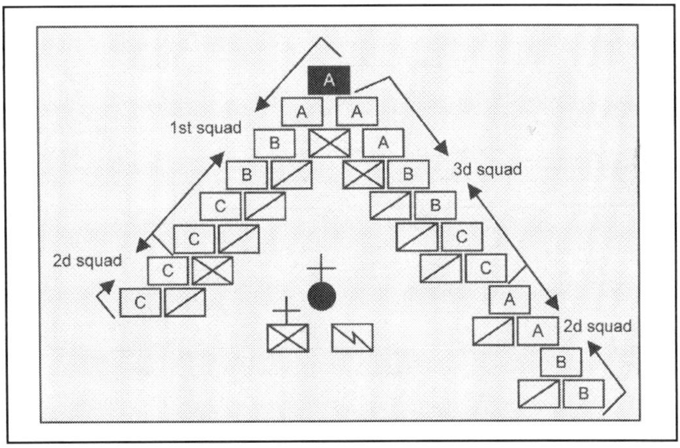

Figure 6-18. Platoon Wedge Formation

may be used to cover one flank by commanding, "2d Squad Right (or Left) Lateral Support" (see *Figures 6-20* and *6-21,* page 6-25).

6-71. To move the 2d squad from general support to lateral support, the platoon leader commands, "2d Squad, Lateral Support, Move." To have the 2d squad join the wedge from general or lateral support, the platoon leader

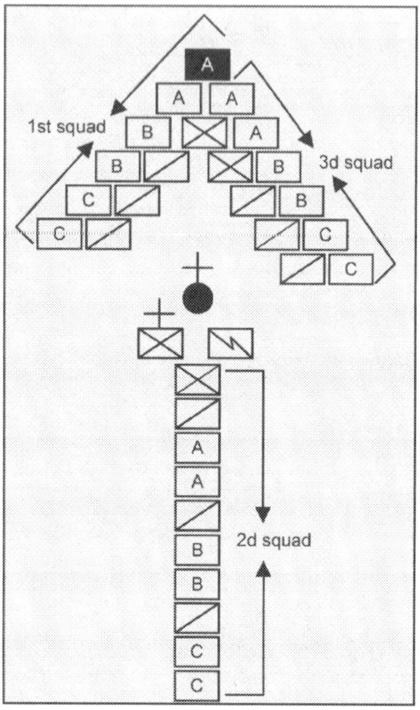

Figure 6-19. Platoon Wedge Formation With 2d Squad in General Support

commands, "2d Squad, Extend the Wedge, Move." To extend the wedge, the 2d squad splits and the teams move to the left and right, respectively.

Platoon Wedge Formation With Direct Support

6-72. The command for forming a platoon wedge formation with the 2d squad in direct support is "Platoon Wedge, 2d Squad in Direct Support, Move." The 1st and 3d squads execute a wedge. The 2d squad waits for the squad and/or team leaders to move to the rear, as in the line formation, before executing a wedge directly behind and centered on the main formation. The individuals in the supporting wedge cover the intervals between individuals in the leading wedge (see *Figure 6-22,* page 6-26).

Platoon Diamond and Circle Formations

6-73. These formations are used in open areas or where the crowd may be able to envelop the formation. The diamond and circle formations are primarily defensive in nature and difficult to maneuver. The decision about which formation to use is based on the ability of the platoon to perform them.

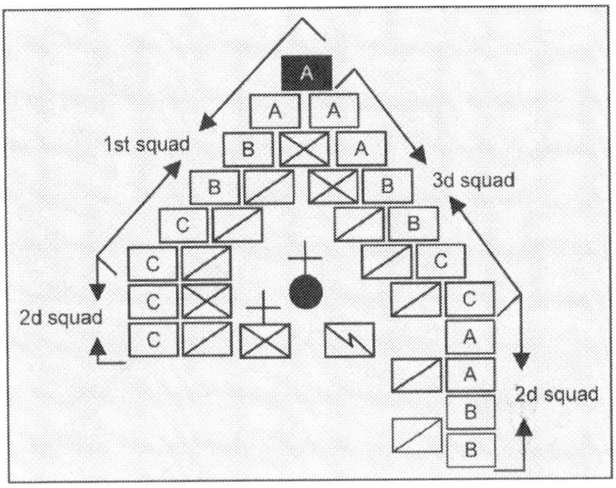

Figure 6-20. Platoon Wedge Formation With 2d Squad in Lateral Support

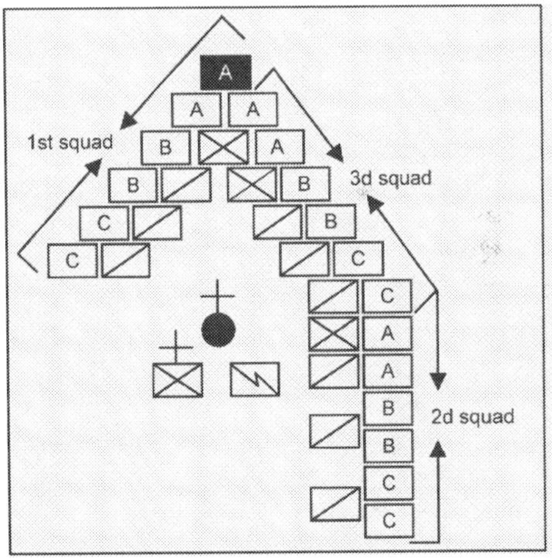

Figure 6-21. Platoon Wedge Formation With 2d Squad in Right Lateral Support

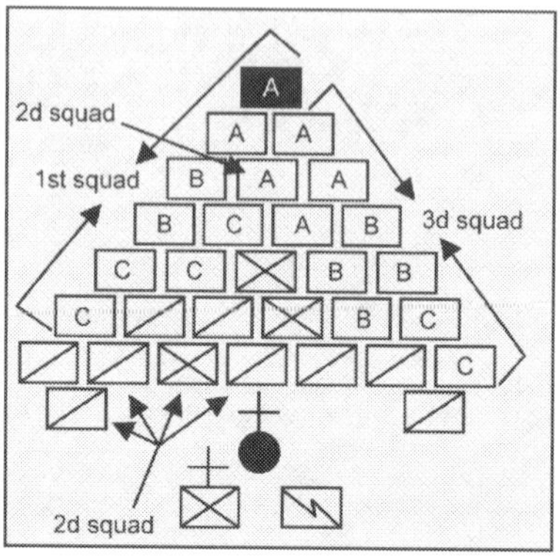

Figure 6-22. Platoon Wedge Formation With 2d Squad in Direct Support

The command for having a platoon form in a diamond or circle formation is "Platoon Diamond (or Circle), Move." On the command of execution, the 1st squad leader moves the baseman to the place indicated by the squad leader. The baseman becomes the 12 o'clock position while the 1st squad lines up to the left of the baseman covering the 8 to 12 o'clock positions. The 3d squad aligns itself to the right of the baseman, covering the 12 to 4 o'clock positions and the 2d squad aligns itself, covering the 4 to 8 o'clock positions (see *Figures 6-23* and *6-24*).

6-74. To assemble from a platoon diamond or circle formation, the HQ element takes the same steps as that of a line formation. The 1st squad leads the platoon back to the AA followed by the 2d and 3d squads. Squads fall into a column formation as they file back to the AA.

RELIEF ELEMENTS IN PLACE

6-75. During crowd control operations, the platoon leader may rotate his squads to give them a rest. He is not limited to using only those squads specified in the preceding examples as the lead elements of the formations. He can replace the base elements by commanding direct support to the 1st squad and then directing the 1st squad into general support. The support element (the 2d squad) then passes through the 1st squad. Once the 2d squad is formed directly behind the 1st squad, the 1st squad leader commands, "Right Face." The members of the 1st squad do a right-face movement and the 2d squad steps quickly though the 1st squad and forms the base formation. The 1st squad leader then commands, "Follow Me," and leads his squad to the general support position. This procedure is often necessary when protective masks are used and chemical irritants employed. The 1st squad then replaces the 3d squad in the same manner. Platoons participating in company operations are relieved in the same manner.

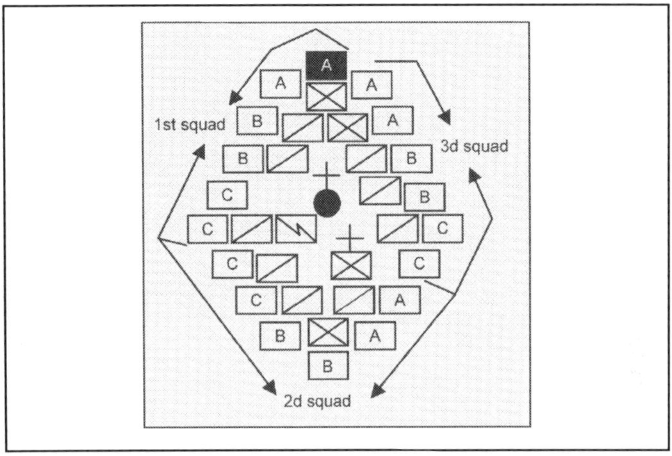

Figure 6-23. Platoon Diamond Formation

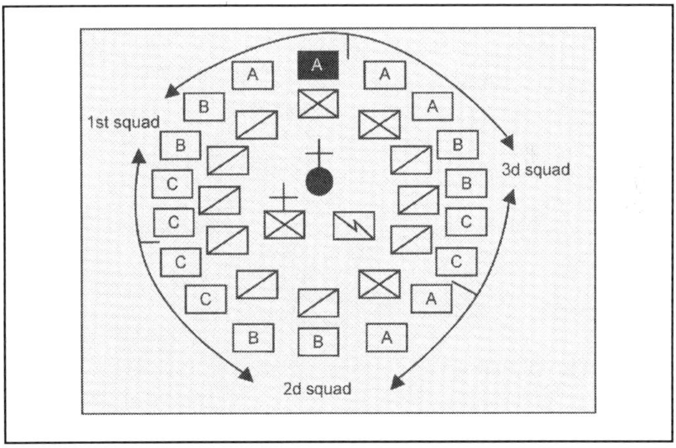

Figure 6-24. Platoon Circle Formation

FOUR-SQUAD PLATOON FORMATIONS

6-76. When forming control force formations from the column, the platoon leader moves to the right of the platoon and faces it when giving his commands. He gives the hand-and-arm signal and the preparatory command. On the command of execution, "Move," he points to the approximate location where he wants the platoon to form. The platoon leader pauses between the preparatory command and the command of execution to allow each squad leader to issue a preparatory command to his squad.

Platoon Line Formation

6-77. The command for forming a platoon line formation is "Platoon Line, Move." Immediately following the platoon leader's preparatory command, the squad leaders of the 1st and 4th squads command, "Follow Me." The squad leaders of the 2d and 3d squads command, "Stand Fast." On the platoon leader's command of execution, the 1st and 4th squads move forward to the point designated by the platoon leader. The 1st A team member of the 1st squad is the baseman for the platoon formation. The 1st squad forms a squad line to the left of the baseman and the 4th squad forms a squad line to the right of the baseman with all lines dressed to the right. After the 1st and 4th squads have cleared the platoon, the 2d and 3d squad leaders command, "Follow Me." The 2d squad forms a squad line to the left of the 1st squad. The 3d squad forms a squad line to the right of the 4th squad. All squads dress with the 4th squad (see *Figure 6-25*).

Platoon Line Formation Assembly

6-78. When assembling from the platoon line formation, the platoon leader and his HQ personnel take a position to the rear of the platoon. He then gives the command, "Platoon Assemble," while raising his right arm and making a circular motion above his head. Platoon members automatically assume the port arms position. The 1st squad leader and the baseman do an about-face movement. All other platoon members do an about-face movement toward the baseman. On the platoon leader's command of execution, he points to where he wants the platoon to assemble. He commands, "Follow Me," and moves his squad toward the AA. The 1st squad files to the baseman position, followed by the 2d squad. As the 2d squad clears the baseman position, the 4th squad leader then commands, "Follow Me," and leads his squad to the AA, followed by the 3d squad. As the 1st squad comes to a halt facing the crowd in the AA, the 2d squad assumes its position to the right of the 1st squad. As the 4th squad assembles, it dresses on the 1st squad, leaving space for the 3d squad to return to its place in the formation. Once the platoon is formed into a column formation, HQ personnel take their position in the formation. The PSG then commands, "Countercolumn, March," and moves the platoon to the rally point.

Platoon Line Formation With General Support

6-79. Whenever the command for establishing a crowd control formation contains the phrase "In Support Without Modification," it means that the support element is to remain in general support. The command for forming a platoon line with two squads in general support is "Platoon Line, 2d and 3d Squads in Support, Move." The 1st and 4th squads execute the line, while the 2d and 3d squads remain in the column (see *Figure 6-26,* page 6-30). To assemble a platoon with two squads in general support, the procedure is the same as that of the platoon line, except that the 1st and 4th squads dress on with the 2d and 3d squads.

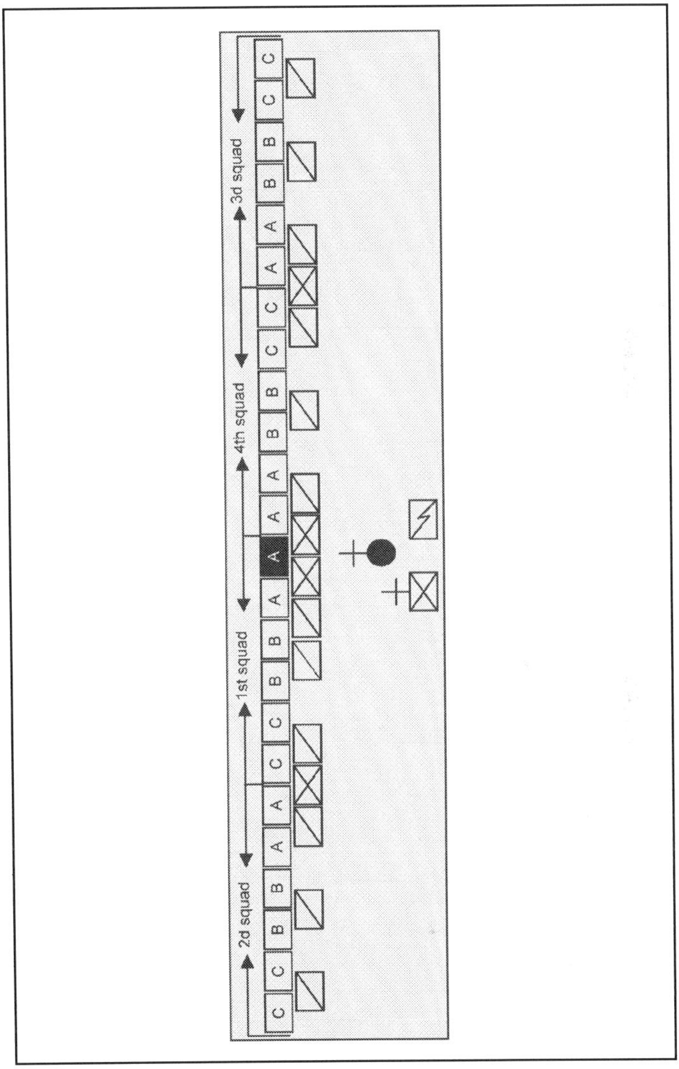

Figure 6-25. Platoon Line Formation With Four Squads

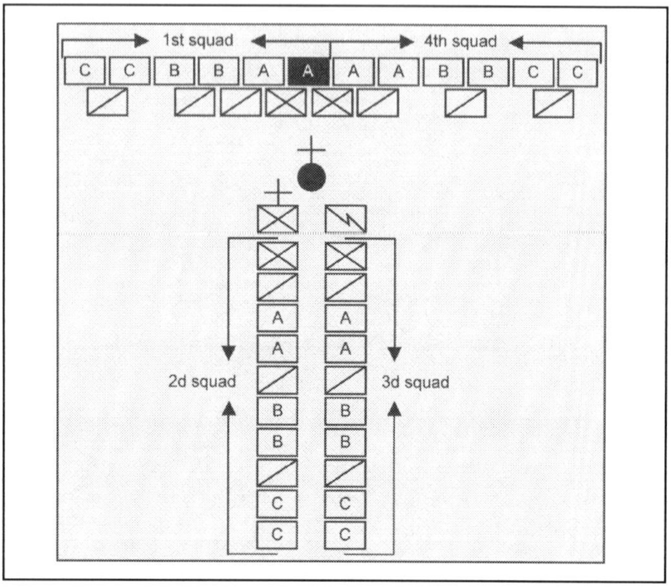

Figure 6-26. Platoon Line Formation With d and 3d Squads in General Support

Platoon Line Formation With Lateral Support

6-80. The command for forming a platoon line formation with two squads in lateral support is "Platoon Line, 2d and 3d Squads in Lateral Support, Move." The 4th and 1st squads execute the line as before, and the 2d and 3d squads stand fast. After the line has been formed, the squad leaders of the 2d and 3d squads command, "Follow Me." The 2d squad forms a column behind the last man of the 1st squad. The 3d squad forms a column behind the last man of the 4th squad (see *Figure 6-27*).

6-81. Squads may be moved from general support to lateral support at any time by the platoon leader. He commands, "2d and 3d Squads, Lateral Support, Move." To have the 2d and 3d squads join the line from either support position, the platoon leader commands, "2d and 3d Squads, Extend the Line, Move." The 2d and 3d squad leaders command, "Follow Me." On the command of execution, the squads set up individual lines, closing and dressing with the existing line. To assemble the platoon from a platoon line with lateral support, the procedure is the same as that of the platoon line, except lateral support squads face forward until they move to the AA.

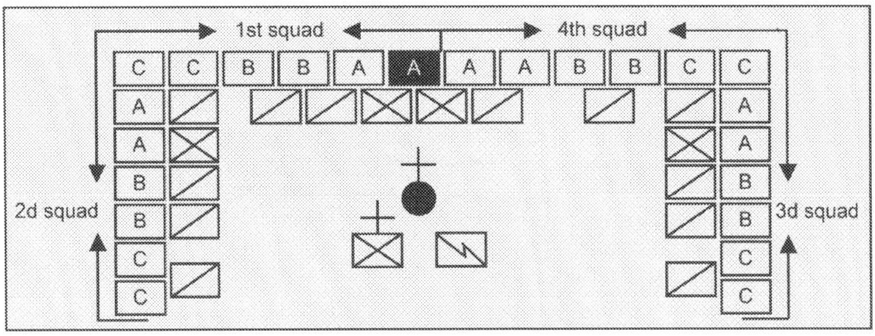

Figure 6-27. Platoon Line Formation With 2d and 3d Squads in Lateral Support

Platoon Line Formation With Direct Support

6-82. The command for forming a platoon line with two squads in direct support is "Platoon Line, 2d and 3d Squads in Direct Support, Move." The 1st and 4th squads execute a line as before. The 2d and 3d squads execute a 2d line directly to the rear and centered on the 1st line. The members in the direct support line cover the intervals between members in the baseline. To do this, the squad and team leaders of the 1st and 4th squads must take two steps to the rear to allow space for the 2d and 3d squads to move into place. After seeing the squad and team leaders move back from the line, 2d and 3d squad leaders command, "Follow Me," and lead their squads into position (see *Figure 6-28*).

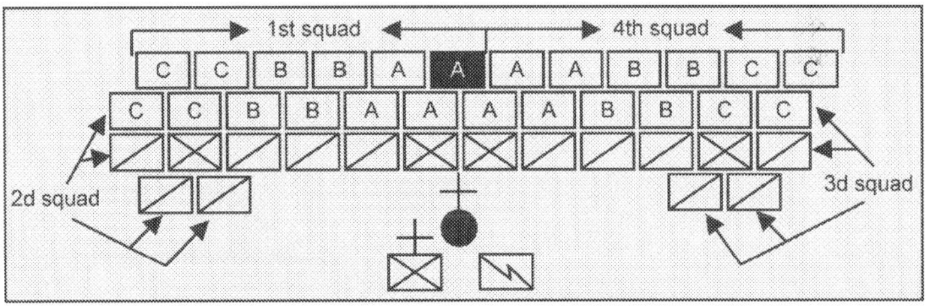

Figure 6-28. Platoon Line Formation With 2d and 3d Squads in Direct Support

6-83. To assemble the support squads from any position to general support, the platoon leader commands, "2d and 3d Squads in Support, Move." The 2d and 3d squads then return to a column behind the line formed by the 1st and 4th squads. To assemble the entire platoon, the procedure is the same as that of a platoon line, except that the 2d squad moves to the AA first, followed by the 3d, 1st, and 4th squads.

Platoon Line Formation With One Support Squad

6-84. The command for forming a platoon line formation with one squad in support is "Platoon Line, 3d Squad in Support, Move." The 1st squad moves out and executes a squad line. The 2d and 4th squads form a line on the side of the left and rights sides of the 1st squad. The 3d squad remains in column formation. The support squad may be used in lateral support on one or both sides of the formation or in direct support of any segment of the formation. To assemble the platoon, the procedure is the same as that of a platoon line, except that all squads dress with the 3d squad.

Platoon Echelon Right and Left Formations

6-85. The command for forming a platoon echelon right formation is "Platoon Echelon Right, Move." After the platoon leader's preparatory command, the squad leader of the 1st squad commands, "Follow Me." The squad leaders of the 2d, 3d, and 4th squads command, "Stand Fast." On the command of execution, the 1st squad moves out and executes an echelon right at the location designated by the platoon leader. As each squad clears the column, the next successive squad moves out and extends the echelon that has been formed. To assemble the platoon, the procedure is the same as that of a platoon line, except that all squads file back in order.

6-86. The command for forming a platoon echelon left formation is "Platoon Echelon Left, Move." The platoon echelon left is formed in the same manner as the echelon right, but in reverse order (see *Figure 6-29*). To assemble the platoon, the procedure is the same as that of an echelon right. See *Figures 6-30,* page 6-34; *6-31,* page 6-35; and *6-32,* page 6-36, for examples of echelon formations with support.

Platoon Wedge Formation

6-87. The command for forming the platoon wedge formation is "Platoon Wedge, Move." After the platoon leader's preparatory command, the 1st and 4th squad leaders command, "Follow Me." At the same time, the 2d and 3d squad leaders command, "Stand Fast." On the platoon leader's command of execution, the 1st and 4th squads move directly to the front. When the last members of the 1st and 4th squads clear the 2d and 3d squads, the 2d and 3d squad leaders command, "Follow Me," and move out to the left and right. The 1st squad executes an echelon left on the baseman and the 4th squad executes an echelon right. The 2d squad forms an echelon left on the last man of the 1st squad. The 3d squad forms an echelon right on the last man of the 4th squad. This completes the wedge formation (see *Figure 6-33,* page 6-36). To assemble the platoon, the procedure is the same as that of a platoon line.

Platoon Wedge Formation With General Support

6-88. The command for forming a platoon wedge formation with two squads in support is "Platoon Wedge, 2d and 3d Squads in Support, Move." The 1st and 4th squads execute the wedge while the 2d and 3d squads remain in column formation (see *Figure 6-34,* page 6-37).

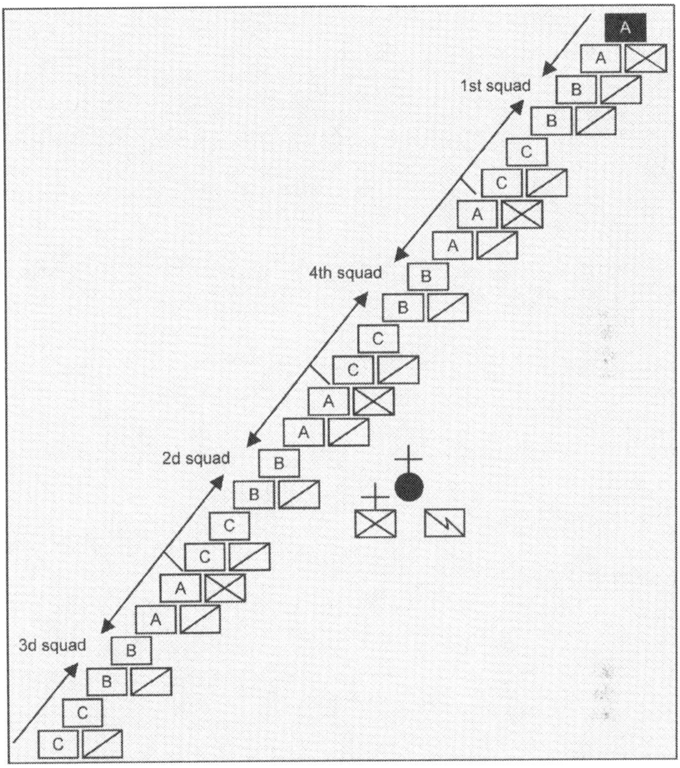

Figure 6-29. Platoon Echelon Left Formation With Four Squads

Platoon Wedge Formation With Lateral Support

6-89. The command for forming a platoon wedge formation with lateral support is "Platoon Wedge, 2d and 3d Squads in Lateral Support, Move." The 1st and 4th squads execute the wedge while the 2d and 3d squads stand fast. After the 1st and 4th squads have cleared the column, the 2d and 3d squad leaders command, "Follow Me," and move out to the left and right. The 2d squad forms a column formation behind the last man of the 1st squad. The 3d squad forms a column formation behind the last man of the 4th squad (see *Figure 6-35,* page 6-38).

6-90. To move the 2d and 3d squads from general to lateral support, the platoon leader commands, "2d and 3d Squads, Lateral Support, Move." To have the 2d and 3d squads join the wedge from general or lateral support, the

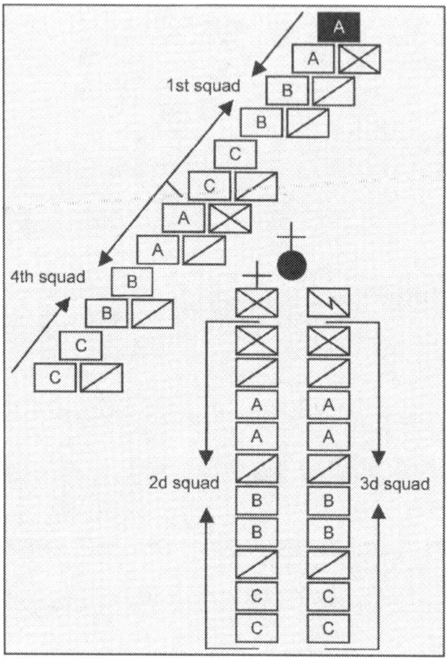

Figure 6-30. Platoon Echelon Left Formation With 2d and 3d Squads in General Support

platoon leader commands, "2d and 3d Squads, Extend the Wedge, Move." The 2d and 3d squad leaders command, "Follow Me," and move out to the left and right to extend the wedge on the 1st and 4th squads.

Platoon Wedge Formation With Direct Support

6-91. The command for forming a platoon wedge with two squads in direct support is "Platoon Wedge, 2d and 3d Squads in Direct Support, Move." The 1st and 4th squads execute a wedge. The 2d and 3d squads execute a wedge and close in on the leading wedge. The members in the supporting wedge cover the intervals between members in the leading wedge. To do this, the squad and team leaders of the 1st and 4th squads must take two steps to the rear to allow space for the 2d and 3d squads to move into place. After seeing the squad and team leaders move back from the line, the 2d and 3d squad leaders command, "Follow Me," and lead their squads into position (See *Figure 6-36,* page 6-38). To move the 2d and 3d squads back to general support, the platoon leader commands, "2d and 3d Squads Support, Move." Then the 2d and 3d squads move to their general support positions. To

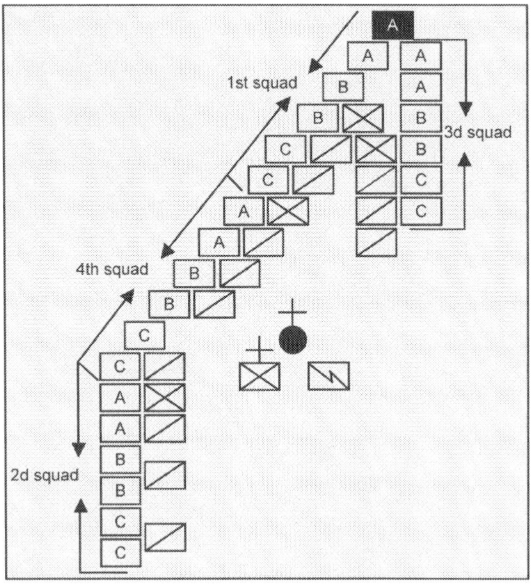

Figure 6-31. Platoon Echelon Left Formation With 2d Squad in Lateral Support

assemble a platoon from a platoon wedge with direct support, the procedure is the same as that of a platoon wedge, except that the 2d squad assembles first, followed by the 3d, 1st, and 4th squads.

Platoon Wedge Formation With One Support Squad

6-92. The command for forming a platoon wedge formation with one squad in support is "Platoon Wedge, 3d Squad in Support, Move." The 1st squad moves out and executes a squad wedge. The 2d and 4th squads form echelons left and right on the 1st squad. The 3d squad remains in the column formation.

Platoon Diamond and Circle Formations

6-93. The platoon diamond and circle formations are used in open areas or where the crowd may be able to envelop the formation. They are primarily defensive in nature and hard to maneuver. The decision concerning which formation to use is based on the ability of the platoon to perform them. The command for having a platoon form in a diamond or circle formation is "Platoon Diamond (or Circle), Move." On the command of execution, the 1st squad leader moves the baseman to the location indicated by the squad leader. The baseman becomes the 12 o'clock position. The 1st squad lines up to the

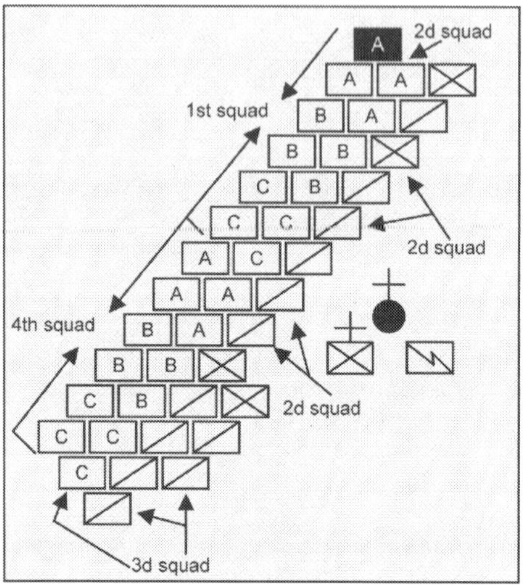

Figure 6-32. Platoon Echelon Left Formation With 2d and 3d Squads in Direct Support

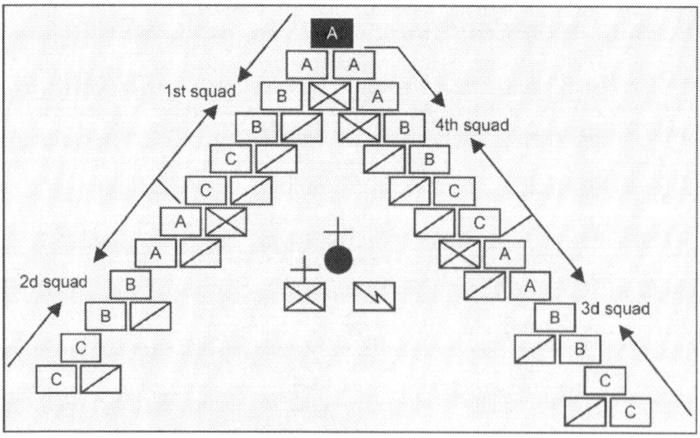

Figure 6-33. Platoon Wedge Formation With Four Squads

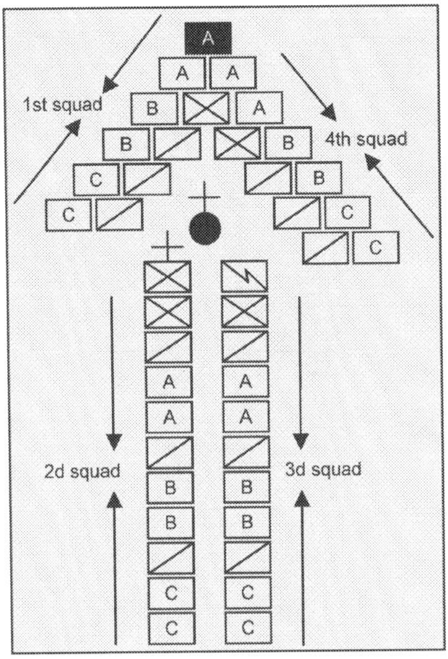

Figure 6-34. Platoon Wedge Formation With 2d and 3d Squads in General Support

left of the baseman covering the 9 to 12 o'clock positions. The 4th squad aligns themselves to the right of the baseman covering the 12 to 3 o'clock positions. The 3d squad aligns itself covering the 3 to 6 o'clock positions. The 2d squad aligns itself covering the 6 to 9 o'clock positions.

6-94. To assemble a platoon from a platoon diamond or circle formation, the HQ element uses the same procedures as that of a line formation. The 1st squad leads the platoon back to the AA followed by the 2d, 3d, and 4th squads. The squads fall into a column formation as they file back to the AA.

COMPANY FORMATIONS

6-95. When a company commander orders his company into control force formations from a column, he moves to the left near the head of the column so he can be seen by the platoon leaders. He then faces the company. As he gives his preparatory command, he points to where he wants the formation. If he does not indicate a location, the company forms immediately in front of the lead platoon.

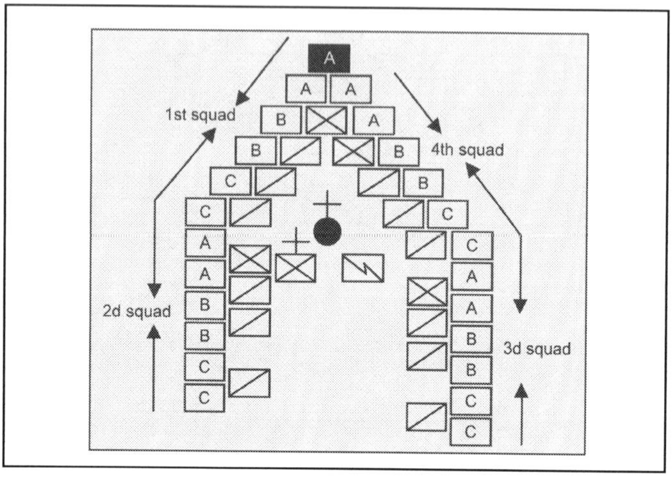

Figure 6-35. Platoon Wedge Formation With 2d and 3d Squads in Lateral Support

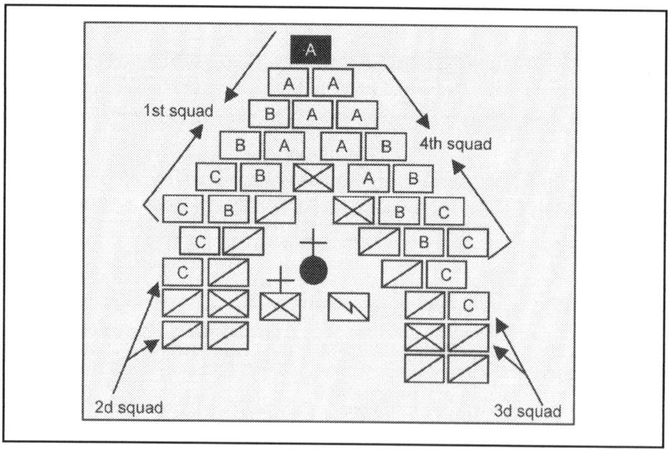

Figure 6-36. Platoon Wedge Formation With 2d and 3d Squads in Direct Support

Relieving Elements in Place

6-96. During crowd control operations, the platoon leader may rotate his squads to give them a rest. See *paragraph 6-76,* page 6-27.

Company Line in Depth Formation

6-97. The command for forming a company line in depth formation (see *Figure 6-37*) is "Company Line in Depth, Move." After the company commander's preparatory command, each platoon leader gives the command, "Platoon Line." The squad leaders follow with their commands to their squads to form the platoon line. On the command of execution, each platoon establishes a platoon line immediately to its front.

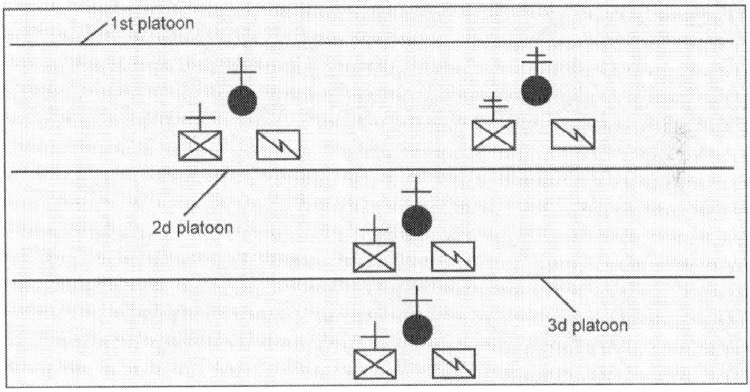

Figure 6-37. Company Line in Depth Formation

6-98. If the commander wants a more formidable formation, he commands, "2d Platoon, Direct Support, Move." The squad and team leaders from the 1st platoon take two steps backward. The 2d platoon moves forward and covers the intervals between the members of the lead platoon. The platoon leader of the 3d platoon then moves his platoon forward to occupy the position formerly held by the 2d platoon.

6-99. If the 3d platoon is called on to support the company line, it moves forward and takes up a position directly behind the 2d platoon in the same manner. Platoon leaders and PSGs help each other in controlling the company.

Company Line in Mass Formation

6-100. The command for forming a company line in mass formation is "Company Line in Mass, Move" (see *Figure 6-38,* page 6-40). The company line in mass is the same as a company line with both platoons in direct support. Instead of forming it in stages, the company commander indicates his desire for this formation in his initial command. The platoons each form a line, and the 2d and 3d platoons close on the 1st platoon without further command.

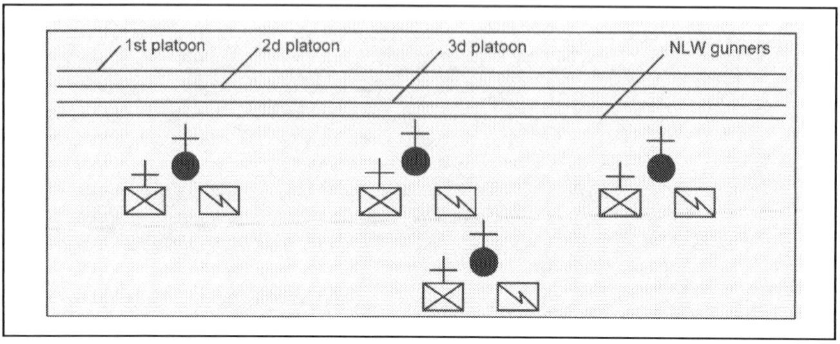

Figure 6-38. Company Line in Mass Formation

Company Line Formation With Support

6-101. In company formations, the 1st platoon usually forms the lead element and the 2d and 3d platoons provide support (see *Figures 6-39; 6-40;* and *6-41*). The support platoons can be employed in the same manner as the support squads in platoon formations. One platoon can be relieved from the lead element by another platoon while in formation by having a support platoon pass through the lead platoon.

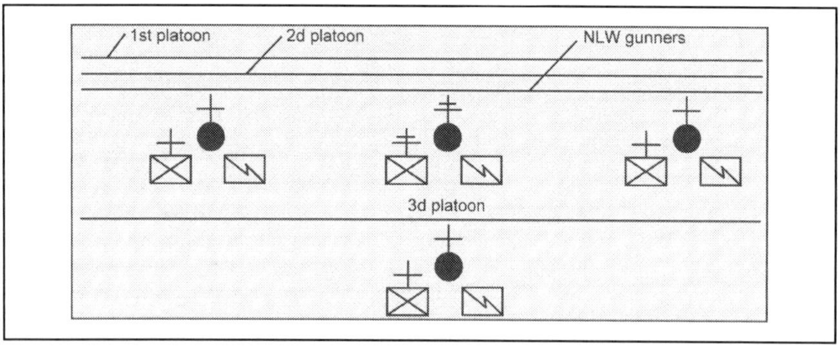

Figure 6-39. Company Line in Mass Formation With One Platoon in Depth

Company Echelons and Wedge Formations

6-102. Company echelons (see *Figures 6-42,* page 6-42; *6-43,* page 6-43; *6-44,* page 6-43; and *6-45,* page 6-44) and company wedge formations (see *Figures 6-46,* page 6-45; *6-47,* page 6-46; *6-48,* page 6-46; and *6-49,* page 6-47) are formed in the same manner and with the same variations as the company line formation. A company assembles from a crowd control formation in the same manner as a platoon or a squad does. However, due to the size of company

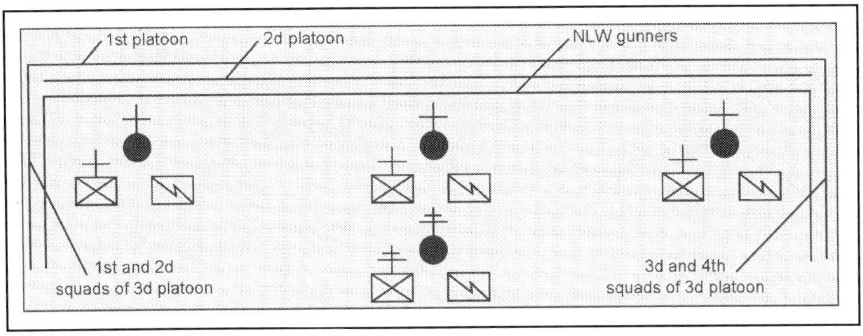

Figure 6-40. Company Line in Mass Formation With One Platoon in Lateral Support

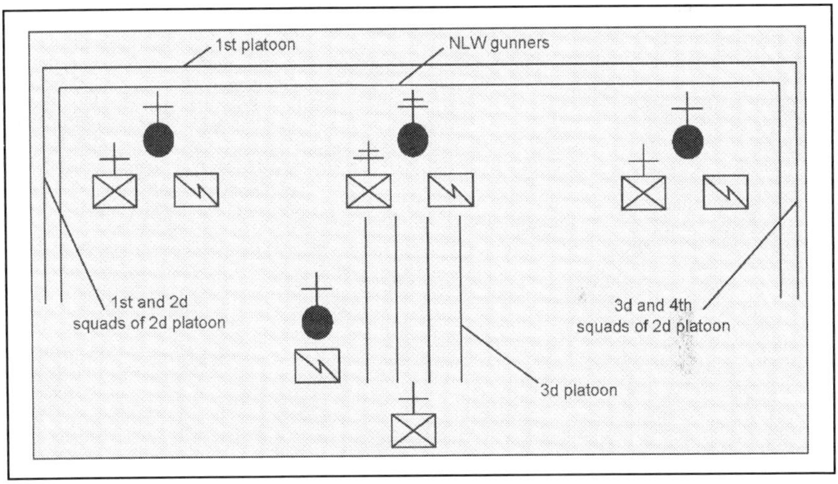

Figure 6-41. Company Line, One Platoon in Lateral Support and One Platoon in General Support

formations, the commander must consider the area that the company is operating in when he assembles the company. If he follows the guidelines of platoon assembly, he will have no difficulty. When assembling the company, the commander takes a position to the rear of the formation and commands, "Company Assemble." The platoon leaders take charge of their respective platoons. Usually, the 3d platoon assembles first, just behind the formation. After the 3d platoon is assembled, the 2d platoon assembles in front of the 3d platoon and the 1st platoon assembles in front of the 2d. All platoons assemble facing the commander. The 2d and 3d platoons then dress with the 1st platoon. As each platoon is assembled, they are ordered to port arms by the platoon leader to await further orders from the commander. If four platoons are used, the 4th platoon assembles first.

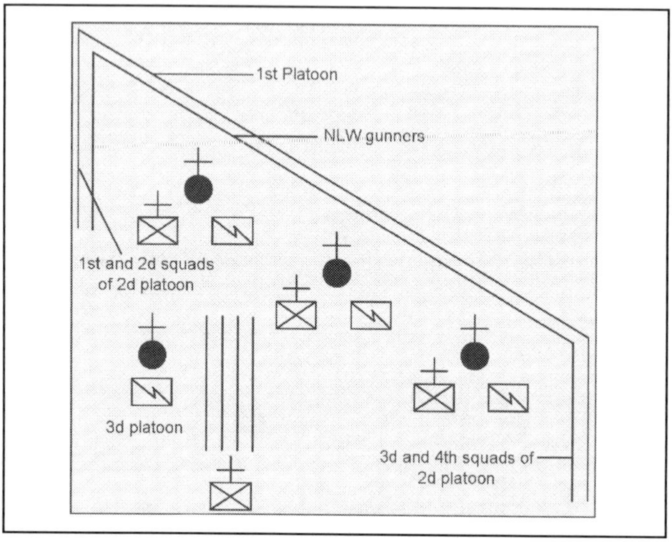

Figure 6-42. Company Echelon Right Formation With One Platoon in Lateral Support and One Platoon in General Support

Company and Platoon Formations With Vehicles

6-103. The commands for vehicles and foot troops are the same as those for foot troops alone. Because of the additional noise and distances involved, the company commander gives the hand-and-arm signal for the line. The command and the signal apply to the lead platoon only. The other platoons remain in general support until further directed (see *Figures 6-50,* page 6-48; *6-51,* page 6-49; *6-52,* page 50; *6-53,* page 6-50; *6-54,* page 6-51; *6-55,* page 6-52; and *6-56,* page 6-53).

6-104. The motor section moves out with vehicle number two moving to the right. Vehicle number two passes the lead vehicle and establishes the position for the center of the line. At the same time, the 3d and 4th vehicles swing out to the left and right. They form to the left and right and slightly to the rear of vehicle number two. The 1st and 4th squads of the lead platoon then move forward and form lines to the left and right, on the front end of vehicle number two. As soon as the troops are in position, the 3d and 4th vehicles close in on the squads and tighten the formation. The 2d and 3d squads of the lead platoon then form a line to the left and right of the 3d and 4th vehicles to complete the company line. The command vehicle (vehicle number one) takes

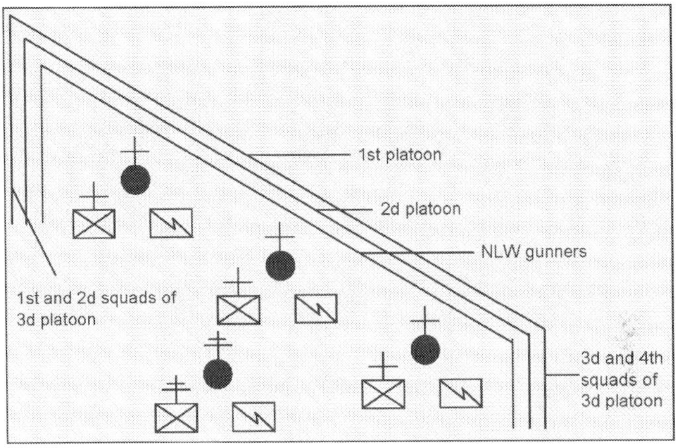

Figure 6-43. Company Echelon Right in Mass Formation With One Platoon in Lateral Support and One Platoon in General Support

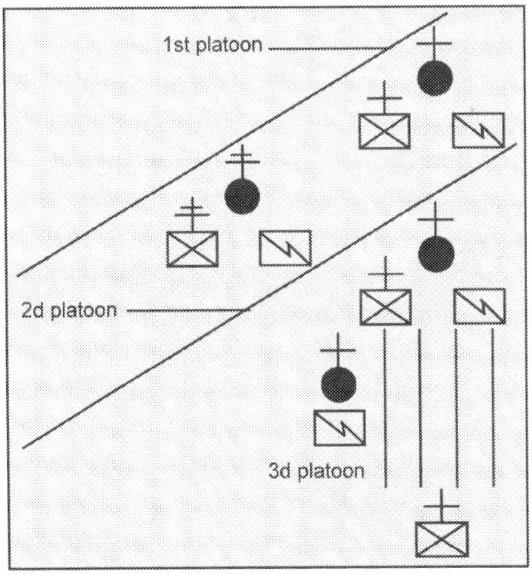

Figure 6-44. Company Echelon Right in Mass Formation With One Platoon in Lateral Support

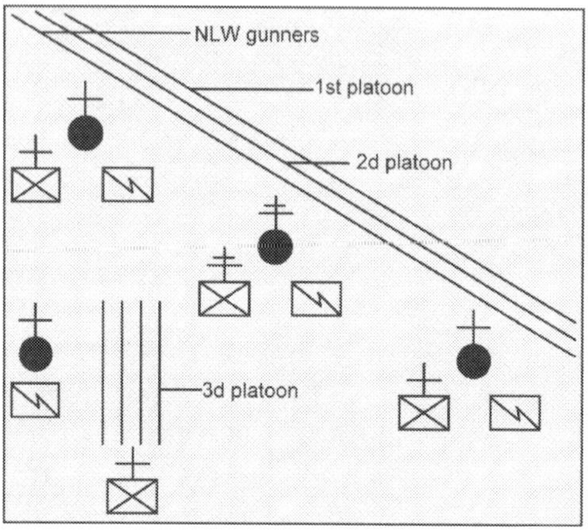

Figure 6-45. Company Echelon Right in Mass Formation With One Platoon in General Support

up a position behind the line where the commander can direct and control the unit. The 2d and 3d platoons move forward and are used in general, lateral, or direct support. In some instances, foot troops move into position first and vehicles join the formation, as needed.

6-105. To form a company echelon or a company wedge with vehicles, the company follows the same procedures used for company formations without vehicles. Vehicles are moved into the appropriate positions using procedures similar to those used for a company line.

6-106. To assemble troops and vehicles, the command vehicle moves to a position behind the formation. The commander faces the formation and gives the vehicle section the signal for assembly. Immediately, the vehicles return to their proper positions in column formation behind the command vehicle while the troops stand fast. The second in command then assembles the troops in the usual manner.

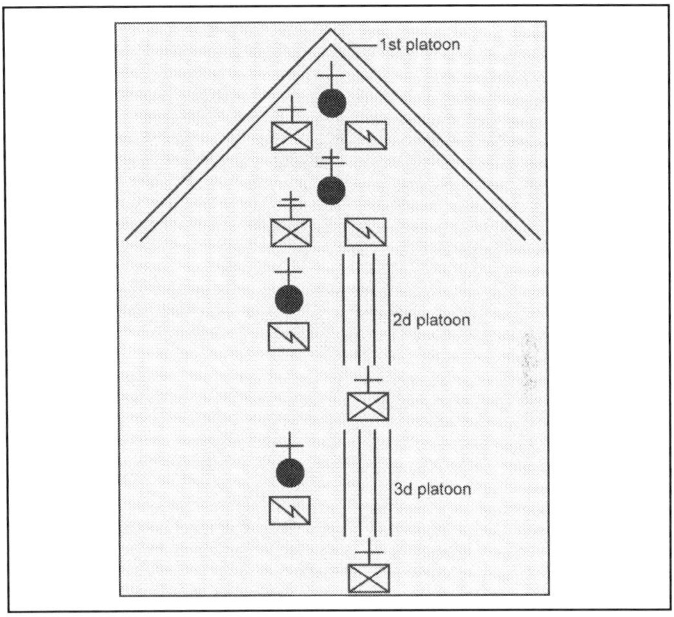

Figure 6-46. Company Wedge Formation With Two Platoons in General Support

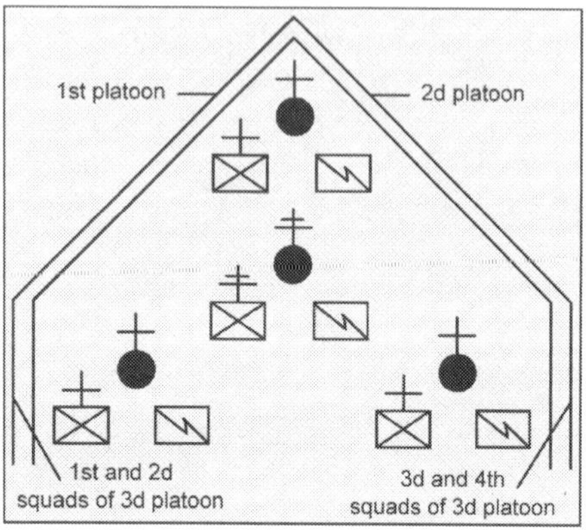

Figure 6-47. Company Wedge Formation With One Platoon in Lateral Support

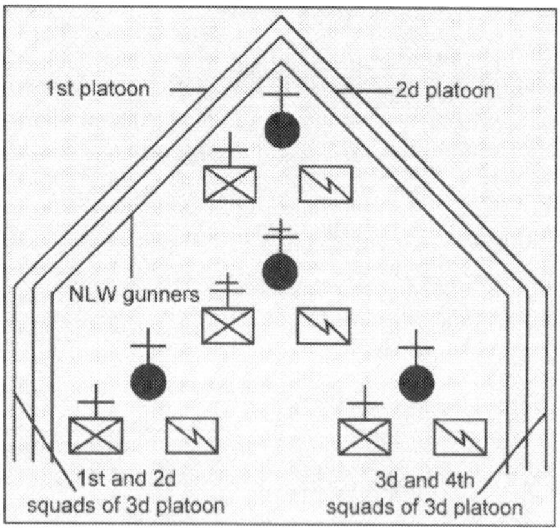

Figure 6-48. Company Wedge Formation With One Platoon in Lateral Support and One Platoon in General Support

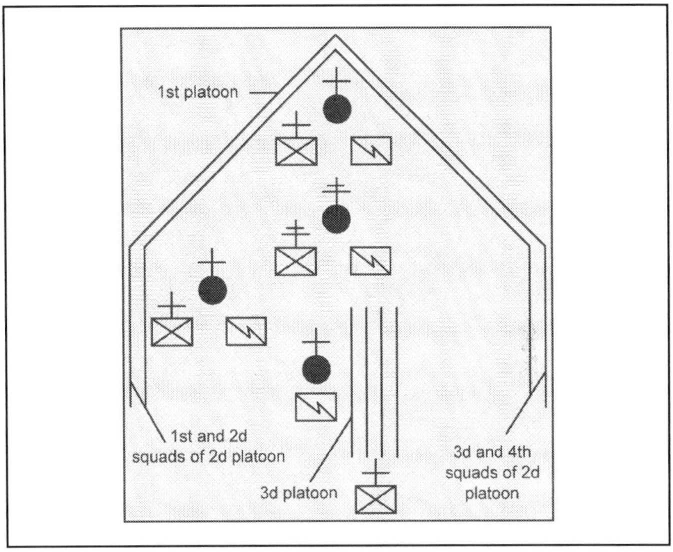

Figure 6-49. Company Wedge Formation With One Platoon in Lateral Support and One Platoon in General Support

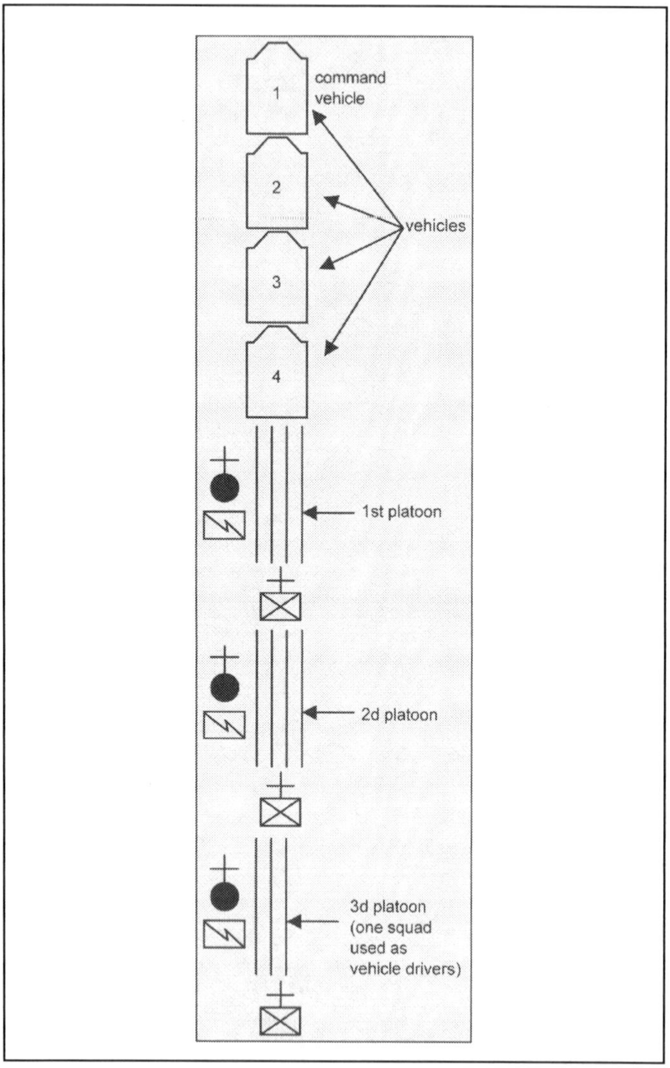

Figure 6-50. Company Column Formation With Vehicles

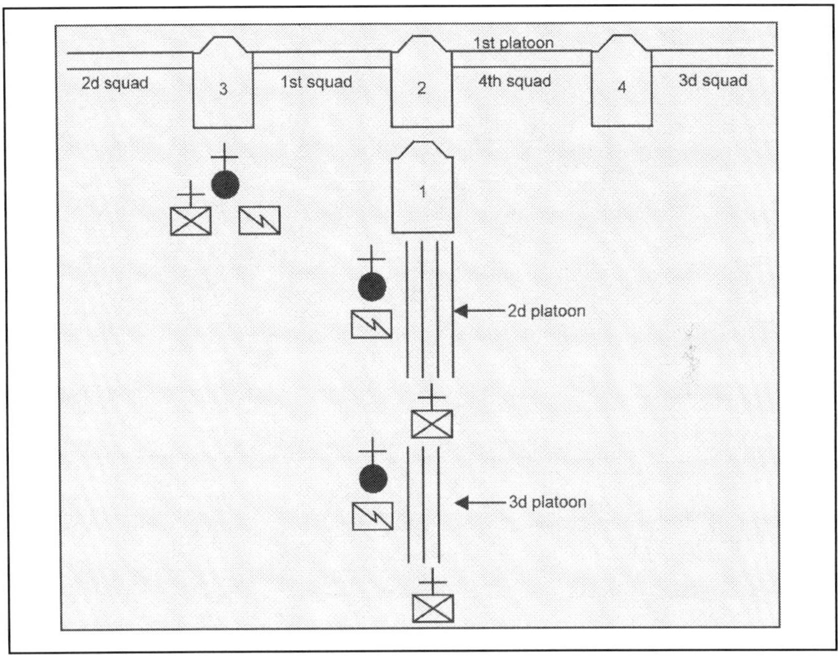

Figure 6-51. Company Line Formation With Vehicles and Two Platoons in General Support

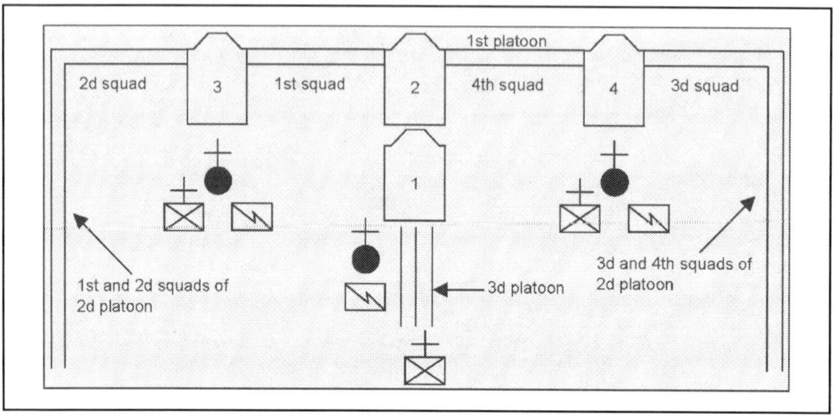

Figure 6-52. Company Line Formation With Vehicles, One Platoon in Lateral Support, and One Platoon in General Support

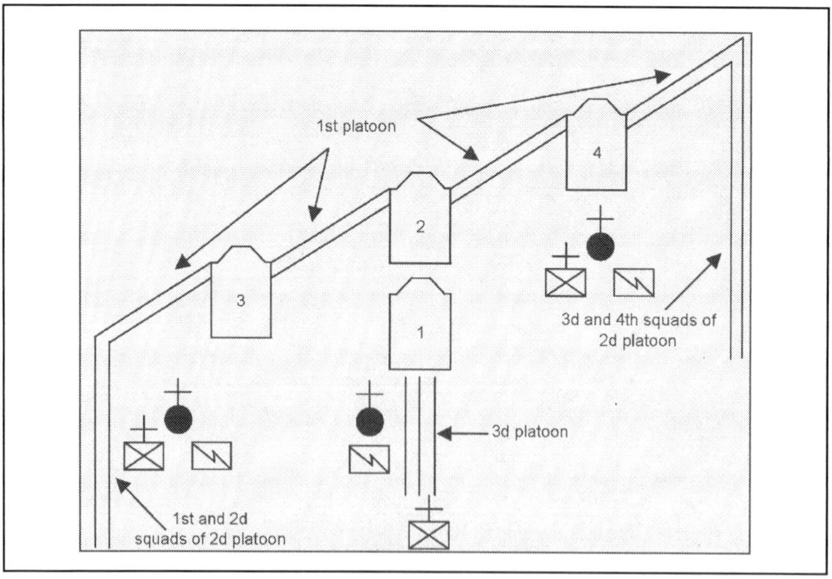

Figure 6-53. Company Echelon Right Formation With Vehicles, One Platoon in Lateral Support, and One Platoon in General Support

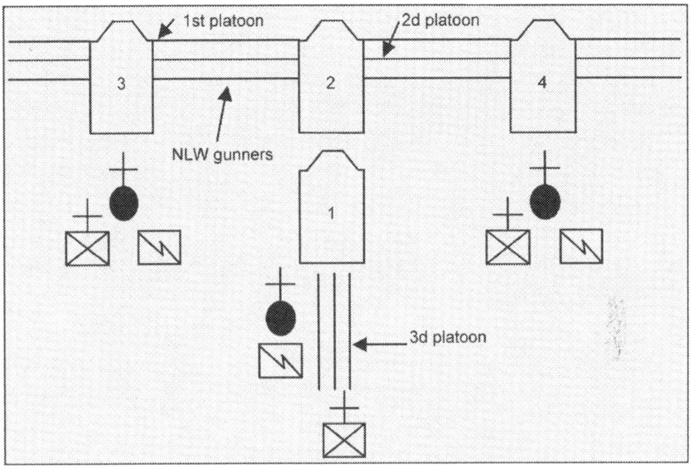

Figure 6-54. Company Line in Mass Formation With Vehicles and One Platoon in General Support

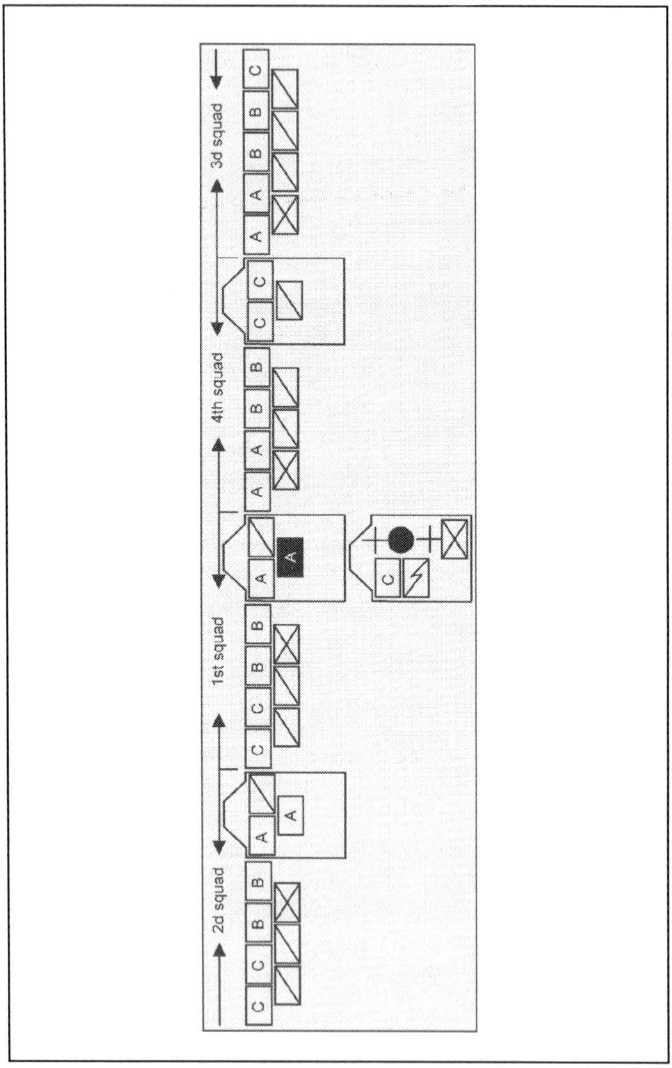

Figure 6-55. Platoon Line Formation With Vehicles

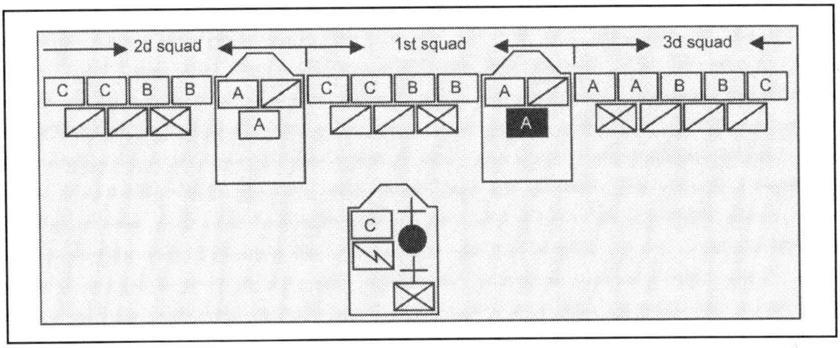

Figure 6-56. Platoon Line Formation With Direct Support and Vehicles

Chapter 7

Civil Disturbance Operations in Confinement Facilities

Civil disturbance doctrine is usually written for quelling disturbances that occur in large open areas where the size of the responding force or the type of formation employed is not hampered by a lack of space. However, in the military confinement environment, inmate disturbances often occur in small, closed areas, such as an inmate's housing unit, a gymnasium, a dining facility, or a chapel where maneuver space and equipment limit the size of the element that may be employed. The layout of each facility is different and presents its own unique problems when employing forces. This chapter covers the use of small military police teams, forced cell move teams (FCMTs), and the formations used to move an unruly inmate from one cell to another. Small generic riot control formations are discussed to provide the commander with a basic employment option that he can tailor to his facility to enable him to regain control of the affected area. The use of NL munitions and RCAs are incorporated into this discussion.

CROWD DYNAMICS WITHIN THE CONFINEMENT FACILITY

7-1. In confinement facility disturbances, inmates employ a number of tactics to resist control or to achieve their goals. Tactics may be unplanned or planned and nonviolent or violent. When a disturbance carries many purposes, it is likely that their tactics are well planned.

NONVIOLENT TACTICS

7-2. Nonviolent tactics range from name-calling to building barricades. Inmates may attempt to distract control force members by shouting at or ridiculing them or by using abusive language, obscene remarks, taunts, and jeers. The inmates' goals are to anger and demoralize the control force. They also want authorities to take actions that may later be exploited as acts of brutality.

VIOLENT TACTICS

7-3. Violent crowd tactics used by inmates are often extremely destructive and can include physical attacks on cadre, other inmates, and property. Use of violent tactics is limited only by the attitudes and ingenuity of the inmates, the training of their leaders, and the materials available to them. Inmates often commit violence with crude homemade weapons. If unplanned violence occurs, inmates will use mops, brooms, chairs, beds, or whatever else is on hand as weapons of violence. During planned violent disturbances, inmates can easily conceal makeshift weapons or tools to use against the control force. Inmates often erect barricades to impede troop movement and/or to prevent the control force from entering certain areas or buildings. Inmates can be expected to vent their emotions on individuals (cadre or inmates), the control force, their equipment, and/or the institution's property. Inmates direct dangerous objects like carts, barrels, liquids, and burning shoe polish at the control force.

GRIEVANCE PROTEST

7-4. A grievance protest may be organized as a riot. However, under normal circumstances, a riot of this type is not of an extremely violent nature. It may turn violent when leaders attempt to exploit any successes or weaknesses found in the control force.

UNORGANIZED RIOTS

7-5. Unorganized riots are characterized as being spontaneous in nature. However, they could be exploited and diverted by leaders at any subsequent stage of the riot, turning it into a different type. They may begin as an isolated assault against authority figures, or acts of civil disobedience. Under determined leadership, an unorganized riot could change to an organized riot.

MULTIPLE RIOTS

7-6. Multiple riots may happen when a disturbance occurs in one area of the confinement facility, causing others to riot in their areas. To control this situation with only one riot control force available, the force commander should first subdue the most violent riot. At the same time, he should employ some of his control forces to contain the other riots until the main control force is able to move to those locations to subdue them.

CADRE BEHAVIOR

7-7. The cadre is susceptible to crowd behavior. They become emotionally stimulated during tense confrontation with unruly and violent inmates. Commanders must be aware of the mood and attitude of the crowd and its effect on the control force cadre. To counteract the effects of crowd behavior on the cadre, the commanders must institute rigorous training and firm and effective leadership. This training must include a complete awareness and understanding of the use of force and the commander's intent. These are all necessary to offset the effect of crowd contagion upon the control force. Even with the best training and leadership, control force members must exercise individual and collective restraint.

PLANNING

7-8. Preplanning is the preparation and training conducted before a crisis occurs. The purpose of preplanning is to have plans and standing operating procedures (SOPs) in place so the cadre may react to an emergency and successfully contain and neutralize the situation. The preplanning process includes plan development, intelligence gathering, and training.

NOTE: Intelligence gathering is essential for the successful containment, neutralization, and/or prevention of a riot.

7-9. Once the control force team has been alerted to a disturbance within the confinement facility, leaders further develop preplans to fit the situation much the same way that DOD develops war plans.

7-10. During the planning process, leaders must be aware of the potential risks involved in quelling disturbances within a confinement facility. Careful planning minimizes collateral damage and risks to control force members and hostages.

7-11. The use of NLW and NL munitions must also be considered when developing plans. There must be strict accountability and control of RCAs, their employment, and other appropriate NL means (such as high-pressure water hoses). When using RCAs, plans must be flexible enough to accommodate changes in the situation and weather because it impacts the effectiveness of some agents. Planners must also consider the—

- Location of the disturbance.
- Estimated number of rioters.
- Accessibility of weapons, tools, or cleaning supplies.
- Inmate's military training.
- Hostage situation.

7-12. Based on an analysis of these factors, the commander makes an estimate of the situation. The estimate must be as thorough as time permits. Using the estimate, the commander considers COAs, selects RCAs, and determines the need for engineer support. In choosing a COA, consider the—

- Number of hostages.
- Number of inmates holding the hostages.
- Leader.
- Attitude and demeanor of the rioters.
- Location of the riot.
  - Inside areas of the confinement facility, such as the dining facility, work site, or housing wing.
  - Outside areas of the confinement facility, such as in exercise or work areas.
- Accessibility of weapons or material to make weapons.
- Accessibility of chemicals, such as cleaning supplies, cooking oils, and shoe polish.

USE OF CHEMICAL IRRITANTS

7-13. The use of chemical irritants can be a valuable NL tool for control force leadership to consider during the planning phase. These chemical irritants can drive a threat from an established, enclosed position or deny the rioters access to a certain area without long-lasting effects to those involved. The proper use of chemical irritants may prevent the control force from having to enter a dangerous area; however, improper use can cause injury, death, or property damage.

USE-OF-FORCE GUIDELINES

7-14. The installation commander authorizes the use of chemical irritants after the control force leader determines the type and dosage. The SOP outlines the procedures for securing authorization and provides guidance to help the control force leader make his decisions. The confinement facility commandant should inform the installation commander of the situation, since the installation commander is ultimately responsible. Team members must be aware of how chemical irritants affect personnel and plan accordingly. Inmates not involved must be removed from the area and the chemical cloud's path. The fire department and emergency medical services (EMS) should be on standby during operations involving chemicals.

PEPPER IRRITANT

7-15. The pepper irritant receives its name from varieties of the capsicum plant that is it made from. The active ingredient, Oleoresin Capsicum (OC), is extracted from the plant and micropulverized to make the irritant. It is neither a chemical nor a gas, but an all-natural organic substance. It is effective against individuals who are emotionally disturbed or under the influence of drugs or alcohol. It is effective against domestic and wild animals. OC is often used to divert the threat from making or continuing an assault. OC does not cause permanent damage. Its effects last 30 to 40 minutes after individuals are placed in fresh air.

7-16. OC can have various effects on people. Some effects are—

- Swelling of mucous membranes and the upper respiratory system; however, it does not shut down the system.
- Burning (intense) and discoloration (bright red) of exposed skin.
- Dilating the capillaries.
- Swelling of the eyelids.
- Burning and involuntary closing of the eyes.
- Coughing uncontrollably.
- Gagging.
- Gasping for air.
- Losing strength and coordination (temporarily).

OC has some disadvantages. They are that—

- The canister depressurizes over time if not checked regularly.
- Inmates may gain access to OC and use it against SRT members.
- The canister must be shaken on a regular basis.

DISSEMINATION OF CHEMICAL IRRITANTS

7-17. There are six methods for disseminating chemical irritants.

Pyrotechnic Method

7-18. In the pyrotechnic method of dissemination, the chemical irritant is placed in a canister with an inert material that is ignited when the device is activated. The agent is then carried into the air on the smoke particles of the inert material.

7-19. There are some disadvantages when using a pyrotechnic device indoors. These disadvantages are that the device—

- May cause a fire.
- Has a very slow saturation time (40 to 45 seconds to burn completely).
- Can be thrown back by the threat, even while burning.

Bursting Method

7-20. In the bursting method of dissemination, the irritant and an inert powder are placed in a saw-toothed canister. After a small detonation, the canister splits and expels the irritant in a cloud.

7-21. The bursting method has its advantages and disadvantages. Some advantages are that—

- There is no risk of fire.
- There is no risk of the canister being thrown back.
- The contents disseminate quickly.
- The canister is effective, easy to carry, and easy to control.

Some disadvantages are that—

- There is a slight possibility of fragmentation when detonated.
- The fuse head may separate from the canister.
- Only 50 to 90 percent of the irritant may be disseminated.
- The direction of the cloud formation depends on where the device is detonated.

Aerosol Method

7-22. In the aerosol method of dissemination, the irritant is suspended in an inert liquid that is located in the rear portion of the device. When the projectile penetrates the structure, the rear portion splits open and the irritant is disseminated in a mist. The aerosol method is most appropriate for tactical operations when chemical irritants are used before entry.

7-23. The aerosol method has its advantages and disadvantages. Some advantages are that it—

- Is easily carried.
- Is an excellent deployment system (uses 40-millimeter, 37-millimeter, or 12-gauge ammunition).
- Disperses the contents quickly.
- Is easily controlled and accurate (fin-stabilized).

NOTE: Practice firing is always necessary.

Some disadvantages are the possibility of—

- Injury to individuals if it is fired from close range.
- Damage to the structure if it strikes a weak portion.

Projectiles Method

7-24. Chemical-containing projectiles can be very effective when used correctly. A soldier must know which projectile to use in each unique situation.

7-25. **Thirty-Seven-Millimeter Projectile.** A 37-millimeter projectile can incapacitate anyone in an enclosed area of 4,500 cubic feet. It can penetrate the following:

- A sheet of ¾-inch-thick plywood at 25 yards.
- An automobile windshield at 33 yards.
- A hollow-core door at 58 yards.

7-26. **Forty-Millimeter Projectile.** A 40-millimeter projectile is more effective at greater ranges than a 37-millimeter projectile because of its rifled bore. It can incapacitate anyone in an enclosed area of 4,500 cubic feet. It can penetrate the following:

- A 1-inch-thick plywood at 10 yards.
- An automobile windshield at 55 yards.
- A hollow-core door at 109 yards.

7-27. **Twelve-Gauge Projectile.** A 12-gauge projectile can be loaded into a shotgun with no modifications to the weapon. It can be fired as a single round, or it can be magazine-fed. The 12-gauge projectile can incapacitate anyone in an enclosed area of 1,000 cubic feet. The angle at which the projectile strikes the barrier can affect its penetration capability. It can penetrate the following:

- A sheet of ¾-inch plywood at 33 yards.
- An automobile windshield at 33 yards.
- A hollow-core door at 100 yards.

Fogging Method

7-28. The irritant is suspended in a fogging solution and dispersed with a fogging machine. It is not recommended for tactical operations.

7-29. The fogging method has its advantages and disadvantages. Some advantages are that it—

- Produces enough chemical irritant to cover an enclosed area of 100,000 cubic feet in 26 seconds.
- Disseminates quickly.
- Provides high-volume capabilities.
- Is effective for riot control or other crowd dispersal situations.

Some disadvantages are that it is—

- Noisy.
- Cumbersome.

Multipurpose Grenade Method

7-30. A multipurpose grenade can be hand-thrown or launched from a shotgun with an adapter. It has an extended shelf life of 6 years and an adjustable fuse delay of 2 to 5 seconds. The irritant is located in the cylinder portion of the device and is forced out the bottom of the grenade upon detonation.

> **WARNING**
> **A multipurpose grenade deployed incorrectly could result in injury. Safety rules should be observed at all times.**

RECORD OF EVENTS

7-31. Reporting procedures for serious incidents should be included in the plans. A record of events must be initiated to provide a basis for the preparation and submission of a formal report to higher HQ. As a minimum, include the—

- Time the incident was reported and by whom.
- Time the incident was reported to the facility commander.
- Time the control force was assembled.
- Time the control force entered the facility.
- Weather conditions as they relate to the use of RCAs.
- Number of cadre and inmates injured or killed, including how they were injured or killed and the medical attention given to them.
- Time the operation was completed.
- Time the riot control force restored order.

TRAINING

7-32. Guard units and associated teams must train on a regular basis in the five basic riot control formations. There must be a continuous training program established, to include—

- Interpersonal communication.
- Self-defense.
- Use of force.
- Use of a riot baton.
- Fatal areas of the body to avoid when using the riot baton.
- First aid.
- Emergency plans.
- Intelligence and counterintelligence techniques.
- The use of RCAs and the various methods of dispersing them.
- The use of NLW.
- The use and effects of high-pressure water.
- Riot control formations.

EQUIPMENT

7-33. The recommended basic riot gear used by the confinement riot control force is as follows:

- Riot baton.
- Riot shield.
- Helmet with face shield.
- Groin protector.
- Flak vest.
- Leather gloves.
- Shin protection.

RIOT BATON

7-34. The riot baton is an invaluable weapon in crowd control situations. A riot baton in the hands of a well-trained control force member is the most appropriate weapon, except with extremely violent crowds. The riot baton is not meant to replace NL munitions, but is instead used in situations in which NL munitions are not needed. Losing a riot baton to the crowd does not present a serious threat.

7-35. Use of the riot baton is based on the commander's appraisal of the situation and his choice of force option. Control forces committed with the riot baton must have teams equipped with RCAs, NLW, and lethal coverage. If the confrontation increases in intensity, the commander has these force options available and may rapidly engage the control force.

7-36. A riot baton is a versatile offensive weapon that varies in length from 26 to 42 inches with the optimum length being 36 inches. It is made of hickory and does not shatter or break easily. Both ends of the riot baton are rounded to prevent unnecessary injury. It is approximately 1¼ inches in diameter. Each riot baton has a hole drilled at the grip end with a leather thong threaded through the hole. The thong helps secure the riot baton to the user's hand.

RIOT SHIELD

7-37. The riot shield (see *Figure 7-1*) is an excellent piece of defensive equipment. It is used by control force personnel in the front and on the flanks of the formation.

7-38. The riot shield has its advantages and disadvantages. Some advantages are that—

- OC pepper spray or high-pressure water can be used to back the control force.
- It provides protection from thrown objects.
- It absorbs some of the impact of rioters hurling themselves at the control force.

The disadvantages of the riot shield are that it—

- Severely limits the use of the riot baton.
- May crack if hit with a heavy object.

Figure 7-1. Riot Shields

RIOT CONTROL FORMATIONS

7-39. The five basic formations for civil disturbance operations are the line, wedge, echelon, diamond, and circle. In the correctional environment the basic formations for disturbances are the line, wedge, or echelon with apprehension teams in support. Generally, the diamond and circle formations are not used in the correctional environment. The minimum size force that the commander should consider is an augmented platoon, with a two-to-one advantage over the rioters.

7-40. Correctional facility cadre must be proficient in all riot control formations. These are usually trained at the squad level and above, but performed by a platoon or company-sized unit. Squad line, echelon left, echelon right, and wedge formations form the basis for platoon and company formations. Each squad must be adept in the basic formations before practicing in platoon-sized or larger formations. For more detailed information on formations, see *STPs 19-95B1-SM* and *19-95B24-SM-TG*. In the correctional environment, the support platoon plays a key role by supporting the control formation with apprehension and equipment teams. See *Chapter 6* for an in-depth discussion on civil disturbance formations.

APPREHENSION TEAMS

7-41. Apprehension teams give the control force the ability to extract inmates. Inmates identified as leaders or agitators and inmates that are injured or no longer want to resist can be extracted from the crowd. Apprehension teams can also be used to enter confined areas where the formation cannot operate.

7-42. The apprehension team will consist of eight members and one leader. Each team will have four shield holders, two nonlethal gunners and two apprehension/search members. The squad leader controls the movement of the team and assists the team in exiting the control formation. He maintains communications with the formation and coordinates the team's operations with the formation. The squad leader also ensures that the team uses the minimum amount of force necessary to extract inmates.

7-43. The apprehension team may deploy through the middle of the formation or from around either of the flanks.

7-44. The apprehension team should be deployed within a distance of approximately 10 meters so that the formation can still provide immediate support. If the apprehension team is deployed farther than 10 meters, the formation must advance to close the gap.

7-45. The team will exit the formation in two columns; shield holders in the first and second ranks, nonlethal gunners in the third rank, apprehension/search members in the fourth rank and the leader last. Team members grip the shoulder or equipment of the member in front of them. The shield holders will envelop the inmate that is to be extracted and form a protective barrier. Once the shield holders have enveloped the inmate, the nonlethal gunners will cover their sides with their weapons while the apprehension/search members control and restrain the inmate. When the team has the inmate under control, the squad leader will give the order for the team to withdraw. The team withdraws in reverse order (apprehension/search members with the inmate, followed by nonlethal gunners and shield holders). Team members then move backwards, face the crowd and grip the shoulder or equipment of the member in front of them.

TEAM EQUIPMENT

7-46. The apprehension team uses the following equipment:

- Flex-Cufs.
- Restraint cutters.
- Pepper spray.
- Helmets with face shields.
- Groin protectors.
- Flak vests.
- Elbow pads.
- Leather gloves.
- Shin protectors.
- Knee pads.
- A video camera.

TEAM COMPOSITION

7-47. The apprehension team is composed of five team members. This team consists of four members and a squad leader. Team member one is the team leader.

Squad Leader

7-48. The squad leader assists the team in exiting the control formation and ensures that the team uses the minimum force necessary to subdue inmates. He is also responsible for signaling the PSG if more help is needed.

Team Member One (Team Leader)

7-49. Team member one's responsibility is to make initial contact with the inmate and use pepper spray and apply Flex-Cufs, as needed. He also carries a set of restraint cutters. Once the apprehension team has returned to the relative safety of the control formation, team member one searches the inmate.

7-50. Team member one receives the command to move forward and secure the inmate from the squad leader. This command is usually given by a signal and verbal command. Team member one then notifies the main force commander where his team exits the control force. Once the team is in position to exit the main formation, team member one signals the member of the formation in front of him. This is usually a prearranged signal, such as a tap on the right shoulder. The control force opens and the team exits the formation and makes contact with the assigned inmate. Team members then subdue and control the inmate. If pepper spray is needed, team member five shouts, "Pepper." This gives the other team members time to react to the command and spray the inmate. Team member one then applies Flex-Cufs to the inmate's wrists and ankles. Depending on the mission, the inmate may be left for follow-on security forces or carried behind the control force to be secured by security forces.

7-51. Should the mission require that the inmate be moved, team member one shouts, "Prepare to Move," this allows each member to grab an arm and position the inmate toward the control force formation. When the team is ready to move, team member two shouts, "Ready," alerting team member one that the team is ready for the last command. Team member one then commands, "Move." The team enters the control formation with the inmate. The inmate is searched by team member one and turned over to the security team.

Team Member Two

7-52. Team member two stands directly behind team member five with his hands on team member five's waist. He protects the left side of team member five by moving to team member five's left and helps to secure the inmate after team member one makes initial contact. He repeats team member five's commands and controls the inmate's head after the inmate is on the ground.

Team Member Three

7-53. Team member three stands behind team member two with his hands on team member two's waist. He protects the right side of team member five by moving to team member five's right and helps to secure the inmate after initial contact.

Team Member Four

7-54. Team member four stands behind team member three with his hands on team member three's waist. His task is to help subdue the inmate in any way possible. He usually concentrates on the inmate's body below the knees. He acts as the eyes and ears for the team during control of the inmate.

Team Member Five

7-55. Team member five gives commands to control the inmate.

FORCED CELL MOVE TEAMS

7-56. The intent of a forced cell move is to move an unruly and/or uncooperative inmate from one cell to another. This is a difficult task and must not be taken lightly. A forced cell move must be thoroughly planned, rehearsed, and resourced to ensure the safety of the team members and the inmate.

7-57. A number of reasons may compel the commander to consider a forced cell move. The inmate may refuse to eat, take medication, accept medical attention, get a haircut, keep himself clean, work, or move or rotate cells when ordered to. These are some of the reasons that the commander may order a forced cell move.

TEAM COMPOSITION AND EQUIPMENT

7-58. The FCMT is composed of six corrections cadre and two support cadre from military police and medical sections specially trained in forced cell movement. Primarily, the team consists of the following personnel:

- One officer in charge (OIC) (E7 or above).
- Five corrections cadre.
- One military police investigator.
- One medic.

7-59. The FCMT uses the following standard equipment:

- Helmets with face shields.
- Groin protectors.
- Flak vests.
- Elbow pads.
- Leather gloves.
- Shin protectors.
- Knee pads.
- A pinning shield.
- Restraints (hand or leg irons or Flex-Cufs).
- A video camera.

ACTIONS PRIOR TO TEAM EMPLOYMENT

7-60. The OIC gives an operations order (OPORD) to team member five (the team leader) of the FCMT. As a minimum, the OPORD should answer the following questions about the incident leading up to the move:

- Did the inmate fail to comply with orders?
- Did the inmate assault cadre or other inmates?
- Is the objective of the move to force a shower or haircut?
- Is OC pepper spray authorized for use?
- Has the inmate been moved before?

- What is the condition of the area where the inmate is currently located?
- Has the inmate tested positive for human immunodeficiency virus (HIV)?
- Does the inmate have a weapon or the materials needed to make one?
- Is the area barricaded?
- Has he smeared himself with body fluids or other slippery substances?
- Can you approach the inmate without risk to him and/or others?

7-61. The team is assembled for inspection by the OIC, and the team members are required to brief the OIC on their individual duties and responsibilities. A line formation is the customary formation for this inspection. It should take place out of site of the inmates.

7-62. Each soldier takes one step forward and briefs the OIC as shown in the following example: *"I am (rank and name), team member one. My mission is to pin the inmate using the minimum amount of force necessary. I carry the pinning shield."* He then faces so the camera can record the number printed on the back of his flak jacket and holds up the pinning shield [see *Figure 7-2*]).

Figure 7-2. Recording Team Members' Duty Position

INDIVIDUAL TEAM MEMBER RESPONSIBILITIES

7-63. Each individual team member has his own responsibilities during a forced cell move. Each team member must do his designated job while working as a team.

Team Member One

7-64. Team member one is the eyes and ears of the team. His responsibility is to carry the pinning shield up to the cell door and describe the scene with as much detail as possible (especially including anything that may cause harm to the team). The pinning shield is used to protect the team from body fluids thrown at the them by the inmate.

7-65. Team member one describes the scene in the cell to his teammates by turning his head to the right and speaking in a loud voice so that all team members can hear him. A slap on the right shoulder by team member two alerts him that all team members have heard him. If he receives a slap on the left shoulder from team member two, he repeats the situation report (SITREP) in a loud, clear voice. The decision to enter the inmate's area with or without the pinning shield is team member one's decision, unless the inmate is known to have a weapon. Should he decide to enter without the pinning shield, team member one then enters with both hands held approximately chest high. When team member one is unable to determine if the inmate has a weapon, the pinning shield is used. Charging through the open cell door, team member one drives the inmate to the bunk or ground with the shield. Once pinned, team member one secures the inmate's head and ensures that it is not twisted or turned in relation to its natural form. He may have to use the various pressure points on the inmate's head to help subdue him.

Team Member Two

7-66. Team member two steps forward and briefs the OIC as follows: *"I am (rank and name), team member two. My mission is to secure the inmate's right arm using the minimum amount of force necessary. I carry one set of hand irons."* He then holds the hand irons up for video camera taping (see *Figure 7-3*) and says, *"I also carry one canister of OC pepper spray to be used at the direction of the team OIC."* He then holds up the canister for video camera taping and faces so the camera can record the number on the back of his flak jacket (see *Figure 7-2,* page 7-13.)

Figure 7-3. Recording Team Members' Equipment

7-67. Team member two lines up directly behind team member one. He keeps contact with team member one until he can secure the inmate's strong arm. Team member two taps team member one on his right or left shoulder after being tapped by team member three. Once inside the cell, team member two secures the prisoners strong hand with wrist restraints and announces, *"Hands secure,"* to team member five.

Team Member Three

7-68. Team member three steps forward and briefs the OIC as follows: *"I am (rank and name), team member three. My mission is to secure the inmate's left arm using the minimum amount of force necessary."* He then faces so the camera can record the number on the back of his flak jacket (see *Figure 7-4*).

Team member three lines up directly behind team member two. Team member three keeps contact with team member two until he can grasp a portion of the inmate's body. He then concentrates on securing the inmate's weak hand and assisting team member two in applying hand restraints. Team member three taps team member two on his right or left shoulder after being tapped by team member four.

Figure 7-4. Recording Number on Flak Jacket

Team Member Four

7-69. Team member four steps forward and briefs the OIC as follows: "*I am (rank and name), team member four. My mission is to secure the inmate's right leg using the minimum amount of force necessary. I carry one set of leg irons.*" He holds the leg irons up for the camera to record (see *Figure 7-2*, page 7-13). He then faces so the camera can record the number on the back of his flak jacket (see *Figure 7-2*). Team member four lines up directly behind team member three. He taps team member three on his right or left shoulder after being tapped by team member five. Team member four keeps contact with team member three until he can secure a portion of the inmate's body. Team member four concentrates on securing the right leg of the inmate. It is his responsibility to apply the leg irons. Once the legs are secure, he announces, "*Legs secure*," to team member five.

Team Member Five (Team Leader)

7-70. Team member five steps forward and briefs the OIC as follows: "*I am (rank and name), team member five. My mission is to secure the inmate's left leg using the minimum amount of force necessary. I carry a restraint key (hold up for the camera to view). I am also the team leader for this mission and it is my responsibility to ensure that the team uses the minimum amount of force necessary to subdue the inmate.*" He then faces so the camera can record the number on the back of his flak jacket (see *Figure 7-4*).

7-71. Team member five taps team member four on the right shoulder when he has heard the SITREP from team member one. If he does not hear the report or needs to hear it again, he taps team member four on the left shoulder. Team member five keeps in contact with team member four until he can secure a portion of the inmate's body. He secures the left leg and assists team member four in applying the leg irons. Once he has heard that the hands and legs are secure, team member five signals for the team to conduct an equipment check. The team members check the equipment in place.

NOTE: At no time is the inmate to be released by anyone.

Medic

7-72. The medic briefs the OIC in the following manner: *"I am (rank and name), the medic. My mission is to treat all injuries that may occur during the move."* He follows the instructions given to him by the OIC/NCOIC. His mission is to observe the inmate for signs of injury before the forced cell move and assess the inmate's condition during and after the move.

Military Police Investigator

7-73. The military police investigator (MPI) is responsible for videotaping the entire forced cell move. This includes the OIC/NCOIC briefing the OPORD, each team member briefing his responsibilities to the OIC, and any medical treatment rendered to the inmate.

ACTIONS DURING THE FORCED CELL MOVE

7-74. There are many actions performed during a forced cell move. Each member of the FCMT must complete his task successfully to accomplish the forced cell move.

Officer in Charge or Noncommissioned Officer in Charge

7-75. The OIC, an E7 or above, is responsible for ensuring that the team uses the minimum amount of force necessary to subdue the inmate. He is also responsible for ensuring that all necessary precautions have been taken to ensure the safety of the team and the inmate. The OIC gives the inmate the initial orders to comply and informs him of the consequences of noncompliance. He does this by placing himself at the front of the inmate's cell (see *Figure 7-5*). The orders are not given until it can be enforced by the FCMT. The following prepared statement is read to the inmate by the OIC: *"I am (rank and name), OIC of this FCMT. At this time I am ordering you to lie face down on your bunk (or other designated location) with your head to the rear of the cell. Cross your hands behind your back and cross your feet as the team enters the cell and uses the minimum amount of force necessary to place you in restraints and move you to (give location). Failure on your part to comply with these instructions or to resist the FCMT will result in the use of force (OC pepper spray if applicable). Do you understand these instructions? If so, lie down on the bunk."*

7-76. After reading the instructions to the inmate, the OIC faces the camera and states whether the inmate complied. He then waits 60 seconds and motions for the team to position itself in front of the inmate's cell (see *Figure 7-6*). The OIC directs the opening of the cell when the team is positioned at the entrance to the cell and prepared to enter. He positions himself so he can observe team entry to and departure from the cell. If OC spray is authorized, the OIC tells the team when to spray the inmate with OC spray. This task is done in silence. Communication between team members should be done through hand-and-arm signals, unless noted otherwise.

Figure 7-5. OIC Reading Compliance Statement to Inmate

Figure 7-6. OIC and FCMT

Team Member One

7-77. Once the team is positioned in front of the inmate's cell, team member one starts his report. At a minimum, he must answer the following questions and report them to the team:

- Does the inmate have a weapon, if so, what kind?
- Does the inmate appear dry?
- Is the inmate clothed or wearing extra clothing?
- Has the inmate smeared anything on his body?
- Can you see the inmate's hands?
- Is the cell barricaded or visibly booby-trapped?

- Is the cell dry?
- Are cell furnishings intact?

7-78. Based on the answers to the above questions, team member one must decide whether to enter the cell with or without the shield (see *Figure 7-7*). Team member one's ultimate goal is to pin the inmate on the bunk or floor.

Figure 7-7. FCMT Preparing to Enter a Cell

Team Member Two

7-79. Team member two keeps contact with team member one until he can secure the inmate's strong arm. If he cannot immediately secure the inmate's arm, he secures a portion of the inmate's body and progresses to the inmate's strong arm. Once the inmate's hands are secure, he announces, "*Hands secure*," to team member five (see *Figure 7-8)*.

Team Member Three

7-80. Team member three keeps contact with team member two until he can grasp a portion of the inmate's body. Team member three concentrates on securing the inmate's weak hand and assists team member two in applying Flex-Cufs. Team member three maintains control of the inmate's hands and looks at team member five for further instructions (see *Figure 7-8*).

Team Member Four

7-81. Team member four keeps contact with team member three until he can secure a portion of the inmate's body. Team member four concentrates on securing the inmate's right leg. He applies leg irons and when the inmate's legs are secure, he announces, "*Legs secure*," (see *Figure 7-8*).

Team Member Five (Team Leader)

7-82. Team member five keeps contact with team member four until he secures a portion of the inmate's body. Team member five concentrates on securing the inmate's left leg and assists with the application of leg irons. When he is informed that the hands and legs are secure, the team searches the inmate for weapons and contraband. Once the search is complete, team member five commands, "Prepare to Lift."

Figure 7-8. FCMT Securing an Inmate

Movement of the Inmate

7-83. Team member one positions his hands to control the inmate's head during the move, turning the inmate's head toward the wall. Team members two and three position themselves on the inmate's right and left sides. They grasp the wrist or forearm of the team member opposite them to form a platform for the inmate to lie on. Team member four positions himself on the inmate's right side even with the inmate's knees. This places him in position to control the inmate's legs. Team member five's position is on the inmate's left side and enables him to control the inmate's lower legs and feet. When in position, team member five ensures that all team members are in place before commanding, "Ready, Lift."

7-84. On the command "Lift," all team members simultaneously stand with the inmate. They should sandwich the inmate between them to facilitate control. Team member five then commands, "Prepare to Turn, Ready—Turn." The inmate is always turned toward the bunk. If a bunk is not available, team member five should specify which direction to turn the inmate. The inmate is then removed from the cell (see *Figure 7-9*, page 7-20).

7-85. Once the inmate is on the tier, team member five commands, "Prepare to Lower, Team—Lower." The team simultaneously lowers the inmate to the floor with the team keeping positive control of the inmate. The medic asks the inmate, "*Do you have any injuries at this time?*" (see *Figure 7-10,* page 7-20). If the inmate has no injuries, the medic looks at the camera and states, "*Inmate (inmate name) appears to have sustained no injuries.*" If the inmate has sustained injuries, the medic treats those injuries and completes a sworn statement. If OC was employed, the medic decontaminates the inmate using water and paper towels. The medic instructs the inmate to tilt his head to the side and flush his eyes with water for approximately 5 minutes. The medic then dries the inmate's face.

Figure 7-9. Moving the Inmate

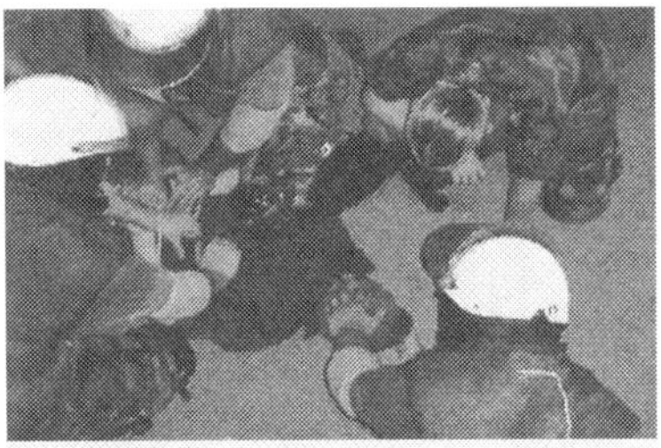

Figure 7-10. Treating Injuries

Placement of the Inmate in a Cell

7-86. Team member five orders an equipment check before returning the inmate to a cell. He then commands, "Prepare to Lift, Team—Lift." Team members position themselves exactly where they were when they removed the inmate from the cell. Next, team members simultaneously lift the inmate and return him to the predetermined cell. They move the inmate into the cell head first with his face toward the bunk. Once inside the cell, team member five commands the team to lower the inmate onto the bunk (see *Figure 7-11,* page 7-20). Team members then remove the inmate's restraints, keeping positive control of the inmate at all times. Team members use a nonverbal signal to notify team member five when the restraints are removed. He signals another equipment check. Team members then account for all equipment in place and prepare to exit the cell. The team exits the cell in the same manner as they entered, but in reverse order.

Figure 7-11. Placing Inmate in Cell

7-87. Team member four moves into position where he can control both of the inmate's legs while team member five moves toward the cell door. When team member four is ready to exit the cell, team member three moves to control the inmate's legs, team member two moves to where he can control both of the inmates arms, and team member one maintains control of the inmate's head. Team member four maintains contact with team member three at all times. When team member three is ready to exit the cell, team member two moves into a position where he can control the inmate's lower body and team member one moves into position to control the inmate's upper body. Using a prearranged, nonverbal signal, the team pulls team members one and two from the cell. The FCMT is prepared to reenter the cell if the inmate attempts to attack team members one and two as they exit the cell. The FCMT remains ready to reenter until the cell door is secured.

Chapter 8

Civil Disturbance Training

Training requires planning and preparation. Unit commanders structure their training program according to the mission essential task list (METL) (see *FM 7-1)* and expected contingencies. These training plans include prior training and the experience and proficiency of their unit. Leaders agree that the operational tempo and expected contingencies for the next 10 to 15 years will likely involve low-intensity military operations other than war (MOOTW). Therefore, civil disturbance training must be included in the commander's annual training program and must cover all aspects of that type of operation. To effectively function as a member of a control force, individual soldiers must be trained in civil disturbance operations. Training must be intensive, realistic, and sustained. Any lack of training (from basic soldier skills to leadership skills) can have an adverse effect to the point of endangering lives, especially the lives of soldiers. Leader involvement in the training process is essential for mission success in crowd control operations.

UNIT TRAINING STRATEGY DEVELOPMENT

8-1. Developing a unit training strategy begins with the mission analysis portion of the MDMP and identifying the GRM. Missions that require soldiers and units to enforce treaties or accords, protect the lives of civilians in uncertain or hostile environments (such as NEOs), or provide large-scale humanitarian assistance demand some sort of graduated-response criteria to maintain order and prevent uncertain environments from becoming hostile.

8-2. When developing training strategies, planners should focus training and evaluation on the echelon level above the target element (for example TF) or the unit identified as needing training.

8-3. Planners must agree on the intent of the GRM. The GRM is for most a training and rehearsal tool. It provides leaders with likely vignettes that can be incorporated into COA analysis, predeployment training, and rehearsals. The GRM becomes a handy reference during situations that require graduated responses.

TRAINING PLAN DEVELOPMENT

8-4. In developing a training plan for civil disturbances, many considerations should be addressed in terms of available resources. Considerations include the number of personnel to train, the equipment needed and its availability, the time available to train to standard, the location of where training will be executed, and the incorporation of SOPs (based on the theater of operations

[CONUS or OCONUS]). A personnel assessment is necessary to determine the training and experience level of soldiers. It is critical to identify and use soldiers with previous real-world experience and leaders that have graduated from the Interservice Nonlethal Individual Weapons Instructor Course (INIWIC) as primary instructors.

8-5. Commanders should structure the unit training program based on the METL, expected contingencies, prior training, and the proficiency of the unit. In developing a training program, use the seven-step approach. For in-depth instructions on training development, see *FMs 7-0* and *7-1*.

8-6. The objective is to train and evaluate units in the performance of civil disturbance procedures. Proficiency in performing civil disturbance procedures is reflected by a unit's ability to quickly and deliberately execute the disturbance control procedures required to prevent loss of life and property or to regain control of a civil disturbance in its assigned area of responsibility.

8-7. Training should focus on, but is not limited to—

- Instructors.
- Individual and collective drills.
- Leaders.
- Equipment.
- Weapons.
- Munitions.
- Live-fire range exercises.
- Qualification and familiarization.
- Soldier training exercises (STXs).
- Culminating exercises.

8-8. There are several successful methods to evaluate trained leader skills and trained soldier skills using scenarios, vignettes, lanes, or STXs. Flexibility is an important key in training plan development due to time restraints.

RANGE SETUP

8-9. The key to setting up a successful range for NL munitions is to ensure that they are in compliance with installation safety regulations. Kinetic energy is the force that pushes the projectiles down range. In most cases, kinetic energy will only carry these NL munitions projectiles 75 meters or less. This gives trainers flexibility as to the locations where a live-fire exercise can be conducted, especially in a field-expedient situation. *Figure 8-1* is an example for conducting a NL live fire range setup.

8-10. As illustrated in *Figure 8-1*, mark off 5-, 10-, 20-, 30-, 40-, and 50-meter distances from the targets with markers. Each of the NL munitions has minimum and maximum effective ranges within the distances of 5 to 50 meters. Impact power and accuracy are greater at close distances and depleted at farther distances. Targets can be anything from the standard E-type to more realistic three-dimensional manikins. Evenly divide your shooters into three groups as shown in *Figure 8-1*. Group one will initially be shield holders, group two will be shooters, and group three will be grenadiers.

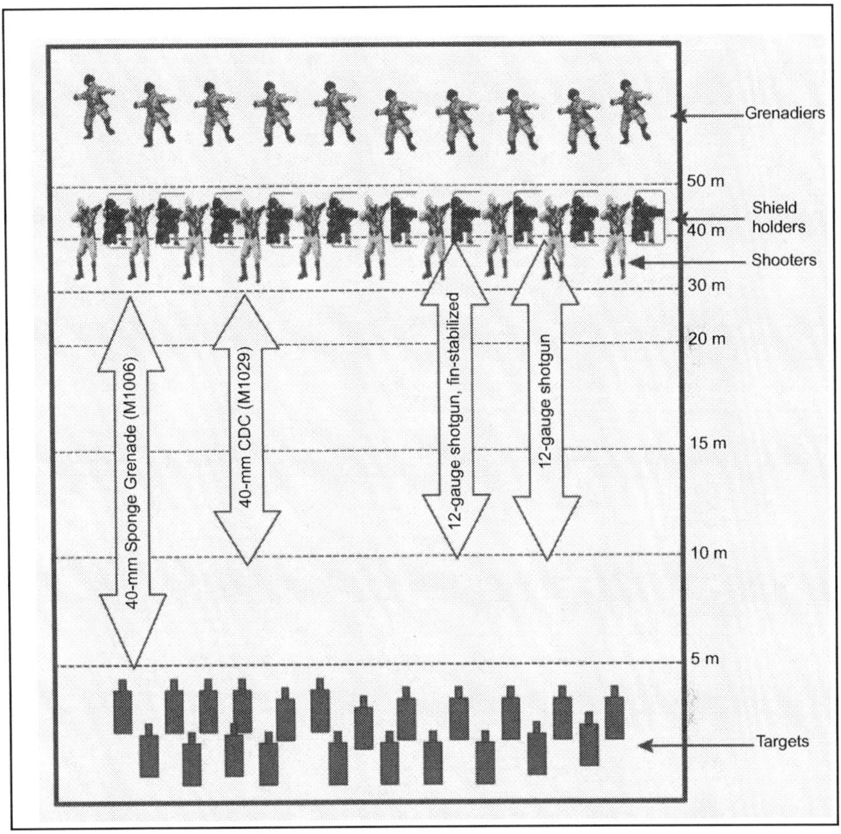

Figure 8-1. NL Munitions Range Setup

8-11. The formation should begin at the farthest effective range firing line (50 meters). After each volley of fire, the formation moves closer to the targets, but stays outside the closest effective range firing line (5 meters). For safety reasons, ensure that the formation stays online throughout each iteration.

8-12. When moving forward to each firing point, it is recommended that the formation always come to a halt before the fire command is given. It is also important to use the tap down drill method (see *Figure 8-1*) rather than having the shooters move forward and fire between the shield holders.

8-13. Range instructors can use verbal commands, such as, "Ready," or "Fire," for each volley. Shooters can be briefed to move forward, tap down the shield holder, and engage targets at the first explosion of distraction grenades thrown by grenadiers. In that instance, the command could simply be "Grenadiers Prepare to Throw, Throw."

8-14. The success of any civil disturbance operation is through effective communication, especially when incorporating the use of NL munitions. For effective training, use verbal commands in conjunction with effective, predetermined hand-and-arm signals during live-fire exercises.

8-15. Each iteration requires two phases of fire. During the first phase, half of the shooters will be firing NL, 40-millimeter rounds using an M203. The others will be firing NL, 12-gauge rounds using a 12-gauge shotgun. In the second phase, shooters switch weapons (safely). It is critical that line safeties are alert during this transition between phases.

8-16. Upon completion of each iteration, clear the weapons and the range. Establish a safe method and return all groups back to the farthest firing line. Once that is accomplished, group one rotates to become group three, group three rotates to become group two, and group two rotates to become group one for the next iteration. Using this method affords a safe and efficient way to get all shooters through the proficiency course. *Figure 8-2* is an example of how to fire each volley based on its characteristics, effectiveness, and capabilities.

Distance	Command	Subject	Actions
10 Meters	12-Gauge Shotgun, Two Point	Shotgun shooters	Load two M1013 rounds.
	M203, Two Point	M203 shooters	Load one M1006 round and have one readily available.
	Ready, Fire	Shotgun shooters	Tap down shields and engage designated targets.
		M203 shooters	Tap down shields, engage designated targets, reload, and reengage.
	Clear Range	Range safeties	Ensure that weapons are clear and safe.
	Safeties, Clear All Weapons	All shooters	Face to right, keep weapons pointed up and downrange.
	Shooters, Right Face, Move	All personnel	Step off shield holder, then shooters, and grenadiers on the command, "Move." Grenadiers reform at 50 meters.

Figure 8-2. Sample Live-Fire Exercise

Distance	Command	Subject	Actions
15 Meters	Shotgun, Two Point	Shotgun shooters	Load two M1013 rounds.
	M203, Two Point	M203 shooters	Load one M1006 round and have another available.
	Grenadiers, Prepare to Throw	Grenadiers	Make distraction grenade ready.
	Throw	Grenadiers	Throw grenade.
		Grenadiers	Throw distraction grenade.
	Ready, Fire	Shot gun shooters	Tap down shields and engage designated targets.
		M203 shooters	Tap down shields, fire at designated targets, reload with M1006 rounds, and engage designated target.
	Clear Range	Range safeties	Ensure that weapons are clear and safe.
	Three Steps Forward	All shooters and grenadiers	Echo preparatory command.
	Move	All shooters and grenadiers	Stay on line and step off three steps, counting each step aloud.
	Clear Range	Range safeties	Ensure that weapons are clear and safe.
	Seven Steps Forward	All shooters and grenadiers	Echo preparatory command.
	Move	All shooters and grenadiers	Stay on line and step off seven steps, counting each step aloud.

Figure 8-2. Sample Live-Fire Exercise (Continued)

Distance	Command	Subject	Actions
20 Meters	Shotgun, Three Point	Shotgun shooters	Load three M1013 rounds.
	M203, One Area, Two Point	M203 shooters	Load one M1026 round and have two M1006 rounds available.
	Grenadiers, Prepare to Throw	Grenadiers	Make distraction grenade ready.
	Throw	Grenadiers	Throw grenade.
		Grenadiers	Throw distraction grenade.
	Ready, Fire	Shotgun shooters	Tap down shields and engage designated targets.
		M203 shooters	Tap down shields, fire area at targets, reload with M1006 rounds, and engage designated target.
	Clear Range	Range safeties	Ensure that weapons are clear and safe.
	Shotgun, Two Point	Shotgun shooters	Load two M1013 rounds.
	M203, Two Point	M203 shooters	Load one M1006 round and have another available.
	Grenadiers, Prepare to Throw	Grenadiers	Make distraction grenade ready.
	Clear Range	Range safeties	Ensure that weapons are clear and safe.
	Three Steps Forward	All shooters and grenadiers	Echo preparatory command.
	Move	All shooters and grenadiers	Stay on line and step off three steps, counting each step aloud.

Figure 8-2. Sample Live-Fire Exercise (Continued)

Distance	Command	Subject	Actions
30 Meters	Shotgun, Two Area, Two Point	Shotgun shooters	Load two M1012 rounds and two M1013 rounds.
	M203, Two Area, Two Point	M203 shooters	Load one round M1026 and have one M1026 and two rounds M1006 available.
	Grenadiers, Prepare to Throw	Grenadiers	Make distraction grenade ready.
	Throw	Grenadiers	Throw grenade.
		Grenadiers	Throw distraction grenade.
	Ready, Fire	Shotgun shooters	Tap down shields, engage targets, and engage designated target with M1013 rounds.
		M203 shooters	Tap down shields, fire area at target, reload with M1006 rounds, and engage designated target.
	Clear Range	Range safeties	Ensure that weapons are clear and safe.
	Seven Steps Forward	All shooters and grenadiers	Echo preparatory command.
	Move	All shooters and grenadiers	Stay on line and step off seven steps, counting each step aloud.

Figure 8-2. Sample Live-Fire Exercise (Continued)

Distance	Command	Subject	Actions
40 Meters	Shotgun, Two Area, One Point	Shotgun shooters	Load two M1012 rounds and one round M1013.
	M203, One Area, One Point	M203 shooters	Load one M1026 round and have one M1006 round available.
	Grenadiers, Prepare to Throw	Grenadiers	Make distraction grenade ready.
	Throw	Grenadiers	Throw grenade.
		Grenadiers	Throw distraction grenade.
	Ready, Fire	Shotgun shooters	Tap down shields and engage targets with M1013 rounds.
		M203 shooters	Tap down shields, fire area at target, reload with M1006 rounds, and engage designated target.
	Clear Range	Range safeties	Ensure that weapons are clear and safe.
	Seven Steps Forward	All shooters and grenadiers	Echo preparatory command.
	Move	All shooters and grenadiers	Stay on line and step off seven steps, counting each step aloud.
50 Meters	Shotgun, One Area	Shotgun shooters	Load one M1012 round.
	M203, One Area, One Point	M203 shooters	Load one M1026 round and have one M1006 round available.
	Grenadiers, Prepare to Throw	Grenadiers	Make distraction grenade ready.
	Throw	Grenadiers	Throw grenade
		Grenadiers	Throw distraction grenade.
	Ready, Fire	Shotgun shooters	Tap down shields, engage targets, and engage designated target with M1013 rounds.
		M203 shooters	Tap down shields, fire area at the target, reload with M1006 rounds, and engage the designated target.
	Clear Range	Range safeties	Ensure that weapons are clear and safe.
	Seven Steps Forward	All shooters and grenadiers	Echo preparatory command.
	Move	All shooters and grenadiers	Stay on line and step off seven steps, counting each step aloud.

Figure 8-2. Sample Live-Fire Exercise (Continued)

NOTE: Shield holders are referred to as group one, shooters are referred to as group two, grenadiers are referred to as group three.

8-17. Currently, there are no qualification standards established in the Army for NL munitions. Area rounds, which contain multiple projectiles designed to affect multiple targets when engaged, only require approximately five rounds. This affords shooters an opportunity to realize the characteristics and capabilities of these rounds.

8-18. Single projectile rounds like the M1013 and the M1006 require a greater degree of proficiency. If situations call for the engagement of targets with these rounds, it is essential that trained shooters be able to hit what they are aiming at. For example, firing 10 rounds from various ranges can provide soldiers with the confidence to be effective shooters. A standard can be established that requires shooters to hit targets with at least 6 out of the 10 rounds used in the example to be considered proficient.

8-19. *Figures 8-3; 8-4,* page 8-11; *8-5,* page 8-21; and *8-6,* page 8-22, give examples of plans to follow if in a situation similar to the following scenario (see *Army Training and Evaluation Program (ARTEP) 19-313-10-MTP*):

A company commander is providing security patrols in support of a joint task force (JTF). Besides providing security patrols for a sector of responsibility, one platoon must always be designated as a quick reaction force (QRF) for civil disturbance operations. For the last 2 weeks the sector has been quiet and the commander has ordered all platoons to conduct civil disturbance training on a rotational basis.

A new order has been sent down to provide a squad-sized element to provide security for the future polling site for an upcoming election. This polling site has yet to open, but intelligence indicates potential problems as the election approaches.

On the day prior to the scheduled election, a crowd of approximately 25 to 30 demonstrators aggressively move toward the polling site, chanting slogans and yelling toward the security forces posted there. While some demonstrators begin throwing whatever they can pick up, the majority are simply yelling, locking arms, and blocking roadways. Hundreds of additional people, not part of the crowd, begin to gather to watch the activities as they unfold.

In an assessment of the demonstrators, it is determined that they intend to stay through the night into the morning of the election with the intention to block access to the polls. It is essential that the crowd is dispersed and that access to the polls is unrestricted in time for the election to begin the next morning.

Evaluation Scheme

The evaluation will consist of five lanes conducted over a 3-day period. Platoons will execute 24-hour operations during the evaluation. Any retraining will be done immediately following the execution of the lane if time permits.

The following tasks will be evaluated:

Lane 1	Cordon and search operations
Lane 2	Civil disturbance QRF
Lane 3	Security and defense of a critical site (polling site)
Lane 4	Night civil disturbance QRF
Lane 5	Night LP/OP (PIO)

Evaluation Schedule

Date and Time	1st Platoon	2d Platoon	3d Platoon	4th Platoon
	Lane 1	Lane 2	Lane 3	Lane 4
	Lane 2	Lane 3	Lane 4	Lane 5
	Lane 3	Lane 4	Lane 5	Lane 1
	Lane 4	Lane 5	Lane 1	Lane 2
	Lane 5	Lane 1	Lane 2	Lane 3

Figure 8-3. Evaluation Scheme

Unit Training Schedule and Firing Tables Unit Training Schedule 1					
Training Day 1					
Time	**Subject**	**Location**	**Uniform**	**Instructor**	**Notes**
0800	Administrative time	Classroom	U1		
0830	GRM	Classroom	U1		
1000	ROE and use of force	Classroom	U1		
1200	Chow				
1300	MOOTW	Classroom	U1		
1500	Effective communication	Classroom	U1		

Administration:

Uniforms:

U1: BDU

U2: BDU and LBE with harness and canteens

U3: BDU and LBE with harness, canteens, flak jacket, and Kevlar helmet

U4: BDU and LBE with harness, canteens, flak jacket, Kevlar helmet, and assigned weapon

Transportation:

Weapons:

Ammunition:

Required Equipment for Training:

1. Projection screen
2. Video projector with CPU
3. Television
4. VCR

Miscellaneous:

1. All personnel are required to have a notebook and a pencil or pen.
2. Handouts must be prepared and passed out.

Figure 8-4. Sample Two-Week Plan

Training Day 2					
Time	**Subject**	**Location**	**Uniform**	**Instructor**	**Notes**
0800	NLCS	Classroom	U1		
0900	PPCT and empty hand	Classroom	U1		
1000	PPCT and empty hand (Practical application)	PT field	U2		
1200	Chow				
1300	PPCT and empty hand (Practical application)	PT field	U2		
1700	COB				

Administration:

Uniforms:

U1: BDU

U2: BDU and LBE with harness and canteens

U3: BDU and LBE with harness, canteens, flak jacket, and Kevlar helmet

U4: BDU and LBE with harness, canteens, flak jacket, Kevlar helmet, and assigned weapon

Transportation:

Weapons:

Ammunition:

Required Equipment for Training:

1. Projection screen
2. Video projector with CPU
3. Training strike bags
4. Flex-Cufs

Miscellaneous:

1. All personnel are required to have a notebook and a pencil or pen.
2. Handouts must be prepared and passed out.
3. Medical personnel must be on location.

Figure 8-4. Sample Two-Week Plan (Continued)

Training Day 3

Time	Subject	Location	Uniform	Instructor	Notes
0800	PPCT and empty hand (Practical application)	PT field	U2		
1200	Chow				
1300	PPCT and empty hand (Practical application)	PT field	U2		
1700	COB				

Administration:

Uniforms:

U1: BDU

U2: BDU and LBE with harness and canteens

U3: BDU and LBE with harness, canteens, flak jacket, and Kevlar helmet

U4: BDU and LBE with harness, canteens, flak jacket, Kevlar helmet, and assigned weapon

Transportation:

Weapons:

Ammunition:

Required Equipment for Training:

1. Projection screen
2. Video projector with CPU
3. Training strike bags
4. Flex-Cufs

Miscellaneous:

1. All personnel are required to have a notebook and a pencil or pen.
2. Handouts must be prepared and passed out.
3. Medical personnel must be on location.

Training Day 4

Time	Subject	Location	Uniform	Instructor	Notes
0800	Riot baton training	Classroom	U1		
0900	Riot baton training (Practical application)	PT field	U2		
1200	Chow				

Figure 8-4. Sample Two-Week Plan (Continued)

1300	Riot baton training (Practical application)	PT field	U2		
1700	COB				

Administration:

Uniforms:

U1: BDU

U2: BDU and LBE with harness and canteens

U3: BDU and LBE with harness, canteens, flak jacket, and Kevlar helmet

U4: BDU and LBE with harness, canteens, flak jacket, Kevlar helmet, and assigned weapon

Transportation:

Weapons:

1. Expandable riot baton with carrier
2. Training riot batons

Ammunition:

Required Equipment for Training:

1. Projection screen
2. Video projector with CPU
3. Training strike bags
4. Flex-Cufs

Miscellaneous:

1. All personnel are required to have a notebook and a pencil or pen.
2. Handouts must be prepared and passed out.
3. Medical personnel must be on location.

Training Day 5

Time	Subject	Location	Uniform	Instructor	Notes
0800	Riot baton training (Practical application)	PT field	U2		
1000	Crowd dynamics	Classroom	U1		
1130	Chow				
1230	Crowd dynamics	Classroom	U1		
1330	CD formations	Classroom	U1		

Figure 8-4. Sample Two-Week Plan (Continued)

1430	RCAs and OC	Classroom	U1
1700	COB		

Administration:

Uniforms:

U1: BDU

U2: BDU and LBE with harness and canteens

U3: BDU and LBE with harness, canteens, flak jacket, and Kevlar helmet

U4: BDU and LBE with harness, canteens, flak jacket, Kevlar helmet, and assigned weapon

Transportation:

Weapons:

1. Expandable riot baton with carrier
2. Training riot baton

Ammunition:

Required Equipment for Training:

1. Projection screen
2. Video projector with CPU
3. Training strike bags
4. Flex-Cufs

Miscellaneous:

1. All personnel are required to have a notebook and a pencil or pen.
2. Handouts must be prepared and passed out.
3. Medical personnel must be on location.

Training Day 6

Time	Subject	Location	Uniform	Instructor	Notes
0800	RCAs and OC (Practical application)	PT field	U2		
1200	Chow				
1300	RCAs and OC (Practical application)	PT field	U2		
1700	COB				

Figure 8-4. Sample Two-Week Plan (Continued)

Administration:

Uniforms:

U1: BDU

U2: BDU and LBE with harness and canteens

U3: BDU and LBE with harness, canteens, flak jacket, and Kevlar helmet

U4: BDU and LBE with harness, canteens, flak jacket, Kevlar helmet, and assigned weapon

Transportation:

Weapons:

Ammunition:

1. MK inert OC
2. MK 4 live OC disperser
3. M36 live CR disperser

Required Equipment for Training:

1. Five large clean trash cans with liners
2. Two water hoses with running water

Miscellaneous:

Medical personnel must be on location.

Training Day 7

Time	Subject	Location	Uniform	Instructor	Notes
0800	NL munitions employment	Classroom	U1		
1030	Chow				
1115	Safety brief	Range	U4		
1130	NL munitions (live fire)	Range	U4		
TBD	Evening chow				
TBD	Safety brief	Range	U4		
TBD	NL munitions night fire exercise	Range	U4		
TBD	Clean up and COB				

Figure 8-4. Sample Two-Week Plan (Continued)

Administration:

Uniforms:

U1: BDU

U2: BDU and LBE with harness and canteens

U3: BDU and LBE with harness, canteens, flak jacket, and Kevlar helmet

U4: BDU and LBE with harness, canteens, flak jacket, Kevlar helmet, and assigned weapon

Transportation:

Yes

Weapons:

1. M16A2 with M203
2. 12-gauge shotgun

Ammunition:

1. Fin-stabilized M1012
2. M1013 multipellet (rubber)
3. M1006 40-millimeter sponge grenade
4. M1026 40-millimeter multiball (rubber)
5. M84 diversionary grenade

Required Equipment for Training:

1. Fifteen to twenty human form targets
2. Eye and hearing protection
3. Fifteen to twenty riot shields
4. Two MREs per student
5. Video recorder
6. Three to five high-intensity lights
7. All students should be issued face shields, shin guards, and ammunition pouches

Miscellaneous:

1. Medical personnel must be on location.
2. Range safety officers must be available.
3. Primary and secondary means of communication with range control must be available.
4. Operational risk assessments must be conducted.

Figure 8-4. Sample Two-Week Plan (Continued)

Training Day 8

Time	Subject	Location	Uniform	Instructor	Notes
0800	CD formations (Practical application)	TA 1	U4		
1200	Chow				
1300	CD formations (Practical application)	TA 1	U4		
1700	COB				

Administration:

Uniforms:

U1: BDU

U2: BDU and LBE with harness and canteens

U3: BDU and LBE with harness, canteens, flak jacket, and Kevlar helmet

U4: BDU and LBE with harness, canteens, flak jacket, Kevlar helmet, and assigned weapon

Transportation:

Yes

Weapons:

1. Training weapons
2. Training riot batons

Ammunition:

Required Equipment for Training:

1. Eye protection
2. Fifteen to twenty riot shields
3. Video recorder
4. Three to five high-intensity lights
5. Face shields, shin guards, and ammunition pouches

Miscellaneous:

1. Medical personnel must be on location.
2. Primary and secondary means of communication must be available.
3. Operational risk assessment must be conducted.
4. Twenty to thirty OPFOR personnel must be available.

Figure 8-4. Sample Two-Week Plan (Continued)

Training Day 9

Time	Subject	Location	Uniform	Instructor	Notes
0800	Introduction to M5, PVAB, and M315	Classroom 1	U1		
1200	Chow				
1300	M5, PVAB, and M315 (Practical application/LFE)	Range	U4		
1700	COB				

Administration:

Uniforms:

U1: BDU

U2: BDU and LBE with harness and canteens

U3: BDU and LBE with harness, canteens, flak jacket, and Kevlar helmet

U4: BDU and LBE with harness, canteens, flak jacket, Kevlar helmet, and assigned weapon

Transportation:

Yes

Weapons:

Two complete M315s and installation kits

Ammunition:

1. Ten live MCCMs
2. Ten live M315 rounds of each

Required Equipment for Training:

1. Eye protection
2. Ten inert MCCMs
3. Two complete PVAB systems

Miscellaneous:

1. Medical personnel must be on location.
2. Primary and secondary means of communication must be available.
3. Operational risk assessment must be conducted.

Figure 8-4. Sample Two-Week Plan (Continued)

Training Day 10					
Time	**Subject**	**Location**	**Uniform**	**Instructor**	**Notes**
0800	Tactical consideration	MOUT classroom	U1		
0930	Safety brief	MOUT classroom	U4		
1000	Tactical exercise	MOUT Facility	U4		

Administration:

Uniforms:

U1: BDU

U2: BDU and LBE with harness and canteens

U3: BDU and LBE with harness, canteens, flak jacket, and Kevlar helmet

U4: BDU and LBE with harness, canteens, flak jacket, Kevlar helmet, and assigned weapon

Transportation:

Yes

Weapons:

1. Assigned weapons
2. Inert RCA/OC individual dispersers
3. Training riot batons
4. Fifty practice grenades

Ammunition:

Required Equipment for Training:

1. Eye protection
2. Ten inert MCCMs
3. Face shields, shin guards, and ammunition pouches
4. Fifteen to twenty full-length riot shields

Miscellaneous:

1. Medical personnel must be on location.
2. Primary and secondary means of communication must be available.
3. Operational risk assessment must be conducted.
4. A video recorder must be available.

Figure 8-4. Sample Two-Week Plan (Continued)

Sample Unit Training Schedule				
Training Day 1				
Time	**Event**	**Location**	**Instructor**	**Notes**
0730-0830	Force continuum	Classroom		L
0830-0930	Management of aggressive behavior	Classroom		L
0930-1030	ROE	Classroom		L
1030-1130	Crowd control	Classroom		L
1130-1230	Chow			
1230-1700	PPCT and empty hand	PT field		L, PA
Training Day 2				
Time	**Event**	**Location**	**Instructor**	**Notes**
0730-1130	Riot baton training	PT field		L, PA
1130-1230	Chow			
1230-1530	Riot baton training	PT field		PA
1530-1700	CD formations	Classroom		L
Training Day 3				
Time	**Event**	**Location**	**Instructor**	**Notes**
0730-1130	CD formations	PT field		PA
1130-1230	Chow			
1230-1700	NL munitions, NLW	Range		L, PA
Training Day 4				
Time	**Event**	**Location**	**Instructor**	**Notes**
0730-1130	RCAs and OC	Classroom and PT field		L, PA
1130-1230	Chow			
1230-1700	RCAs and OC	PT field		PA

Figure 8-5. Sample One-Week Plan

Training Day 5				
Time	**Event**	**Location**	**Instructor**	**Notes**
0730-1700	Tactical exercise	MOUT facility		OPS
NOTE: Appointed student leaders ensure that troops are fed and rotated for rest.				

Figure 8-5. Sample One-Week Plan (Continued)

Sample Unit Training Schedule				
Training Day 1				
Time	**Event**	**Location**	**Instructor**	**Notes**
	Shield and riot baton techniques	MOUT site		Individual and collective tasks
	NLCS and munitions	Range Classroom		Leader training
	Formation training	MOUT site		Individual and collective tasks
	Chow			
	Formation training	MOUT site		Individual and collective tasks
Training Day 2				
Time	**Event**	**Location**	**Instructor**	**Notes**
0730-1930, 12-hour operations	Live-fire exercise	Range		M203/12-gauge
	QRF exercise	MOUT site		Individual and collective tasks
NOTE: Once weapons and munitions familiarization is complete, shooters join the QRF exercise to be used. For live-fire range setup, see *Figure 8-1*.				

Figure 8-6. Sample Two-Day Plan

Appendix A

Metric Conversion Chart

This appendix complies with current Army directives which state that the metric system will be incorporated into all new publications. *Table A-1* is a metric conversion chart.

Table A-1. Metric Conversion Chart

US Units	Multiplied By	Metric Units
Cubic feet	0.0283	Cubic meters
Feet	0.3048	Meters
Feet per second	0.3048	Meters per second
Inches	0.0254	Meters
Inches	25.4001	Millimeters
Miles per hour	1.6093	Kilometers per hour
Pounds	0.4536	Kilograms
Yards	0.9144	Meters
Metric Units	**Multiplied By**	**US Units**
Cubic meters	35.3146	Cubic feet
Kilograms	2.2046	Pounds
Kilometers per hour	0.6214	Miles per hour
Meters	3.2808	Feet
Meters	39.3701	Inches
Meters	1.0936	Yards
Meters per second	3.2808	Feet per second
Millimeters	0.0394	Inches

Appendix B

Operations and Legal Considerations in the Continental United States

Providing military assistance to state and local governments to assist them in quelling a civil disturbance or riot requires close coordination through a host of state and federal agencies. It requires a thorough briefing of soldiers at all levels on what they can and cannot do with respect to law enforcement. Civil authorities must be briefed on the restrictions placed on federal forces by the *Constitution of the United States* and federal statutes and laws. The national guard (NG), when not in a federal status, operates under the control of the state governor and the adjutant general (AG). The NG has historically been the first military responder during emergencies.

FEDERAL INTERVENTION AND AID

B-1. Under the *Constitution of the United States* and *United States Code*, the President is empowered to direct federal intervention in civil disturbances to—

- Respond to state requests for aid in restoring order.
- Enforce the laws of the United States.
- Protect the civil rights of citizens.
- Protect federal property and functions.

B-2. Under the *Constitution of the United States*, each state is responsible for protecting life and property within its boundaries. State and local governments use their LEA forces to maintain law and order and to quell civil disturbances. However, if civil disturbance support requirements exceed the resources of a state, federal troops may be requested to help restore and maintain law and order.

B-3. The *Constitution of the United States* and federal statutes authorize the President to direct the use of federal armed troops within the 50 states and territories and their political subdivisions. The President is also empowered to federalize the NG of any state to suppress rebellion and enforce federal laws.

B-4. Federal assistance is provided to a state when the state has used all of its resources, including its NG, to quell a disorder and finds the resources insufficient. Usually, active duty federal forces are used to augment the NG of the requesting state. However, the President may choose to federalize the NG of another state and use them alone or with other forces to restore order.

B-5. The President may employ federal troops to ensure the execution of US law when a state opposes or obstructs US law or impedes the course of justice under those laws. The President may employ armed federal troops to suppress insurrection, domestic violence, unlawful assemblies, and conspiracy if such acts deprive the people of their constitutional rights and privileges and the civil authorities of a state cannot or will not provide adequate protection.

B-6. The president is also authorized to use armed federal troops to protect federal property and functions when the need for protection exists and local civil authorities cannot or will not give adequate protection. The right of the United States to protect all federal property and functions regardless of their locations is an accepted principle of our government.

B-7. As a temporary measure, federal military equipment and facilities may be loaned to law enforcement agencies of state and local governmental bodies for use during civil disturbances. These resources may also be loaned to state NG and non-DOD federal agencies. The requesting agencies are expected to provide enough resources of their own to minimize the need for US military resources. In addition, the loan of the resources must not conflict with US military needs.

ROLES AND RESPONSIBILITIES OF VARIOUS AGENCIES

B-8. Several agencies within the federal and state governments play a key role in supporting the request of local authorities for support during a civil disturbance.

SECRETARY OF DEFENSE

B-9. The Secretary of Defense retains approval for federal support to civil authorities involving the use of DOD forces, personnel, and equipment. Approval may also involve DOD support that will result in a planned event with the potential for confrontation with specifically identified individuals and groups or will result in the use of lethal force.

SECRETARY OF THE ARMY

B-10. The Secretary of the Army, as an executive agent for the Secretary of Defense, is the approval authority for federal emergency support in response to natural and man-made disasters (except weapons of mass destruction [WMD]). Military forces respond to directions and guidance from the director of military support (DOMS) (the action agent for the Secretary of the Army).

DIRECTOR OF MILITARY SUPPORT

B-11. The DOMS plans and executes the DOD domestic support mission to civil authorities. He is a general officer appointed by the Secretary of the Army. The DOMS is the DOD primary contact for all federal departments and agencies during DOD involvement in most domestic support operations.

STATE COORDINATING OFFICER

B-12. The governor appoints the state coordinating officer (SCO) to coordinate disaster operations for the state. The SCO is the primary point of contact for the federal coordinating officer (FCO) in facilitating disaster assistance.

DIRECTOR OF THE FEDERAL EMERGENCY MANAGEMENT AGENCY

B-13. The director of the Federal Emergency Management Agency (FEMA), on behalf of the President, appoints the FCO. The FCO coordinates the timely delivery of federal disaster assistance to the affected state and local governments and disaster victims. He works closely with the SCO to determine state requirements and to coordinate national level issues with the catastrophic disaster response group. There must be close and continuous coordination with FEMA to ensure the proper use of military assets in support of civil authorities.

DEFENSE COORDINATING OFFICER

B-14. The defense coordinating officer (DCO) operates under the authority of the President and is the DOD representative designated to coordinate on-the-scene activities with the FCO. FEMA and other federal agencies that request support from DOD go through the DCO for validation and resourcing.

NATIONAL GUARD

B-15. The NG, in state status (see *Title 32, United States Code {USC}),* is the governor's primary military response organization for emergencies and disasters. The NG responds and operates under state laws unless they are federalized, when they operate under the same constraints and laws as federal units.

UNITED STATES ARMY RESERVE

B-16. The US Army Reserve is capable of providing military support. This assistance and support may include the use of equipment and other resources, including units and individuals.

STATE AND LOCAL GOVERNMENT RESPONSIBILITIES

B-17. State and local government officials, operating under the authority granted by state constitutions and the *Constitution of the United States*, are responsible for daily safety and security issues that impact the quality of life of their citizens. State and local officials have primary responsibility for emergency preparedness.

STATE RESPONSIBILITIES

B-18. Many state government agencies are responsible for coming to the aid of state citizens.

Governor

B-19. A state governor is empowered by each state constitution and the *Constitution of the United States* to execute the laws of the state and to command the state NG while it is serving in state status. Governors are also responsible for issuing executive orders or emergency proclamations, such as declaring states of emergency and ensuring that state agencies plan for potential disasters and civil disorder.

B-20. Once a man-made or natural disaster occurs, the governor assesses its extent and determines if local government requests for assistance should be honored. If appropriate, the governor declares a state of emergency, activates the state emergency response plan, and may order the NG into state active duty. The governor assigns the NG its mission through the office of emergency service (OES)/emergency medical dispatching (EMD).

Plans, Operations, and Military Support Officer

B-21. Each state NG has a plans, operations, and military support officer (POMSO) who prepares and coordinates contingency plans for military support to civil authorities (MSCA) during disaster response and recovery operations. The POMSO coordinates training plans and exercises between the state NG and federal, state, and local emergency management agencies. The POMSO serves as the NG point of contact with DOD officials during a declared federal emergency or disaster.

Office of Emergency Services

B-22. Each state has a specific agency and/or office that coordinates emergency preparedness and serves as the governor's coordinating agency during an emergency. The titles of these offices vary from state to state. Some examples are the Division of Emergency Government, Emergency Management Agency, Department of Public Safety, or Office of Emergency Preparedness. This manual refers to this office using the generic term OES/EMD.

B-23. Generally, the OES/EMD is organized as a stand-alone office under the governor or aligned under the AG or the state police, and in some instances, the AG is the OES/EMD director. It operates the state emergency operations center and coordinates with state agencies and federal officials for assistance.

Adjutant General

B-24. The state NG is the governor's primary military response force in an emergency. The AG, through the state area command (STARC) (specifically the POMSO), develops and coordinates emergency response plans for disasters and emergencies. The AG is in command of state NG forces ordered into state active duty.

State Government Agencies

B-25. State government departments and agencies prepare emergency preparedness plans for their areas of specialization in support of the governor's plan. They also participate in emergency preparedness exercises.

LOCAL RESPONSIBILITIES

B-26. Mayors, city managers, local police, fire protection officials, county executives, sheriffs, prosecuting attorneys, and public health officials are some of the people responsible for law enforcement, safety, health, and fire protection on a daily basis. They are responsible for developing appropriate emergency response plans and responding to emergencies within their jurisdictions. Most local jurisdictions have an OES/EMD to plan and coordinate actions in an emergency. In many cases, local jurisdictions have mutual aid agreements with other jurisdictions that allow for firefighting, medical, and police assistance across jurisdictional boundaries. Once local officials determine that an emergency is beyond the scope of their resources or ability to respond, the senior local official is responsible for requesting additional assistance from the next agency in the assistance request process.

LEGAL CONSIDERATIONS AND CONSTRAINTS

B-27. The *Constitution of the United States*, laws, regulations, policies, and other legal issues limit the use of federal military personnel in domestic support operations. Any Army involvement in civil disturbance operations involves many legal issues requiring comprehensive legal reviews.

B-28. Under the *Constitution of the United States*, Congress has the authority to raise and support an army, provide and maintain a navy, and make rules for governing and regulating the land and naval forces. The *Constitution of the United States* places the military under civilian control and designates the President as commander-in-chief. Statutes provide for civilian leadership in the office of the secretary of defense, service secretaries, and various other civil authorities. Under the *Constitution of the United States*, the civilian government must preserve public order and carry out governmental operations within its territorial limits. The *Constitution of the United States* allows the use of the federal military to execute or enforce the law when necessary to protect federal or civilian property and functions. However, significant restrictions exist on employing federal military forces within the United States.

B-29. The unique capabilities of the military enable it to support federal, state, or local civilian agencies. In most circumstances, the DOD is one of many federal agencies reacting to a domestic emergency or crisis, playing a subordinate, supporting role to a lead civilian agency.

LAW

B-30. In the United States, civilian agencies provide for the needs of citizens. Civilian, federal, state, and local government and law enforcement agencies execute US laws. Laws governing the use of the military in domestic operations are complex, subtle, and constantly changing. Commanders involved in civil disturbance operations should staff plans, policies, programs, exercises, funding, operations, constraints, and limitations with their local SJA to ensure conformity with legal requirements. They should scrutinize each request for aid (whether it is for equipment, individuals, or training) to ensure that it conforms to statutory requirements. The following references are a snapshot of the law on military support to civilian authorities:

- *Section 331, Chapter 15, Part 1, Subtitle A, Title 10 USC.*
- *Section 332, Chapter 15, Part 1, Subtitle A, Title 10 USC.*
- *Section 333, Chapter 15, Part 1, Subtitle A, Title 10 USC.*
- *Section 334, Chapter 15, Part 1, Subtitle A, Title 10 USC.*
- *Section 1385, Chapter 67, Part 1, Title 18 USC.*
- *DOD Directive 3025.12.*

- *DOD Directive 3025.15*
- *DOD Directive 5525.5*
- *DOD Civil Disturbance Plan 55-2*
- *NGR 500-1/ANGI 10-8101*
- *FM 3-19.15*

INFORMATION OPERATIONS

B-31. Information superiority helps forces anticipate problems and requirements. It allows commanders to control situations earlier and with less force, creating the conditions necessary to achieve the end state. Public affairs, PSYOP, and civil military operations are activities that will allow the commander to control situations earlier and with less force.

B-32. Information is available from a multitude of sources, the primary being open sources, law enforcement, and the military. A diversity of sources is the best approach because it prevents biased behavior.

Open sources include—

- Libraries.
- Newspapers and news periodicals.
- Radio and television.

Law enforcement sources include—

- Local law enforcement agencies.
- National law enforcement agencies.

Military sources include—

- DOD intelligence community (most restrictive source).
- Local MI field offices.

B-33. The DOD intelligence community operates under limitations imposed by regulations and executive orders. Attempts to skirt these restrictions may violate regulations or federal statutes.

RESTRICTIONS

B-34. Strict limits are placed on collecting information related to a civil disturbance in order to protect the civil rights of people and organizations not affiliated with the DOD. Civil disturbance plans and materials must not include lists of groups or people not affiliated with the DOD. Information on civilians and civilian organizations can be collected only with specific authorization from the Secretary or the Under Secretary of the Army. Conditions for collecting information include the existence of threats against Army personnel, functions, or property (see *ARs 380-10* and *380-13)*. Civil disturbance information (available in public documents) or open-source information may be collected. However, specific rules regarding its storage must be followed. Commanders must coordinate with SJA, MI, and USACIDC before collecting any such information.

B-35. Civilian law enforcement agencies may limit access to law enforcement intelligence. In such circumstances, the military may request that agencies release the unclassified version.

B-36. If the Department of Justice (DOJ) determines that federal intervention into a civil disturbance is likely, then information relating to the disturbance is provided to the Army Assistant Chief of Staff for Intelligence. The information is analyzed and then passed through channels to the TF commander for planning purposes.

B-37. Military intelligence collection efforts, except liaison, may begin only when DA orders. The orders will come through the CSA's personal liaison officer and the TF commander. Covert operations to gather information on non-DOD individuals and groups must be approved on an operation-by-operation basis by the Undersecretary of the Army.

B-38. Once DA approves the collection effort, MI elements establish and maintain liaison with the appropriate local, state, and federal authorities. Then they collect information pertaining to the incident and general situation. Their estimate of the situation is passed to the TF commander and DA.

POSSE COMITATUS

B-39. Generally, federal military forces may not give law enforcement assistance to civil authorities without conflicting with the *Posse Comitatus Act.* However, constitutional and statutory exceptions to this prohibition do exist. The recent emphasis on drug interdiction has led to an increase in those exceptions.

B-40. The *Judiciary Act of 1789* allowed US marshals to call upon the military as a posse comitatus. This continued until after the Civil War, when the federal government used the Army to execute reconstruction era policies. The southern states regarded the use of the military for this purpose as abusive and repressive, and in 1878, President Rutherford B. Hayes signed the original bill ending the practice.

B-41. The *Posse Comitatus Act* prescribes criminal penalties for the use of the US Army or Air Force to execute the laws of or to perform civilian law enforcement functions within the United States. DOD policy extends this prohibition to the US Navy and Marine Corps. Prohibiting the military from executing the laws means that military personnel may not participate directly in—

- An arrest; a search and seizure; a stop and frisk; or an interdiction of vessels, aircraft, or vehicles.
- A surveillance or pursuit.
- A civilian legal case or any other civilian law enforcement activity as informants, undercover agents, or investigators.

B-42. The *Posse Comitatus Act* does not apply to the following:

- Members of the NG when in state service.
- Members of a Reserve Component when not on active duty or active duty for training.
- DOD civilians, unless under the direct command and control of an active duty officer.

- A soldier when off duty and acting only in a private capacity.
- Soldiers taking action for the primary purpose of furthering a military or foreign affairs function of the United States; for example, enforcing military justice, maintaining law and order on military installations, and protecting classified materials.

CONSTITUTIONAL EXCEPTIONS TO THE POSSE COMITATUS ACT

B-43. Under its inherent authority, the US government is responsible for preserving public order and carrying out governmental operations within its territorial limits, by force if necessary. Under the *Constitution of the United States*, exceptions allow the use of military force to execute or enforce the law. Some of the exceptions are to—

- Protect civilian property and functions.
- Ensure preservation of public order and carry out government operations, by force if necessary.
- Protect civil rights where local authorities do not or cannot protect them.
- Protect federal property and functions.
- Quell civil disturbances and labor strife that rises to the level of civil disorder.

B-44. The President may order the armed forces to support state civil authorities suffering from an insurrection or civil disturbance. He must act personally by first issuing a proclamation calling on insurgents to disperse and retire peacefully within a limited time. This occurred when federal forces were called in to counter the Los Angeles riots in 1992.

Protect Civilian Property and Functions

B-45. A sudden and unexpected civil disturbance, disaster, or calamity may seriously endanger life and property and disrupt normal governmental functions to such an extent that local authorities cannot control the situation. At such times, the federal government may use military force to prevent loss of life or wanton destruction of property and to restore government functions and public order. This exception has rarely been used.

Protect Federal Property and Functions

B-46. The federal government may use military force to protect federal property and federal government functions when local authorities cannot or decline to provide adequate protection.

ARMED FORCES

B-47. Basic guidance for the interaction of the military, the Reserve Component, and civilian LEAs is given in *10 USC Chapter 18*. Guidelines on reimbursement and restrictions, direct participation in law enforcement activities, the use of information collected during military operations, and the use of military equipment and facilities are some of the topics covered. Title 10 prohibits the military from directly participating in arrests, searches, seizures, or other similar activity unless authorized by law (such as arrests on military property). The *Fiscal Year 1989 Act* and subsequent *National Defense Authorization Acts* have authorized the DOD to provide more support to LEAs in the counterdrug effort.

NATIONAL GUARD

B-48. For a description of how the secretary of defense may provide funds to state governments (including the District of Columbia, Commonwealth of Puerto Rico, and US territories) for civil disturbance operations performed by the National Guard when not in federal service, see 32 *USC*. The *Posse Comitatus Act* does not apply to the NG when not in federal service. Each state specifies how NG may be employed.

SUPPORT TO CIVIL LAW ENFORCEMENT

B-49. It is DOD policy to cooperate with civilian law enforcement officials to the greatest practical extent. However, cooperation must consistently meet the needs of national security and military preparedness, the historic tradition of limiting direct military involvement in civilian LEAs, and the requirements of applicable law. US military forces are never placed under the command of civilian law enforcement officers or nonfederalized NG. DOD policy concerning the provision of military support to LEAs, including personnel and equipment, are contained in *DOD Directive 5525.5.*

B-50. The Army assists civilian law enforcement by providing personnel, equipment, training, and expert advice within the limits of applicable law. Army NG units in state status (see 32 *USC*) provide the primary source of military assistance. They may be able to provide assistance to civil authorities in instances when federal units are precluded due to the restrictions of the *Posse Comitatus Act.*

ROLES OF THE NATIONAL GUARD AND FEDERAL FORCES

B-51. The preservation of law and order in the civilian community is the responsibility of state and local governments and law enforcement authorities. When a civil disturbance in a civilian community turns to widespread rioting that includes arson, looting, and acts of violence, the civil authorities may decide that they do not have the resources to quell the riot. They may then turn to the NG and federal forces to support the civil authorities in restoring law and order. Civil disturbances in any form are prejudicial to public law and order.

NATIONAL GUARD FORCES

B-52. The NG (as a state organization) responds to the governor according to state law for civil disturbance operations. NG regulations direct planning and training for the civil disturbance mission. During most civil disturbance situations, the NG will be the first military responder and will usually remain in state active duty status (according to 32 *USC*) throughout the operation. In extreme circumstances, the NG can be brought on federal service for civil disturbance operations when ordered to under the appropriate federal statute by the President. During the 1992 Los Angeles civil disturbance, the President federalized the California NG. Once the NG has been deployed to assist the civil authorities with a civil disturbance, it is too far into the deployment cycle to start determining what tasks the NG can do as a state or as a federally controlled force. It is the commander's responsibility not to accept missions that the unit is not adequately trained to do. The following are examples of appropriate missions for the NG:

- Manning traffic control points.
- Providing building security.
- Providing area security and patrols.
- Providing security at custody facilities.
- Providing security and escort for emergency personnel and equipment.
- Protecting sensitive sites.
- Transporting law enforcement personnel.
- Providing show of force.
- Dispersing crowds.
- Employing riot control munitions.
- Providing very important person (VIP) protection and escort.
- Providing quick reaction and reserve force.

B-53. Examples of inappropriate taskings:

- Negotiating hostage situations.
- Barricading suspects.
- Participating in criminal investigations.
- Acting as federal Army forces.

B-54. The commitment of federal troops to deal with domestic civil disturbances must be viewed as a drastic measure of last resort. Their role should never be greater than what is absolutely necessary under the circumstances. Commanders should take every measure to avoid the perception of an invading force. A JTF designated to respond to a civil disturbance should project the image of a restrained and well-disciplined force whose sole purpose is to help restore law and order with minimal harm to the people and property and with due respect for all law-abiding citizens.

B-55. The role of federal Army forces is to assist civil authorities in restoring law and order when the magnitude of the disturbance exceeds the capabilities of local and state LEA, including the NG. Under the provisions of the *Constitution of the United States* and selected federal statutes, the President may order the employment of the federal armed forces to aid local and state civil authorities to protect the Constitutional rights of citizens. Federal military forces may also protect federal facilities and installations in any state, territory, or possession.

B-56. The *DA Civil Disturbance Plan 55-2* gives guidance for Army forces directed to quell civil disturbances. In addition, *AR 500-50* outlines responsibilities, prescribes policies, and provides guidance for the use of Army resources in response to civil disturbances. It outlines statutory restrictions and designates staff and command responsibilities for planning and executing civil disturbance operations.

B-57. The JTF commander exercises control of all federal military forces (including NG in federal status) committed to assist civil authorities. Federal military forces remain under the military chain of command during civil disturbance operations. Federal forces will not be placed under the command of either civil officials or NG commanders in a nonfederal status. Civil authorities retain control of their state and local law enforcement agencies. The JTF commander establishes liaison with the senior representative of the attorney general (SCRAG) and other appropriate federal, state, and local civil authorities.

B-58. When a JTF is being assembled it must deploy with a provost marshal (PM) who will be the POC for situation assessment, operational information, and security matters. He also serves as a liaison with civilian and law enforcement agencies.

B-59. Federal military forces must be tailored to the specific civil disturbance situation. Sufficient combat support and combat service support units will be required to sustain the force throughout the deployment. Coordination with civil officials may allow the force to draw on resources available from state and local agencies. Close and continuous coordination between federal military forces and LEAs will provide the commander with the detailed and timely information required to employ and protect the force effectively.

B-60. In supporting *DA Civil Disturbance Plan 55-2,* intelligence personnel may conduct close and continuous liaison with LEAs and military police to ensure that their units receive the information needed to allow the commander to adequately protect the force. The JTF commander should staff intelligence support missions with his senior intelligence officer and legal counsel before approving the mission.

B-61. Federal military forces must be employed in tasks or missions appropriate to their organization and training; they must not be employed in ways that violate the legal restrictions in effect. Military forces may be used to disperse unlawful assemblies and to patrol disturbed areas to prevent unlawful acts. They may be used to assist in the distribution of essential goods and the maintenance of essential services. Forces may also establish traffic control points, cordon off areas, release smoke and obscurants, and serve as security or quick reaction forces. Certain types of missions (such as gathering intelligence on civilians) are always inappropriate for military forces during civil disturbance operations.

B-62. Requests for specific military missions are typically passed through a single state or federal law enforcement coordinating officer, as approved by the SCRAG. Validated requests are transmitted to the JTF commander and his HQ for staffing and coordination. Approved missions are assigned through the military chain of command to the appropriate element or unit for execution. Units and soldiers will not accept taskings or missions directly from law enforcement or civilian officials, except in a direct support relationship as approved and ordered through the military chain of command.

B-63. It is important that a military liaison be provided to each LEA HQ generating requests for support. This liaison can assist LEA officials in determining the types and quantities of military support to be requested. JTF HQ can facilitate this mission assignment process by providing LEAs with a detailed listing of the types of missions military forces may conduct.

B-64. The AO for a deployed unit should coincide with the jurisdiction or subdivision boundaries of the LEA it supports. This arrangement facilitates liaison and coordination between law enforcement and military chains of command.

LEAD AGENCY CONCEPT AND ROLE OF MILITARY

B-65. The DOJ is the lead federal agency for civil disturbance operations. The Attorney General's on-scene representative is the SCRAG. The DOD has designated DA as their executive agent for military assistance for civil disturbance (MACDIS) (see *FM 27-100.*)

JOINT TASK FORCE

B-66. Military support to civil authorities in disasters and domestic emergencies is a DOD (not a service component) responsibility. The supported commander may decide that the severity and scope of a disaster require a joint response. In civil assistance, as in operational level warfighting, the commander-in-chief (CINC) uses the different and complementary capabilities of each service to accomplish the mission. Such use requires knowledge of both the capability and the availability of all service component assets, to include their agencies and installation.

B-67. The CINC may establish a JTF to provide comprehensive military support. Establishing a JTF may provide the best DOD response to civil emergencies. A JTF is established to execute a specific mission limited in scope and duration. The JTF objective in civil emergencies is to deploy forces to the area rapidly to assist immediately in saving lives and safeguarding property and to continue providing assistance required by the federal response plan (FRP) and the particular situation.

B-68. The JTF is configured for each specific mission. In some civil emergencies, the JTF may require a greater proportion of combat service support type units and capabilities than in typical warfighting deployments. The JTF must be able to provide emergency assistance across all lines of support.

JOINT TASK FORCE CONTROL ELEMENT

B-69. Responsibility for controlling the civil disturbance resides with the civil authorities. The JTF is organized to support and lend assistance to them in their effort to return civil order to the community.

B-70. The control element consists of the command group and the crisis management team (CMT). The command group is made up of city, county, state, and military command personnel. They set policy and issue directives, coordinate the activities of civil and military support agencies, supervise the CMT, and coordinate with outside agencies. The CMT is made up of representatives of civilian and military staff sections. The CMT advises the command group and coordinates operations and support for the action element of the TF. The control elements are located with the emergency operations center (EOC) to facilitate information processing, resource management, and operational control. The JTF command element should colocate with the civilian command element at the EOC. This will expedite the exchange of information. If an EOC has not been set up, the commander may establish one while making provisions for including civil authorities to ensure a unity of effort.

B-71. Not all CMT members are located at the EOC. Some key people may use liaisons to represent them at the EOC. Key personnel can then research and discuss ideas freely, away from the confusion associated with crisis management. The liaison can transmit guidance and answers to the EOC. Some agencies may not be needed in the EOC at first, but they must be prepared for inclusion. The decision to include an agency in the EOC is based on the likelihood that they may have to take an action or a support role on the importance of the agency to the mission.

Appendix C

Urban Terrain Analysis

Of all the environments in which the Army may need to conduct operations, the urban environment confronts commanders with a combination of difficulties rarely found in other environments. The distinct characteristics of the environment result from the combination of man-made features and supporting infrastructure superimposed on the existing natural terrain and the density of civilians in close proximity to combat forces. The human dimension it represents is potentially the most important and perplexing for commanders and their staffs to understand and evaluate. Commanders at all levels must make extraordinary efforts to assess and understand their particular urban environment to plan, prepare for, and execute effective civil disturbance operations.

SIGNIFICANT FACTORS

C-1. Although urban areas throughout the world possess similar characteristics, no two are alike. The sprawl of Los Angeles, for example, bears little physical resemblance to New Delhi. The most significant factors affecting the uniqueness and complexity of each area are its societal characteristics. While the terrain can be extremely complex, information about this aspect, its potential effects on operations, and how it changes over time may be determined with some degree of certainty. However, the human dimension is more difficult to understand and assess, particularly its effect on military operations. The urban environment is not neutral—it helps or hinders. The commander who can best understand and exploit the effects of the urban environment has the best chance of success.

C-2. Whether a metropolitan area or a village, each urban environment has an identifiable system of urban components that constantly change and interact. This system of components consists of the terrain, the society, and the infrastructure that provides the critical link between the two (see *Figure C-1,* page C-2).

C-3. These systems are not separate and distinct categories but instead are overlapping and interdependent. A thorough analysis of these systems contributes to the commander's situational understanding and allows him to develop COAs that apply appropriate resources to control a crowd and prevent a major civil disturbance. This understanding allows commanders to engage decisive points critical to maintaining peace or restoring normalcy to the urban environment.

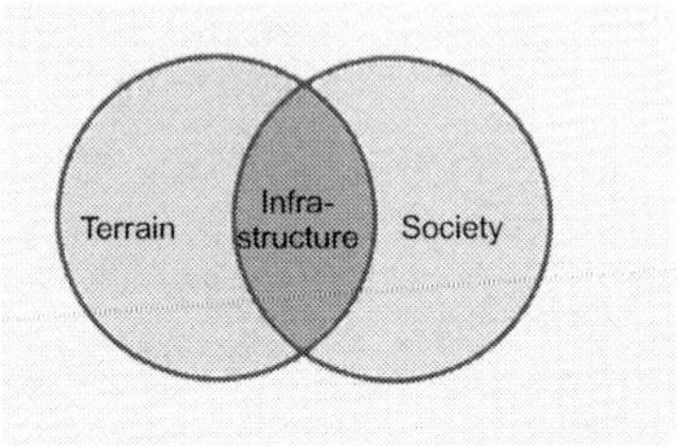

Figure C-1. Keys to Understanding the Urban Environment

C-4. The social aspects and the integrating infrastructure are the most difficult for commanders to understand and assess. These issues require greater dependence on LEAs and local governmental agencies for their information, knowledge, and expertise.

URBAN TERRAIN TYPES

C-5. The physical terrain is the most basic and easily recognizable aspect of the urban area; however, truly understanding it requires comprehending its multidimensional nature, urban and street patterns, urban form and function, and size. The infinite ways in which these factors (see *paragraph C-15)* can intertwine make it impossible to describe a typical urban area. However, these factors provide a framework for understanding the complex terrain within an urban environment. Although the terrain consists of both natural and man-made features, the latter dominates the physical characteristics of the urban environment and has significant effects on military systems and soldiers, tactics, and operations.

URBAN PATTERNS

C-6. The combinations of urban areas, lines of communications (LOCs), and natural terrain result in the identification of four urban patterns (see *Figure C-2*) that will influence civil disturbance operations. An integral part of each urban pattern is the hub or the central urban area around which outlying urban areas radiate. In SOSO, the hub is important, as it is often the economic, political, or cultural center of the surrounding area.

Satellite Pattern

C-7. This pattern consists of a central hub surrounded by smaller, dependent urban areas. LOCs tend to converge on the hub. The natural terrain throughout this pattern is relatively homogenous. Outlying or satellite urban areas often support the principal urban area at the hub.

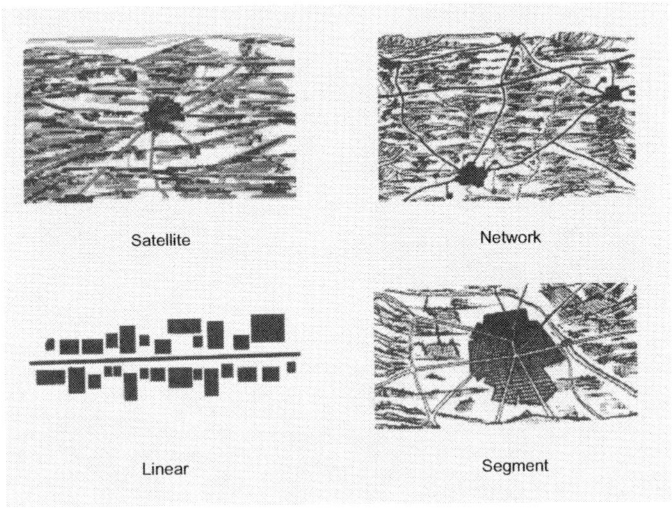

Figure C-2. Urban Patterns

Network Pattern

C-8. This pattern represents the interlocking of the primary hubs of subordinate satellite patterns. Its elements are more self-sufficient and less supportive of each other, although a dominant hub may exist. Major LOCs within a network are more extensive than in a satellite pattern and take a rectangular rather than a convergent form. The natural terrain within a network may vary more than in a single satellite array.

Linear Pattern

C-9. This pattern is a potential subelement of the previous two patterns. It may form one ray of the satellite pattern or be found along connecting links between the hubs of a network. Most frequently, however, this pattern results from the string of minor urban areas aligning a confined natural terrain corridor such as an elongated valley, a body of water, or a man-made communications route.

Segment Pattern

C-10. When dominant natural terrain (such as a river or man-made features [canals, major highways, or railways] divides an urban area, it creates a segmented pattern). This pattern makes it easier to assign areas of responsibility to units. However, if operations are in support of the city LEA, then units should be assigned to a police precinct to avoid crossing several precinct lines, as would happen when assigning units by dominant terrain features.

STREET PATTERNS

C-11. Patterns within the urban area result from the arrangement of the streets, roads, highways, and other thoroughfares. They are a result of the influences of natural terrain, the original designer's personal prejudices, and the changing needs of the inhabitants. The commander must understand the street patterns in the area in which his control force will operate. This will offer force protection, preventing the control formation from being outmaneuvered, flanked, or surrounded by a hostile crowd. Urban areas can display any of the three basic patterns or any combination of the three (see *Figure C-3).*

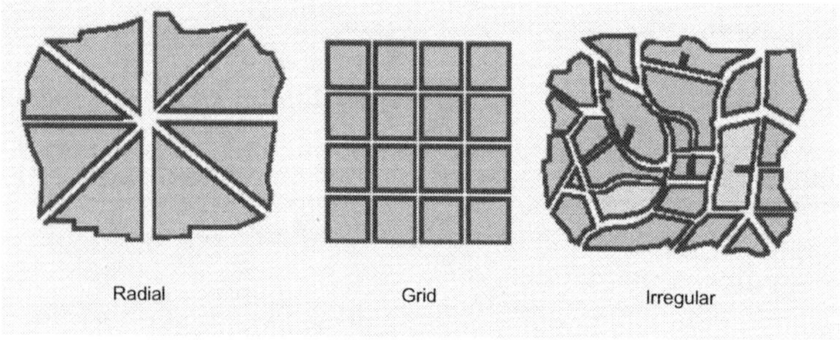

Figure C-3. Basic Internal Street Patterns

Radial

C-12. Societies of highly concentrated religious or secular power often construct urban areas with a radial design with all primary thoroughfares radiating out from the center of power. Terrain permitting, these streets may extend outward in a complete circle or may form a semicircle or arc when a focal point abuts a natural barrier such as a coastline or mountain. To increase mobility and traffic flow, societies often add concentric loops or rings to larger radial patterns.

Grid

C-13. The most adaptable and universal form for urban areas is the grid pattern (lines of streets at right angles to one another forming blocks similar to the pattern of a chessboard). A grid pattern can fill in and eventually take over an original radial pattern. Most urban areas, regardless of the original intent, plan, or vision, emerge from successive plans overlaid on one another.

Areas are well planned to fit with previous plans and others a haphazard response to explosive urban growth. The results may mix different patterns, blend patterns in symmetric combinations, or have no discernable geometric pattern (the irregular pattern).

Irregular

C-14. Irregular patterns can be specifically designed for aesthetic reasons (as in many suburban housing developments). They are sometimes used to conform to marked terrain relief (see *FMs 3-06* and *3-06.11*).

URBAN FORM AND FUNCTION

C-15. Throughout the world, urban areas have similar form and function. In form, urban areas contain like characteristics, readily divisible into distinct sections or areas. Functionally, they tend to be the center of population, finance, politics, transportation, industry, and culture.

- **Core.** The core is the heart of the urban area, the downtown, or central business district. Relatively small and compact, it contains a large percentage of the urban area's shops, offices, and public institutions.
- **Core Periphery.** As the name implies, the core periphery forms the outer edges of the core of the urban area. Generally, it has undergone less change than the core, resulting in buildings of uniform height (generally two to three stories in towns and five to ten stories in larger urban areas).
- **Industrial Area.** Industrial areas generally develop on the outskirts of the urban areas, where commercial transportation is easiest (along airfields and major sea, river, rail, and highway routes). The open/ dispersed pattern of the buildings provides sufficient space for large cargoes, trucks, and materials-handling equipment.
- **High-Rise Area**. High-rise areas consist of multistoried apartments, commercial offices, and businesses separated by large open areas such as parking lots, parks, and individual one-story buildings. The automobile, mass transit systems, and improved road networks encourage these areas to grow and function further from the urban core.
- **Residential Area.** Residential areas can be found dispersed throughout the urban area; however, large suburban areas (or sprawl) normally form on the outskirts. The combined population of surrounding suburban areas often outnumbers that of the urban area. Proper and specific suburbs tend toward homogeneity based on ethnicity, religion, economics, or some other social aspect.
- **Commercial Ribbon Area**. Commercial ribbon areas are rows of stores, shops, and restaurants built along both sides of major streets that run through urban areas. These same types of areas often develop along the roads that connect one urban area to another (strip areas).

URBAN SOCIETY

C-16. Although complex, understanding the urban terrain is relatively straightforward in comparison to comprehending the multifaceted nature of urban society. As urban areas increase in size, they generally lose homogeneity; therefore, commanders must understand and account for the characteristics of a diverse population whose beliefs may vary. The behavior of civilian populations within an urban area is dynamic and poses a special challenge to commanders.

C-17. The center of gravity during an urban operation may be the civilian inhabitants themselves. The side that enjoys the support of the population has many advantages. To gain and retain this support, commanders must first understand the complex nature and character of the urban society. Second, they must understand and accept that every military action (or inaction) may influence the relationship between the urban population and Army forces, and by extension, mission success.

C-18. The identification of groups helps commanders focus on specific segments of the urban society to determine their beliefs and needs and how those beliefs and needs motivate them to future action (or inaction). Groups may be categorized based on race, religion, national origin, tribe, clan, class, party affiliation, education level, union memberships, occupation, age, or any other significant social grouping.

C-19. Commanders should consider political, economic, and historical factors in this analysis. These factors usually affect all groups to some extent and often provide the basis for many of their beliefs, needs (actual or perceived), and subsequent behavior. In most cases, training and discipline, grounded in cultural understanding and sensitivity, will help mitigate many of the potential adverse effects resulting from military and civilian interaction.

URBAN INFRASTRUCTURE

C-20. Urban infrastructures are those systems that support the inhabitants, their economy, and their government. They provide the link between the physical terrain and the urban society. Hundreds of systems may exist. Each system has a critical role in the smooth functioning of the urban area. All systems fit into the following five broad categories: communications and information, transportation and distribution, energy, commerce, and human services.

Appendix D

Practical Application

It is always best to derive lessons learned from practical operations, in the field. These are operations where a measure of success was acquired, along with important lessons learned. The following reflects an actual TF application of the doctrine highlighted in the contents of this manual.

OVERVIEW

D-1. The purpose of this manual is to provide TF units and elements with the latest guidance on providing crowd control and preventing civil disturbances.

D-2. Local civilians often gather in towns or along strategic locations throughout an AO to demonstrate grievances or other causes (for example, protesting the apprehension of a fellow citizen, policies of US forces, and policies of government services). Although civilians have the right to assemble, crowd gatherings are predominantly ethnically based, motivated, and influenced. Local civilians can apply for a permit to legally conduct a rally 72 hours in advance as long as their assembly is peaceful and unobtrusive to government services, freedom of movement, or the rights of others. On occasion, intelligence collection can indicate a possibility for violence. In these instances or during spontaneous crowd formations, the TF must provide forces in support of a safe and secure environment or to assist civilian police in preventing unlawful acts. If unlawful acts are committed, detain violators for adjudication under the legal system.

PROCEDURES

D-3. Take precautions to avoid becoming a catalyst to the civil disturbance. During lawful assemblies, the TF must make every effort to treat HN individuals with respect. This means that the TF must engage in peaceful gatherings using interpersonal communication skills. This ensures that demonstrators understand that TF personnel respect their right to lawfully assemble during peaceful demonstrations. When possible, TF engagement of a peaceful assembly should be unobtrusive.

D-4. Do not display civil disturbance equipment or weapons, overtly prepare defenses against a crowd, or aggressively violate crowd space, unless presented with information, intelligence, or negative indicators of pending violence.

NEGATIVE INDICATORS

D-5. Any response to negative indicators should be sequential, measured, and focused on deescalating rather than escalating the crowd.

NEGATIVE INDICATORS OF A POTENTIAL CIVIL DISTURBANCE

D-6. Predictive intelligence is the best indicator of a potential civil disturbance and may provide insight on internal planning, external influences, environmental conditions, or catalyst events that should influence TF planning and response.

D-7. Other indicators of a potential civil disturbance are people beginning to assemble during a scheduled or spontaneous rally. When people begin to assemble, leaders should pay particular attention to crowd dynamics. As the crowd demeanor or dynamics change from peaceful assembly to unruly behavior or unlawful acts, negative indicators should provide an advance warning and prompt a measured response. Negative indicators of a potential change in crowd dynamics and unlawful activities should prompt immediate countermeasures (see *Figure D-1*).

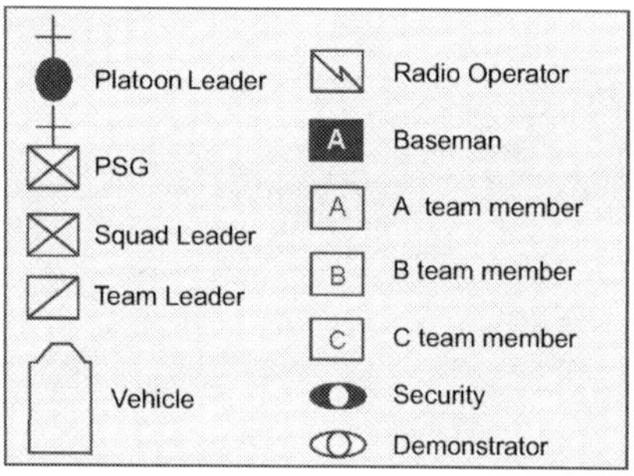

Figure D-1. Symbols for a Formation Element

Orientation

D-8. Crowd dynamics change from people milling around and talking among themselves in isolated groups to a more collective focus concentrated on a single objective (for example, TF personnel, specific agitators or magnets, a government figure, or other activity).

Massing

D-9. The crowd begins to mass or tighten into a large contiguous body from a loose formation to a static congregation located in a more concentrated area. Unchecked, massing can provide anonymity and a collective feeling of invulnerability. This change in group dynamics could provide ideal conditions for a civil disturbance (see *Figure D-2*).

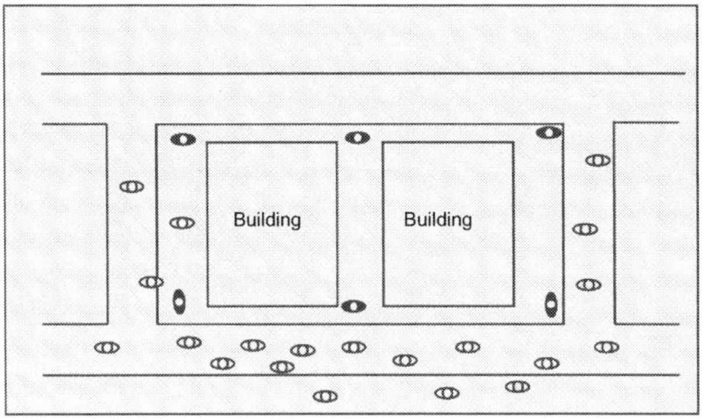

Figure D-2. Crowd Massing

Chanting

D-10. People may begin chanting slogans or yelling ethnically charged rhetoric or obscenities. Because chanting can indicate an increase in crowd intensity, interpreters should be positioned in strategic locations to interpret crowd rhetoric.

Signaling

D-11. Signaling occurs when members of the crowd position themselves to communicate and synchronize crowd actions tied to a plan with specific objectives. Signaling may indicate the intent to escalate the message through the use of violent or unlawful acts.

Blocking

D-12. The crowd may attempt to block public thoroughfares (for example, entrances and exits to public buildings, MSRs, community streets, or other traffic areas that may degrade freedom of movement or similar actions). This escalates the level of attention given to the crowd's message or prevents freedom of movement to opposing persons or groups.

Arming

D-13. Any signs that the crowd is arming itself with objects (for example, bottles, rocks, and sticks) is a clear indication of impending danger to the TF and other civilians. In fact, during a recent civil disturbance, some members of the crowd were seen handing out bottles to others in the crowd. Arming demonstrates intent toward violence is unlawful.

Outside Influence

D-14. The presence of people from outside the community may indicate more sophisticated planning and resources than would otherwise be expected from local community members. Outside people may feel insulated from responsibility for local actions and, therefore, provide a dangerous catalyst for a violent civil disturbance.

Agitators and Instigators

D-15. Agitators and instigators are common elements associated with civil disturbances. They often provide a catalyst to incite the crowd toward violence. Whether spewing rhetoric or committing unlawful acts, their continued presence will fuel potential strife.

Absence of Children

D-16. Recent civil disturbances have shown that even when children were initially present at a rally, they were absent before the civil disturbance erupted with violence. While the presence of children is not a guarantee of a peaceful assembly, the removal of children from a rally may provide early warning for a pending escalation of hostilities or a full-blown riot.

PREVENTION OF CROWD ESCALATION

D-17. If TF personnel believe, based on intelligence or negative indicators, that a crowd is planning to, or may spontaneously, riot they should immediately engage crowd participants using the preventive TTP measures.

D-18. The force executing these preventive TTP can include a number of civil disturbance teams and elements capable of executing the tasks shown in *Figures D-2,* page D-3; *D-3; D-4,* page D-6; and *D-5,* page D-8. These figures demonstrate a full range of crowd control teams and elements, to include camera teams with interpreters, checkpoints, QRF response, riot control formation, NLW teams, snatch and detain teams, blocking teams, and detainee evacuation, as required. *Figures D-2, D-3, D-4,* and *D-5,* illustrate the evolution of a crowd formation and the recommended positioning of teams and elements to engage people as they congregate. Included with each security team is a designated camera or video recorder.

D-19. Depending on the situation and crowd dynamics, it may not be necessary to establish all crowd control teams and elements, but they should be included in planning. Regardless of task organization, teams and elements must be prepared to execute the crowd control and civil disturbance TTP. The order to execute the following TTP should be tailored based on local conditions.

Early Contact

D-20. Engaging people as a crowd forms is essential to prevention. Whether based on intelligence or negative indicators, people should be engaged early, before the crowd masses as shown in *Figure D-2,* rather than waiting until after the crowd masses as shown in *Figure D-3.* Early contact will assist TFs in personalizing their interaction and ensuring that people understand their presence is known, documented, and can be linked to their actions.

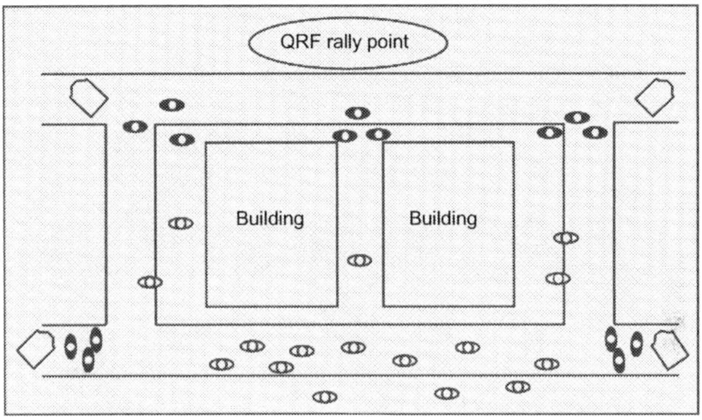

Figure D-3. Early Contact and Unobtrusive Crowd Control Measures

D-21. Cameras are one of the best tools for preventing civil disturbances. The use of cameras to document crowd formations, particularly lens-size groups (small enough to fit in a single photo), will personalize contact and eliminate feelings of anonymity among the crowd. People whose identity has been documented will think twice before committing unlawful acts. Even without documenting names, photo analysis can identify people associated with the civil disturbance, especially those who unlawfully participated as crowd instigators, agitators, or magnets or may commit unlawful acts.

D-22. As people begin to gather in and along streets, camera teams should be deployed as unobtrusively as possible to engage people before they congregate in mass. As depicted in *Figure D-3*, camera teams should consist of two soldiers with an interpreter and be employed in a widely dispersed pattern. Such small, dispersed camera teams are less likely to intrude on lawful rallies or create invasive friction that might become a catalyst to a civil disturbance.

D-23. As the crowd begins to demonstrate negative indicators for a possible civil disturbance (*Figure D-4,* page D-6), camera teams should become more invasive by actively engaging groups to take their pictures. Camera teams should focus cameras on instigators, agitators, or magnets (personnel that garner respect or compliance from the crowd). If a person is purposely avoiding the camera, even if his activities are lawful, the team should take special measures to obtain his picture. Each time a camera team takes a picture of a person or group of people, they must immediately engage them using the warning outlined in *paragraph D-26.*

D-24. Once the crowd has massed and displays negative indicators or begins to riot, camera teams should immediately move behind the riot formation line (see *Figure D-4*). It is essential, however, that camera teams continue to document the event, focusing on people who are committing illegal acts such as arming themselves, throwing objects, and assaulting TF personnel.

D-25. Immediately following a crowd control or civil disturbance operation, photos must be consolidated, processed, and submitted to the TF analysis and control element (ACE) for analysis. A near real-time analysis will provide immediate intelligence for ongoing operations to prevent subsequent civil disturbances or to detain violators from the previous operation.

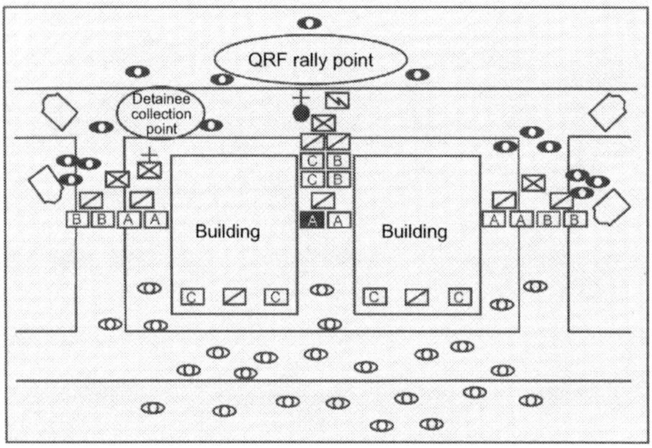

Figure D-4. More Obtrusive Engagement as the Crowd Demonstrates Negative Indicators

Crowd Warning

D-26. Once the team has taken a picture of a person or group of people, they should warn the crowd that as individuals they would be held accountable for their actions. Using an interpreter, the camera team leader will make solid eye contact with the person or group and state the following warning: *"We have your picture. You will be held accountable for your actions. If you commit an unlawful act, you will be detained, if not today, tomorrow. Blocking roads or access to public buildings is unlawful."*

Limit Crowd Size

D-27. Persons seeking to conduct a rally must apply for a permit through civilian authorities at least 72 hours prior to the meeting. If a rally has not been lawfully scheduled through civilian authorities, or if it has, but is demonstrating negative indicators and is turning into a civil disturbance, TF units and elements should attempt to limit the crowd or separate gatherings under 10 people (see *Figure D-3,* page D-5). Walking patrols can assist camera teams in breaking up large groups. Walking patrols should operate in conjunction with camera teams. This will ensure that camera teams can accurately document and provide the aforementioned warning to prevent further escalation of crowd size or behavior.

Clear Routes

D-28. Routes should be considered key terrain and must be controlled by the TF to retain the initiative. Note how a QRF divides in order to clear side streets between protected buildings in *Figure D-4.* Blocking routes or public thoroughfares is not only illegal, but severely restrictive to civil disturbance operations. Routes must remain open to allow movement of TFs and elements, whether it is for camera teams during early contact or for the QRF to respond during an actual civil disturbance. Because blocking routes is illegal, camera teams must document responsible persons for immediate or subsequent detention and legal processing.

Isolate the Area

D-29. As depicted in *Figure D-3,* page D-5, checkpoints should be established to isolate the area. While checkpoints can prevent people from entering the isolated area, they should never prevent people from exiting the area. When trapped, people may escalate the level of violence out of fear, anger, or desperation. When placing checkpoints, leaders should find a location that is both close enough to limit the number of persons being isolated, but far enough away that they will not draw a crowd from people within the isolated area. Checkpoints are essential to preventing others from joining the already gathered crowds. These checkpoints must provide storage room for processing vehicles in both directions; but when limited by space or other considerations, it should simply block traffic trying to enter the isolated area. Improve force protection measures in and around checkpoints as necessary. Strategically placed checkpoints can provide numerous countermeasures to prevent and control a civil disturbance or to identify and detain persons who may have committed unlawful acts during the riot. Use checkpoints to prevent potential participants from joining the crowd or civil disturbance. Checkpoints should also prevent any activity that may commingle ethnic groups. The checkpoint must process all people departing the area by documenting personal and vehicle information with the pictures of all occupants in front of or adjacent to the vehicle.

Establish Lethal Overwatch

D-30. The mentality of "taking the high ground" is just as important in civil disturbance operations as in any other type of operation. Although QRF assets on the ground have both NL and lethal capabilities, it is critical to ensure their safety by deploying marksman/observer teams as shown with the C-team designates on top of the buildings in *Figures D-4* and *D-5,* page D-8. Detain persons committing unlawful acts. The detention and legal processing of persons who commit unlawful acts, against TFs or local laws sends a clear message that all people will be held accountable for their actions. Immediately detain persons who are committing unlawful acts and are isolated or easily accessible, unless the detain operation would create a catalyst event or endanger the apprehension team. If it is not feasible to immediately detain them, a picture with relevant information should be forwarded for analysis and inclusion on the detain list for their detention at a later date and time. Remember, as stated in the warning, they will be detained, if not today, tomorrow. Detainees should be quickly relocated to the forward processing site for detainee evacuation (see *Figure D-5*). Detainee evacuation sites should be behind the control formations to prevent any interference from the local population. Sites should be positioned at two separate locations, such as at the end of a town or located along an MSR. All people who are detained based on

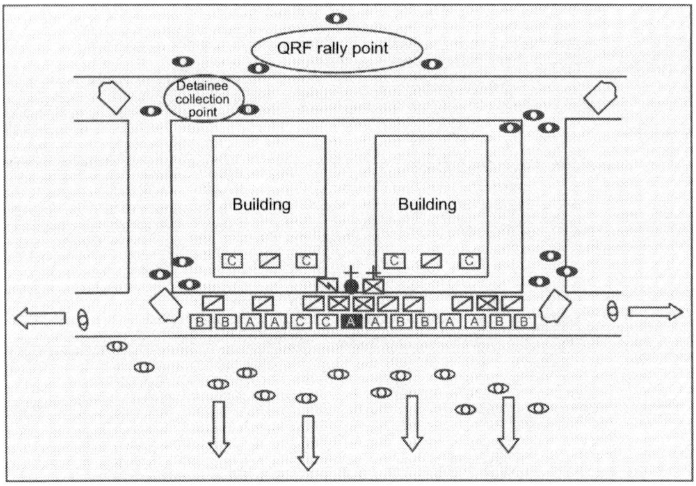

Figure D-5. Full Engagement as Crowd Unlawfully Demonstrates

solid evidence of a crime (for example, pictures of them armed with dangerous objects, blocking a road, assaulting TF members) will be turned over to local authorities for legal processing.

Detain List

D-31. When executing operations based on the detain list, do not show people the list or separate pictures from the list.

Quick Reaction Force Position

D-32. The QRF should be positioned with easy access to decisive points to interdict formations or gathering of crowds, support checkpoints, or provide sustained operations in support of early contact teams. Notice in *Figure D-1*, page D-2, how the QRF is positioned out of direct sight of the crowd until employment becomes necessary. QRF personnel are readily organized, equipped, and easily moved into position to move between buildings and through obstacles (see *Figure D-4,* page D-6).

D-33. At this point, if the crowd ignores verbal orders, NL munitions from appropriate standoff distances could be used to influence and motivate the crowd to comply. Use of NL munitions could prevent the necessity for QRF personnel to come into direct physical contact with the crowd. Early positioning of the QRF could become a catalyst event that may instigate a civil disturbance.

D-34. Assembling in mass (see *Figure D-5*), directly in front of the protected buildings, illustrates to the crowd that there is strength in numbers and the QRF is determined to disperse the crowd from the area. As a reminder, remove the flanking checkpoints that may block the dispersal of the crowd.

Glossary

A2C2	Army airspace command and control
AA	assembly area
AFL-CIO	American Federation of Labor-Congress of Industrial Organizations
AG	Adjutant General
AO	area of operations
ARTEP	Army training and evaluation program
AS	area security
ASV	armored security vehicle
ATTN	attention
BDU	battle dress uniform
CA	civil affairs
CD	combat development
CDC	crowd dispersal cartridge
CFR	Code of Federal Regulations
CINC	commander-in-chief
CMT	crisis management team
COA	course of action
COB	close of business
CONUS	continental United States
COTS	commercial off-the-shelf
CP	candle power
CPU	central processing unit
DA	Department of the Army
DCO	defense coordinating officer
DD	Department of Defense
DM	designated marksmen
DOD	Department of Defense
DOJ	Department of Justice
DOMS	director of military support
EMD	emergency medical dispatching
EMS	emergency medical services
EOC	emergency operations center
FCMT	forced cell move team
FCO	federal coordinating officer
FEMA	Federal Emergency Management Agency
FM	field manual

FRP	federal response plan
FSE	fire support element
FSO	fire support officer
G2	Assistant Chief of Staff, G3 (Intelligence)
G3	Assistant Chief of Staff, G3 (Operations and Plans)
G5	Assistant Chief of Staff, G5 (Civil Affairs)
G6	Assistant Chief of Staff, G6 (Information Management)
GOTS	government off-the-shelf
GRM	graduated response matrix
HIV	human immunodeficiency virus
HMMWV	high-mobility multipurpose wheeled vehicle
HN	host nation
HQ	headquarters
I/R	internment and resettlement
INIWIC	Interservice Nonlethal Individual Weapons Instructor Course
IO	information operations
IPB	intelligence preparation of the battlefield
JMC	joint military commission
JTF	joint task force
L&O	law and order
LBE	load-bearing equipment
LBV	load-bearing vest
LEA	law enforcement agency
LOC	lines of communication
LOS	line of sight
LP	listening post
MACDIS	military assistance for civil disturbance
MANSCEN	Maneuver Support Center
MCCM-GE	modified crowd control munition-ground emplacement
MCM	Manual for Courts-Martial
MDMP	military decision-making process
METL	mission essential task list
METT-TC	mission, enemy, terrain, troops, time available, and civilian considerations
MI	military intelligence
mm	millimeter
MO	Missouri
MOOTW	military operations other than war
MOUT	military operations on urbanized terrain
MPI	military police investigator
MRE	meal, ready to eat

MS	mobility support
MSCA	military support to civil authorities
MTP	mission training plan
MWD	military working dog
NCO	noncommissioned officer
NEO	noncombatant evacuation operations
NG	National Guard
NL	nonlethal
NLCS	nonlethal capabilities set
NLW	nonlethal weapons
OC	Oleoresin Capsicum
OCONUS	outside continental United States
OES	office of emergency services
OIC	officer in charge
OP	observation post
OPFOR	opposing forces
OPORD	operations order
PA	public address
PAO	Public Affairs Office
PIO	police intelligence operations
PM	provost marshal
PLT	platoon
PMO	Provost Marshal Office
POMSO	plans, operations, and military support officer
PPCT	pressure point control tactic
PPE	personal protective equipment
PSG	platoon sergeant
PSYOP	psychological operations
PT	physical training
PVAB	portable vehicle arrest barrier
QRF	quick reaction force
RCA	riot control agent
ROE	rules of engagement
RUF	rule for the use of force
S2	Intelligence Officer (US Army)
SCO	state coordinating officer
SCRAG	senior representative of the attorney general
SITREP	situation report
SJA	Staff Judge Advocate
SOFA	status of forces agreement

SOMA	status of mission agreement
SOP	standing operating procedure
SOSO	stability operations and support operations
SQD	squad
STARC	state area command
STP	soldier training publication
STX	soldier training exercise
TA	Training Area
TF	task force
TLP	troop leading procedure
TM	team
TPT	tactical psychological operations team
TRADOC	United States Army Training and Doctrine Command
TSOP	tactical standing operating procedure
TTP	tactics, techniques, and procedures
UCMJ	Uniform Code of Military Justice
UN	United Nations
UNGA	United Nations General Assembly
UNHCR	United Nations High Commissioner for Refugees
UNSC	United Nations Security Council
US	United States
USAMPS	United States Army Military Police School
VCR	video cassette recorder
VIP	very important person
WMD	weapons of mass destruction
WTO	World Trade Organization

Bibliography

AR 380-10. *Foreign Disclosure and Contacts With Foreign Representatives.* 6 June 2003.

AR 380-13. *Acquisition and Storage of Information Concerning Nonaffiliated Persons and Organizations.* 30 September 1974.

Code of Federal Regulations.

Constitution of the United States.

DA Form 2028. *Recommended Changes to Publications and Blank Forms.*

DA Form 4137. *Evidence/Property Document.*

DD Form 2708. *Receipt For Inmate or Detained Person.*

DOD Civil Disturbance Plan 55-2. *Operation Garden Plot.* 15 February 1991.

DOD Directive 3025.12. *Military Assistance for Civil Disturbances.* 4 February 1994.

DOD Directive 3025.15. *Military Assistance to Civil Authorities.* 18 February 1997.

DOD Directive 5525.5. *DOD Cooperation With Civilian Law Enforcement Officials.* 15 January 1986.

Executive Order 12333. *United States Intelligence Activities.* 4 December 1981.

Fiscal Year 1989 Act.

FM 3-06. *Urban Operations.* 1 June 2003.

FM 3-06.11. *Combined Arms Operations in Urban Terrain.* 28 February 2002.

FM 19.10. *Military Police Law and Order Operations.* 30 September 1987.

FM 3-19.4. *Military Police Leaders' Handbook.* 4 March 2002.

FM 3-19.40. *Military Policy Internment/Resettlement Operations.* 1 August 2001.

FM 3-21.5. *Drills and Ceremonies.* 7 July 2003.

FM 3-22.40. *Multiservice Tactics, Techniques, and Procedures (MTTP) For Tactical Employment of Nonlethal Weapons (NLW).* 15 January 2003.

FM 7-0. *Training the Force.* 22 October 2002.

FM 7-1. *Battle Focused Training.* 15 September 2003.

FM 27-100. *Legal Support to Operations.* 1 March 2000.

Judiciary Act of 1789.

MCM.

National Defense Authorization Acts.

NGR 500-1/ANGI 10-8101. *Military Support to Civil Authorities.* 1 February 1996.

Posse Comitatus Act. 1878.

STP 19-95B1-SM. *Soldier's Manual for MOS 95B Military Police Skill Level 1.* 14 January 2003.

STP 19-95B24-SM-TG. *Soldier's Manual and Trainer's Guide for MOS 95B, Military Police, Skill Levels 2/3/4.* 2 December 2002.

STP 19-95C1-SM. *MOS 95C, Corrections Specialist, Skill Level 1 Soldier's Manual.* 30 October 2003.

STP 19-95C24-SM-TG. *MOS 95C, Corrections Specialist, Skill Levels 2/3/4, Soldier's Manual and Trainer's Guide.* 30 October 2003.

Uniform Code of Military Justice.

USC, Title 32. *National Guard.*

USC, Title 10, Chapter 18. *Military Support for Civilian Law Enforcement Agencies.*

USC, Title 10, Subtitle A, Part I, Chapter 15, Section 331. *Federal Aid for State Governments.*

USC, Title 10, Subtitle A, Part I, Chapter 15, Section 332. *Use of Militia and Armed Forces to Enforce Federal Authority.*

USC, Title 10, Subtitle A, Part I, Chapter 15, Section 333. *Interference With State and Federal Law.*

USC, Title 10, Subtitle A, Part I, Chapter 15, Section 334. *Proclamation to Disperse.*

USC, Title 18, Part I, Chapter 67, Section 1385. *Use of Army and Air Force as a Posse Comitatus.*

Index

Numerics

12-gauge shotgun, 5-6
12-gauge utility pouch, 5-16
12-guage buttstock cuff, 5-17
12-guagc high-intensity light system, 5-15

A

accountability, D-6
agendas, 1-4
air patrols, 2-23
analysis, 7-3, D-6
anarchists, 1-3
annual unit training plan, 6-3
anonymous, 1-4
anti-free-enterprise, 1-3
antiglobalization, 1-3
apprehension teams, 3-3, 6-12, 7-9
  apprehension element, 6-12
  squad leader, 7-10
  team member four, 7-11
  team member one, 7-11
  team member three, 7-11
  team member two, 7-11
area isolation, D-7
arming, D-3
assembly from a platoon line, 6-17

B

ballistic full-length riot body shield, 5-3
ballistic riot face shield, 5-2
barricades, 1-7, 1-9, 2-19, 7-1
behavior, 1-7
blocking, 2-19, D-3
by-the-number commands, 2-22

C

cadre behavior, 7-2
caltrop, 5-17
cameras, D-5
carrying pouch (40 millimeter), 5-15
casual crowds, 1-6
catalyst, D-1
CDC. See crowd dispersal cartridge.
centralized planning, 1-3
changing sides, 3-15
chanting, D-3
checkpoints, D-7
chemical irritants, 7-3, 7-4
chemical-containing projectiles, 7-6
  12-gauge projectile, 7-6
  40-millimeter projectile, 7-6
children, D-4
civil disturbance training, 6-3
civil law enforcement, 3-2, B-9
civil unrest, 1-1
clear routes, D-7
closed-cell shock-absorbent foam, 5-19
COA. See course of action.
coercion dispersal, 1-5
commands, 6-7, 6-9
commercial ribbon, C-5
containment, 2-21
control force team, 7-3
cooperative subjects, 3-2
core, C-5
core capabilities, 5-20
core periphery, C-5
correctional facility cadre, 7-9
counterdemonstration workgroups, 2-6
course of action (COA), 8-1
course of action analysis, 8-1
CR solution, 5-12
crowd assessment, 2-8, 2-9
crowd control, 2-5
crowd control options, 2-17
crowd dispersal cartridge (CDC), 5-9
crowd types
  agitated crowds, 1-7
  sighting crowds, 1-6
crowds
  building of, 1-4
CS1, 5-12

D

deescalating the crowd, D-1
deliberate operation, 2-16
demonstrations, 1-1
Department of Defense (DOD), 2-4
designated marksmen (DM), 2-2
designated marksmen (DM) teams, 6-11
detain list, D-8
detention facilities, 3-18, 3-19
Dibenz, 5-11
director of military support (DOMS), B-2
dispersal process, 1-4
  emergency dispersal, 1-5
  routine dispersal, 1-4
dispersing, 2-19
dissemination of chemical irritants, 7-4
  aerosol method, 7-5
  bursting method, 7-5
  fogging method, 7-6
  pyrotechnic method, 7-4
DM. See designated marksmen.
DOD. See Department of Defense.

E

emergency preparedness, B-4
equipment
  M33A1, 2-3
expandable riot baton, 5-5
explosives, 1-10, 1-11
extraction teams, 6-10
  deployment, 6-10

F

FCMTs. See forced cell move teams.
Federal Emergency Management Agency (FEMA), B-3
FEMA. See Federal Emergency Management Agency.
female subjects, 3-4
fight-or-flight, 1-6, 2-19
final phase, 1-4
final-target areas, 4-5
fire, 1-10
firearms, 1-10
Flex-Cufs, 3-4, 5-11, 7-18
  applying from the prone position, 3-5
  applying from the standing-supported position, 3-4
forced cell move teams (FCMTs), 7-12
  inspection, 7-13
  medic, 7-16
  officer in charge (OIC), 7-16
  team member five, 7-15, 7-18

team member four, 7-15, 7-18
team member one, 7-13, 7-17
team member three, 7-14, 7-18
team member two, 7-18
forced cell moves, 7-12
military police investigator, 7-16
moving the inmate, 7-19
returning the inmate to the cell, 7-20
team member three, 7-18
team member two, 7-14
forced dispersal, 2-22
formations, 7-9
company, 6-37
company echelons and wedge, 6-40
company line in depth, 6-39
company line in mass, 6-39
company line with support, 6-40
company with vehicles, 6-42
control force, 6-1, 6-3
control force base element, 6-5
control force circular, 6-4
control force command element, 6-6
control force diamond formations, 6-4
control force echelon formation, 6-4
control force lethal overwatch element, 6-6
control force line, 6-4
control force support element, 6-5
control force wedge, 6-4
control force with vehicles, 6-5
crowd control, 6-17
platoon, 6-27
platoon diamond and circle, 6-24, 6-35
platoon echelon right and left, 6-20, 6-32
platoon line, 6-16, 6-28, 6-30
platoon line assembly, 6-28
platoon line with direct support, 6-19, 6-31
platoon line with general support, 6-17, 6-28
platoon line with lateral support, 6-18
platoon line with one support squad, 6-32
platoon wedge, 6-32
platoon wedge with direct support, 6-24, 6-34
platoon wedge with general support, 6-22, 6-32
platoon wedge with lateral support, 6-22, 6-33
platoon wedge with one support squad, 6-35
three-squad platoon, 6-16
frisk search, 3-10
performance phase, 3-10
preparation phase, 3-10

G

G8 summit protest, 2-22
gathering
assembly process of, 1-2
impromptu, 1-3
organized, 1-3
GG04, 5-9
goal definition, 2-5
graduated response matrix, 8-1
development, 2-13, 2-14
graduated response matrix (GRM) development, 2-13
grievance protest, 7-2
GRM. See graduated response matrix.
ground-mounted bullhorns, 5-13

H

hand-and-arm signals, 6-7, 6-8, 6-16
handcuffed escort position, 3-14
hard rubber balls (.48 caliber), 5-9
hard rubber pellets (.25 caiber), 5-9
hasty operations, 2-17
high-intensity light, 5-14
high-rise areas, C-5
HN. See host nation.
host nation (HN), 2-10
host nation (HN) law, 2-11
hostages, 7-3
humanitarian assistance, 1-1, 2-13, 8-1

I

impact training suits, 5-19
individual disposable riot control agent dispersers, 5-11
individual high-intensity searchlight, 5-14
industrial areas, C-5
INIWIC. See Interservice Nonlethal Individual Weapons Instructor Course.
international law, 2-10
Interservice Nonlethal Individual Weapons Instructor Course (INIWIC), 8-2
interval and distance, 6-7

J

joint task force (JTF), B-12
JTF. See joint task force.

K

Kevlar helmet, 5-1

L

launch cup, 5-9
law enforcement agencies, B-1
law enforcement sources, B-6
lethal overwatch, D-7
lethal protection, 6-11
load bearing equipment, 5-15
load bearing vest, 5-15
Los Angeles Sheriffs Department, 2-7

M

M1006, 5-8
M1012, 5-7
M1013, 5-8
M33A1, 5-12
M36, 5-11
M37, 5-11
M40, 5-13
M7, 5-13
M7 66-millimeter launcher, 5-6
M84, 5-9
mandates, 2-10
marksmen and observer teams, 6-1
massing, D-2
MCCM-GE. See modified crowd control munition-ground emplacment.
metric conversion chart, A-1
military police functions, 2-4
military police investigator, 7-16
military police investigator (MPI), 7-16
military sources, B-6
military transport vehicles, 2-22
military working dogs (MWDs), 2-3
mobs, 1-7
modified crowd control munition-ground emplacment (MCCM-GE), 5-7

Molotov cocktails, 1-9
monitoring, 2-17
motorized patrols, 2-23
MPI. See military police investigator.
multinational force, 2-12
multiple riots, 7-2
MWDs. See military working dogs.

N

national guard (NG), B-9
negative indicators, D-2
negotiated management, 1-5
negotiations, 2-7
NG. See national guard.
NL. See nonlethal.
NLW. See nonlethal weapons.
nonballistic full-length riot shield, 5-1
nonballistic riot face shield, 5-1
nonballistic shin guard, 5-2
nonlethal (NL), 5-20
nonlethal (NL) counterpersonnel capabilities, 5-21
nonlethal (NL) munitions, 4-21
nonlethal countermateriel capabilities, 5-21
nonlethal engagements, 6-11
nonlethal munitions, 2-2, 7-3
nonlethal weapons, 2-2, 4-1, 5-22, 7-3
nonlethal weapons (NLW), 2-2
nonviolent tactics, 7-1

O

OC. See Oleoresin Capsicum.
officer in charge, 7-16
Oleoresin Capsicum (OC), 7-4
open sources, B-6
opposite-gender searches, 3-13, 3-16
orientation, D-2

P

patrol routes, 2-23
pawns in civil disturbance operations, 1-10
personnel effectors, 5-4
personnel protectors, 5-1
photographic record, 2-3
physical barriers, 1-7
planned violent disturbances, 7-1
planning process, 7-3
plans, 8-9
plans, operations, and military support officer (POMSO), B-4
platoon formations
  vehicles, 6-42
political grievances, 1-1
POMSO. See plans, operations, and military support officer.
portable vehicle arrest barrier, 5-18
posse comitatus, B-7, B-8
practice grenade fuse, 5-20
predictive intelligence, D-2
preparation and training, 7-2
primary-target areas, 4-4
proclamation, 2-20
prone search, 3-15
  performance phase, 3-17
  preparation phase, 3-16
psychological operations (PSYOP), 2-4
PSYOP. See psychological operations.
public disorders, 1-5
public disturbances, 1-5

R

rate of march, 6-7
RCAs. See riot control agents.
records, 7-7
religious-based fighting factions, 1-1
reserve forces, 6-12, B-3
residential areas, C-5
responsibilities, B-3
riot baton, 4-1, 4-3, 7-8
riot baton blocking techniques, 4-10
  high block, 4-10
  low block, 4-11
  middle block, 4-13
  strong-side block, 4-12
  weak-side block, 4-12
riot baton carries
  outside-arm carry, 4-8
  two-hand carry, 4-7
  vertical high-profile carry, 4-8
  vertical low-profile carry, 4-9
riot baton retention, 4-19
riot baton striking techniques, 4-13
  one-hand forward strike, 4-14
  one-hand reverse strike, 4-14
  two-hand front jab, 4-17
  two-hand middle strike, 4-18
  two-hand rear jab, 4-18
  two-hand strong-side horizontal strike, 4-16
  two-hand weak side horizontal strike, 4-16
riot batons, 4-7
  types of batons, 4-6
riot control agent disperser, 2-3
riot control agents, 5-13, 7-3
riot control agents (RCAs), 1-6, 2-14, 2-16, 2-19
riot gear, 7-8
riot shield, 4-1, 7-8
riot shield carries
  at the ready, 4-2
  on guard, 4-2
riots, 1-5, 1-8, 2-2, 2-9, 7-2
rocks, 1-9
ROE. See rules of engagement.
rubber ball grenade pouch, 5-17
rules of engagement (ROE), 2-11

S

same-gender searches, 3-9
sample proclamation, 2-20
scalable effects process, 2-12
search man, 3-7
search principles, 3-8
search teams, 3-9
secondary-target areas, 4-4
Secretary of Defense, B-2
Secretary of the Army, B-2
security man, 3-8
seven-step approach, 8-2
show of force, 2-21
single projectile rounds, 8-9
smoke grenades, 1-9
sniper, 1-10
SOFA. See status of forces agreement.
SOMA. See status of mission agreement.
SOP. See standing operating procedure.
Spectra Shield, 5-3
squad formations, 6-13
  assembly, 6-15
  diamond and circle formations, 6-15
  diamond or circle formation, 6-16
  echelon formation, 6-14
  line formation, 6-14
  wedge formation, 6-14
squad rotation, 6-26, 6-38
staff coordination, 2-16
standing operating procedure (SOP), 7-2
standing operating procedures (SOP), 8-1
standing search, 3-13
  performance phase, 3-14

preparation phase, 3-13
standoff distance, 6-1
status of forces agreement (SOFA), 2-10
status of mission agreement (SOMA), 2-10
street patterns
grid, C-4
irregular, C-5
radial, C-4
symbols for a formation element, 6-2

T

tactical standing operating procedure (TSOP), 2-3
tap-down technique, 4-22
target development, 2-14
targets, 8-2
E-type, 8-2
three-dimensional manikins, 8-2
terrain, C-1
terror organizations, 1-2
The Balkans, 1-10
threat cover man, 3-8
thrown objects, 1-10
tracing-C technique, 4-20
training, 7-7
training strike bags, 5-19
TSOP. See tactical standing operating procedure.

U

unanimous, 1-4
uncooperative subjects, 3-2
unit training program, 8-2
unit training strategy, 8-1
unorganized riots, 7-2
urban environment, C-1
urban infrastructure, C-6
urban operations, C-6
urban pattern
linear, C-3
segment, C-3
urban patterns
network, C-3
satellite, C-2
urban society, C-6
analysis, C-6
segments, C-6
urban terrain, C-2
US law, 2-10

V

vehicles, 1-10
verbal abuse, 1-9
violent tactics, 7-1

W

walking patrols, D-6
war, 1-1
wargame, 2-16
weapon positions, 6-6
high-port arms, 6-7
safe-port arms, 6-7
weapons
12-gauge shotgun, 2-1
M16, 2-1
M203, 2-1
M4, 2-1
M9 pistol, 2-1
wooden riot baton, 5-5
World Trade Organization (WTO), 2-24
WTO. See World Trade Organization.